*Stochastic Mechanics*
*Random Media*
*Signal Processing*
*and Image Synthesis*
*Mathematical Economics and Finance*
*Stochastic Optimization*
*Stochastic Control*
*Stochastic Models in Life Sciences*

# Stochastic Modelling and Applied Probability

(Formerly:
Applications of Mathematics)

# 62

Xianping Guo • Onésimo Hernández-Lerma

# Continuous-Time Markov Decision Processes

## Theory and Applications

Springer

Xianping Guo
School of Mathematics
and Computational Science
Zhongshan University
Guangzhou 510275
People's Republic of China
mcsgxp@mail.sysu.edu.cn

Onésimo Hernández-Lerma
Departamento de Matemáticas
Centro de Investigación y de Estudios
Avanzados del Instituto Politécnico
Nacional (CINVESTAV-IPN)
Apdo Postal 14-740
México D.F. 07000
Mexico
ohernand@math.cinvestav.mx

ISSN 0172-4568
ISBN 978-3-642-02546-4 e-ISBN 978-3-642-02547-1
DOI 10.1007/978-3-642-02547-1
Springer Heidelberg Dordrecht London New York

Library of Congress Control Number: 2009935465

Mathematics Subject Classification (2000): 90C40, 93E20, 90B05, 90B22, 60J27, 60K30



Printed on acid-free paper

Springer is part of Springer Science+Business Media (www.springer.com)

*OHL dedicates this book to:*
*Marina, Adrián, Gerardo, Claudia, Azziza, Rosita, and Lucas*

# Preface

This book is about continuous-time Markov decision processes (MDPs) with a countable state space. These processes are also known as continuous-time controlled Markov chains or stochastic dynamic programs and are widely used for modeling decision-making (or control) problems that arise in many fields, for instance, operations research (inventory, manufacturing, and queueing systems), communications engineering, computer science, population processes (such as fisheries and epidemics), and management science, to name just a few. We develop their structure and main properties and describe their applications. To the best of our knowledge, this is the first book completely devoted to continuous-time MDPs.

The book consists of twelve chapters and three appendices, which are briefly described below. (A more complete description is provided at the end of Chap. 1, after introducing suitable terminology and notation.)

We begin in Chap. 1 with some examples that motivate the different components of a continuous-time MDP, which are then formalized in Chap. 2. Chapter 2 also introduces the basic optimality criteria we are concerned with. The remaining material is organized in an increasing order of difficulty. The simplest case, namely, finite continuous-time MDPs, is studied in Chap. 3. In Chaps. 4 and 5, we move on to continuous-time MDPs with a countable state space and nonnegative cost rates. Then, in Chaps. 6 and 7, we consider general unbounded rewards.

The rest of the book deals with special topics, such as pathwise average rewards (Chap. 8), so-called advanced optimality criteria (Chap. 9), average reward continuous-time MDPs with minimum variance (Chap. 10), and continuous-time MDPs with constraints (Chaps. 11 and 12). Finally, the three appendices summarize important facts from real analysis and Markov chains used in the text.

Most chapters have two final sections, one on examples and one with bibliographical notes.

To conclude, we should mention that although the book focuses on continuous-time MDPs with a countable state space, the presentation and approaches are sufficiently general to facilitate the extension of results to other important problems, such as continuous-time stochastic games and MDPs in general (say, Polish) spaces.

**Acknowledgements** The first author wishes to thank the National Natural Science Foundation of China (NSFC) for generous support to his research work through the grants NSFC-60874004 and NSFC-60736028. The work of the second author (OHL) has been partially supported by the Consejo Nacional de Ciencia y Tecnología (CONACyT) grants 37355E (2002–2004) and 45693F (2005–2008).

*Xianping Guo and Onésimo Hernández-Lerma*

# Contents

# Notation

| | |
|---|---|
| $a$ | A generic action |
| $i$ | A generic state |
| $a(t)$ | Action at time $t$ |
| $A$ | Action space |
| $A(i)$ | Set of actions available at state $i$ |
| $A_n^f(i)$ | Set defined in (3.20) |
| $B(S)$ | The space of all bounded functions on $S$ |
| $\mathcal{B}(X)$ | The Borel $\sigma$-algebra on a Borel space $X$ |
| $B_n^f(i)$ | Sets defined in (3.36), (3.45), and (3.52) |
| $B_w(S)$ | The space of all $w$-bounded functions on $S$ |
| $c(i, a)$ | Cost function of states $i$ and actions $a$ |
| $\delta_{ij}$ | The Kronecker delta |
| $e$ | Vector with all components equal to 1 |
| $E_i^\pi$ | Expectation operator with respect to $P_i^\pi$ |
| $E_\nu^\pi$ | Expectation operator with respect to $P_\nu^\pi$ |
| $f$ | Deterministic stationary policy |
| $f^*$ | Optimal policy for some criterion |
| $f(i)$ | Action taken in state $i$ under policy $f$ |
| $F$ | Set of all deterministic stationary policies |
| $F_{\text{ao}}$ | Set of all average optimal policies |
| $F_{\text{ca}}$ | Set of all canonical policies |
| $F_n^*$ | Set of all $n$-potential optimal deterministic stationary policies |
| $g(f)$ | Average reward of policy $f$, column vector |
| $g_n(f)$ | $n$-potential of policy $f$, column vector |
| $g_n^*$ | The optimal $n$-potential, column vector |
| $H_f$ | Bias operator under $f$ defined in (3.4) and (9.10) |
| $I$ | The identity matrix |
| $J_\alpha^*$ | Optimal discounted-cost function |
| $J_c^*$ | Optimal average-cost function |
| $J_\alpha(i, \pi)$ | Expected discounted cost of a policy $\pi$ |
| $J_c(i, \pi)$ | Expected average cost of a policy $\pi$ |

| | |
|---|---|
| $J_t^c(i, \pi)$ | $t$-horizon expected cost |
| $m(i)$ | Positive number such that $m(i) \geq q^*(i)$ |
| $M(X)$ | The set of all finite signed measures on $X$ |
| $M_w(X)$ | The set of all signed measures with finite $w$-norm |
| $\|\mu\|$ | The total variation of a signed measure $\mu$ |
| $\\|\mu\\|$ | The total variation norm of a signed measure $\mu$ |
| $\\|\mu\\|_w$ | The $w$-norm of a signed measure $\mu$ |
| $\mu_f$ | Invariant probability measure under $f$ |
| $\nu$ | Initial distribution on $S$ |
| $p(j\|i, a)$ | Transition probability defined in (6.12) |
| $\bar{p}(j\|i, a)$ | Transition probability defined in (6.2) |
| $p_\pi(s, i, t, j)$ | Transition probability function under $\pi$ |
| $P(t, f)$ | Homogeneous transition probability matrix under $f \in F$ |
| $P^*(f)$ | Limiting matrix of the transition probability matrix $P(t, f)$ |
| $P(X)$ | The set of all probability measures on $\mathcal{B}(X)$ |
| $P_i^\pi$ | Probability measure determined by a policy $\pi$ and initial state $i$ |
| $P_i^\pi$-a.s. | Almost surely with respect to $P_i^\pi$ |
| $P_\nu^\pi$ | Probability measure determined by a policy $\pi$ and initial distribution $\nu$ |
| $P_{\nu,s}^\pi$ | Probability measure determined by a policy $\pi$ and distribution $\nu$ at time $s$ |
| $\Pi$ | The set of all randomized Markov policies |
| $\Pi^s$ | The set of all randomized stationary policies |
| $\pi$ | Randomized stationary policy |
| $(\pi_t)$ | Randomized Markov policy |
| $q(j\|i, \pi)$ | Transition rate from state $i$ to state $j$ under $\pi$ |
| $q(j\|i, \pi_t)$ | Transition rate from state $i$ to state $j$ under $(\pi_t)$ |
| $q(j\|i, a)$ | Transition rate from state $i$ to state $j$ under action $a$ |
| $q^*(i)$ | $q^*(i) := \sup\{-q(i\|i, a)\|a \in A(i)\}$ for every $i \in S$ |
| $q_i(a)$ | The absolute value of $q(i\|i, a)$ |
| $q_i(f)$ | The absolute value of $q(i\|i, f(i))$ |
| $q(j\|i, f)$ | Transition rate from state $i$ to state $j$ under $f \in F$ |
| $Q(f)$ | Transition rate matrix under $f$ |
| $Q(\pi)$ | Transition rate matrix under a randomized stationary policy $\pi$ |
| $Q(\pi_t)$ | Transition rate matrix under a policy $(\pi_t)$ |
| $r(f)$ | Reward function of policy $f$, column vector |
| $r(i, a)$ | Reward for taking action $a$ at state $i$ |
| $r(i, \pi_t)$ | Reward under a policy $(\pi_t)$ |
| $r(i, \pi)$ | Reward under a randomized stationary policy $\pi$ |
| $S$ | State space |
| $\|S\|$ | Number of elements in $S$ |
| $u^T$ | Transpose of a vector $u$ |
| $u \succeq v$ | $u \geq v$ and $u(i) > v(i)$ for at least one $i \in S$ |
| $u \succeq_{lg} v$ | $u$ is lexicographically greater than or equal to $v$ |
| $u \succ_{lg} v$ | $u \succeq_{lg} v$ and $u \neq v$ |

| | |
|---|---|
| $V_\alpha(i,\pi)$ | Discounted reward under $\pi$ given an initial state $i$ |
| $\bar{V}^*$ | Optimal average reward function |
| $\bar{V}(i,\pi)$ | Expected average reward of a policy $\pi$ |
| $V_t(i,\pi)$ | $t$-horizon expected reward |
| $v_f^h$ | $v_f^h := r(h) + Q(h)g_0(f) - g_{-1}(f)$ |
| $x(t)$ | State at time $t$ |
| $\square$ | Completion of a proof |
| "0" | The matrix or the vector with zero at every component |

# Abbreviations

| | |
|---|---|
| ACOI | Average-cost optimality inequality |
| ACOE | Average-cost optimality equation |
| AR | Average reward |
| AROE | Average-reward optimality equation |
| EAR | Expected average reward |
| PAR | Pathwise average reward |
| a.e. | Almost everywhere with respect to the Lebesgue measure on $[0, \infty)$ |
| a.s. | Almost surely |
| D-LP | Dual linear program (for the ergodic model) |
| D-LP-M | Dual linear program (for the multichain model) |
| DROE | Discounted reward optimality equation |
| MDP | Markov decision process |
| MDPs | Markov decision processes |
| i.p.m. | Invariant probability measure |
| P-LP | Primal linear program (for the ergodic model) |
| P-LP-M | Primal linear program (for the multichain model) |
| w.l.o.g | Without loss of generality |

# Chapter 1
# Introduction and Summary

## 1.1 Introduction

This book concerns continuous-time Markov decision processes (MDPs), also known as continuous-time controlled Markov chains. The main objective is to present a systematic exposition of some recent developments in the theory and applications of continuous-time MDPs with the discounted and the average cost/reward criteria, as well as some advanced criteria. We do not aim at completeness and technical generality, and so we are mainly interested in continuous-time MDPs with a *countable* state space. However, we allow unbounded transition rates and reward functions that may be *unbounded* from below and from above, and the formulation can be applied to the case of Polish spaces.

Continuous-time MDPs appear in many classes of applications, including queueing systems, manufacturing processes, inventory control, population processes, and control of infectious diseases, to name a few. In this chapter, to motivate the problems we will deal with, we briefly discuss some application areas, namely, control of queueing systems, control of epidemic processes, and control of population processes. Finally, we close this chapter with a description of the main contents of the book.

## 1.2 Preliminary Examples

We first introduce the basic elements of a continuous-time Markov decision process (MDP) by means of an example.

*Example 1.1* (Controlled queueing systems) Queueing systems appear in a wide variety of contexts including telecommunication networks, computer systems, manufacturing processes, repair–maintenance shops, supermarket checkout lines, and even dynamic power management—see Qiu et al. (2001) [130]. In the simplest case, that is, a single-server queueing system, jobs (also known as customers) arrive, enter

X. Guo, O. Hernández-Lerma, *Continuous-Time Markov Decision Processes*,
Stochastic Modelling and Applied Probability 62,
DOI 10.1007/978-3-642-02547-1_1, 

the queue (or waiting line or buffer), wait for service, receive service, and then leave the system.

There are three control possibilities and combinations thereof. In an *admission control* problem a controller or decision-maker at the entrance to the queue decides which jobs to be admitted to the queue. Alternatively, in an *arrival control* problem the controller can control the arrival process itself by either increasing or decreasing the arrival rates. Finally, in a *service control* problem a controller selects the rate at which jobs are served.

In each of these problems we can identify the following components at each time $t$ during the period of time, say $[0, T]$, in which the system is controlled, where $T > 0$ is a finite number in a *finite-horizon* problem or $+\infty$ in an *infinite-horizon* problem. First, we need to specify the *state* $x(t)$ of the system at time $t \geq 0$, which is the information known to the controller. Here we will assume that $x(t)$ is the total number of jobs in the system at time $t$, that is to say, the jobs waiting in the queue and the job being served at time $t \geq 0$, if any. Therefore, the *state space*, that is, the set of all possible system states will be either $S = \{0, 1, \ldots\}$, the set of nonnegative integers, if the system has *infinite capacity*, or a finite set $S = \{0, 1, \ldots, C\}$ if the system has *finite capacity* $C$.

Second, we need to specify the set of *actions* available to the controller. This set will typically depend on the current state $i$ and will be denoted by $A(i)$. In turn, these sets will be seen as subsets of a larger set $A$, called the controller's *action space*. For instance, in an admission control problem the action space is $A = \{0, 1\}$, where "0" means "reject" arriving customers, and "1" means "admit" them. Hence, if the current state is $i$, we will usually have the following situation: if $i$ is "large", say $i > i^*$ for some integer $i^* > 0$, then the controller will reject new customers; if $i$ is "small", say $i < i_*$ for some integer $i_* > 0$, then the controller will accept new customers; otherwise, he/she will reject or admit them. Thus the corresponding action sets will be

$$A(i) := \begin{cases} \{0\} & \text{if } i > i^*, \\ \{0, 1\} & \text{if } i_* \leq i \leq i^*, \\ \{1\} & \text{if } i < i_*. \end{cases}$$

As another example, in a service control problem, the action space, that is, the set of service rates available to the controller, can be an interval $A = [\underline{a}, \bar{a}]$ with $0 \leq \underline{a} < \bar{a}$ or a finite set $A = \{a_1, \ldots, a_n\}$ with $0 \leq a_1 < \cdots < a_n$. And, again, if the current state of the system is $i$, the set $A(i)$ of available actions to the controller is a subset of $A$.

Third, we need to specify the random dynamic evolution of the system. Therefore, since we are dealing with continuous-time control problems, for every pair of states $i, j \in S$, we need to introduce the *transition rate* $q(j|i, a)$ from $i$ to $j$ when the controller takes an action $a \in A(i)$ at state $i$. As we will see in Chap. 2, under suitable assumptions on the matrix $[q(j|i, a)]$ of transition rates with $(i, j)$-element $q(j|i, a)$, we can determine the system's *transition probabilities*, as well as the control problem's *state process* $\{x(t), t \geq 0\}$ and *action process* $\{a(t), t \geq 0\}$.

Finally, to complete the specification of a *stochastic optimal control* problem, we need, of course, to introduce an *optimality criterion*. This requires to define the class of *policies* (also known as *control policies*) admissible to the controller, which is done in Sect. 2.2 in Chap. 2 below. In the meantime, however, we will interpret a (control) policy simply as a family $\pi = \{a(t), 0 \le t \le T\}$ of actions $a(t) \in A(x(t))$ for each $t \in [0, T]$. We also require a *reward* (or *cost*) function $r(x(t), a(t))$ to measure (or evaluate) the utility of taking action $a(t)$ at state $x(t)$, where $r(i, a)$ is a real-valued function defined on the set

$$K := \{(i, a) \mid i \in S,\ a \in A(i)\} \subseteq S \times A$$

of so-called feasible state–action pairs. The function $r$ is interpreted as a "net" reward rate, that is, the difference between revenues (or income) and costs for operating the system.

We are mainly—but not exclusively—interested in the following types of optimality criteria: for each policy $\pi = \{a(t)\}$ and each initial state $x(0) = i$,

- *The finite-horizon expected reward*

$$V_T(i, \pi) := E_i^\pi \left[ \int_0^T r(x(t), a(t))\, dt \right],$$

  where $T > 0$ is the so-called *optimization horizon*.
- *The infinite-horizon expected discounted reward*

$$V_\alpha(i, \pi) := E_i^\pi \left[ \int_0^\infty e^{-\alpha t} r(x(t), a(t))\, dt \right], \tag{1.1}$$

  where $\alpha > 0$ is a given *discount factor*.
- *The long-run expected average reward*

$$\bar{V}(i, \pi) := \liminf_{T\to\infty} \frac{1}{T} V_T(i, \pi) \tag{1.2}$$

  with $V_T(i, \pi)$ as above.

In (1.1) and (1.2), $E_i^\pi$ denotes the expectation operator when using the policy $\pi$, given that the initial state of the system is $x(0) = i$.

If $\Psi(i, \pi)$ denotes any of the expected reward functions above, and $\Pi$ stands for the class of admissible policies, then the corresponding *optimal control* problem is defined as follows. Determine a policy $\pi^* \in \Pi$ such that

$$\Psi(i, \pi^*) = \sup_{\pi\in\Pi} \Psi(i, \pi) =: \Psi^*(i) \quad \forall i \in S. \tag{1.3}$$

If such a policy $\pi^* \in \Pi$ exists, then $\pi^*$ is said to be an *optimal policy*, and the function $\Psi^*$ in (1.3) is called the *optimal reward function*, which is also called the (*optimal*) *value function* of the control problem.

For particular problems on the control of queueing systems, see, for instance, Kitaev and Rykov (1995) [98], Sennott (1999) [141], and Tadj and Choudhury (2005) [146].

To summarize, the problems we are concerned with can be expressed in terms of a so-called *control model*

$$\{S, (A(i) \subseteq A), q(j|i,a), r(i,a)\} \tag{1.4}$$

together with an optimality criterion, for instance, as in (1.1) and (1.2). The control problem is then defined as in the previous paragraph. The corresponding stochastic state process $\{x(t)\}$ is called a *continuous-time MDP*, which is also known as a *continuous-time controlled Markov chain.* Actually, by an abuse of terminology, the control problem itself is sometimes referred to a continuous-time MDP.

To further explain the elements in the control model (1.4), we next present some examples.

*Example 1.2* (Control of epidemic processes) The mathematical theory of epidemics was initiated by Kermack and McKendrick (1927) [100]. A detailed account of the literature up to the mid-1970s appeared in Bailey (1975) [7] and Wickwire (1977) [155] for deterministic and stochastic models, respectively.

In addition to their intrinsic interest, the mathematical models for the spread of infectious diseases are closely related to models for the control of populations in general (e.g., fisheries, predator-prey systems, pests) and for the control of rumors—see, for instance, Bartholomew (1973) [8]. Here we present just a couple of models that can be expressed as continuous-time MDPs.

As in (1.4), let $S$ be a finite or countably infinite set denoting the state space of the continuous-time MDP. In epidemic models $S$ is divided into two or more disjoint classes of individuals, such as the *susceptibles*, the *infectives*, the *immunized*, and so forth. The susceptibles are those individuals exposed to a contagious disease but are not yet infected, whereas the infectives are those individuals who carry the infection and can transmit it to the susceptibles. The immunized individuals are those protected against infection, but they can come from different sources. For instance, they are immunized because they were susceptibles subjected to some vaccination programs or because they were infectives subjected to some kind of medical treatment, and after recovery they were immune to the disease. There are some models, however, in which an infective may recover from the disease and then become again susceptible to reinfection—see, e.g. Lefèvre (1979) [103] and Wickwire (1977) [155]. Here, to keep matters simple, we consider a model introduced by Lefèvre (1981) [104], which is quite elementary, but at the same time it contains many of the ingredients that appear in the control of an epidemic process.

Lefèvre (1981) [104] considers a birth and death epidemic process in which the state space $S$ is a closed population of $N$ individuals divided into two disjoint classes, the susceptibles and the infectives. The state process is $\{x(t)\}$, where $x(t)$ denotes the number of infectives at time $t \geq 0$. When not controlled, this is assumed to be a birth and death process with transition rates

$$q(j|i) = \begin{cases} \lambda_i + \hat{\lambda}_i & \text{if } j = i+1 \ (0 \le i \le N-1), \\ -(\lambda_i + \hat{\lambda}_i + \mu_i) & \text{if } j = i, \\ \mu_i & \text{if } j = i-1, \\ 0 & \text{otherwise.} \end{cases}$$

The birth rates $\lambda_i$ and $\hat{\lambda}_i$ represent, respectively, the rates of propagation of infection through the susceptibles due to the presence of $i$ infectives and a source of infection from outside the population. Thus, it can be assumed that $\lambda_0 = 0$ and $\hat{\lambda}_0 > 0$. The death rate $\mu_i$ represents the recovery rate of the $i$ infectives. Lefèvre (1981) [104] mentions several particular expressions for the birth and death rates. These rates will depend of course on the control actions, which are defined as follows.

The action space is the two-dimensional set

$$A := [\underline{a}, \bar{a}] \times [\underline{b}, \bar{b}] \equiv A(i) \quad \forall i \in S.$$

If $a = (a_1, a_2) \in A$ is a control action, the first component, $a_1 \in [\underline{a}, \bar{a}]$, denotes the controllable level of a quarantine program whose purpose is to protect the susceptibles against the external source of infection. Thus $\hat{\lambda}_i(a_1)$ denotes the birth rate $\hat{\lambda}_i$ when a quarantine program of level $a_1$ is chosen. On the other hand, $a_2 \in [\underline{b}, \bar{b}]$ denotes the level of a medical treatment to the infectives to slow the propagation of infection (so that the birth rate $\lambda_i(a_2)$ is a decreasing function of $a_2$) and increase the recovery rate of the infectives (so that the death rate $\mu_i(a_2)$ is an increasing function of $a_2$).

Finally, the continuous-time MDP in Lefèvre (1981) [104] is to *minimize* an infinite-horizon expected discounted cost. Hence, the "reward" function $r(i,a)$ in (1.1) and (1.4) should be replaced with a "cost" function $c(i,a)$, which is taken to be of the form

$$c(i,a) := h_0(i) + h_1(a_1) + h_2(i, a_2) \quad \forall i \in S,\ a = (a_1, a_2) \in A,$$

where $h_0(i)$ is the social cost per unit time due to the existence of $i$ infectives, $h_1(a_1)$ is the cost for adopting a quarantine program of level $a_1$, and $h_2(i, a_2)$ denotes the cost for applying a medical care program of level $a_2$ to the $i$ infectives in the population. This completes the specification of the continuous-time MDP, in the sense that we now have all the components of the control model (1.4), with $c(i,a)$ in lieu of $r(i,a)$. (Furthermore, under mild continuity conditions, it is easily seen that this continuous-time MDP satisfies all our assumptions in Chap. 2.)

As mentioned earlier, epidemic models are related to several classes of "population" models, as well as models of consumer behavior. The interested reader may consult, for instance, Albright and Winston (1979) [1], Allen (2003) [2], Bartholomew (1973) [8], Iosifescu and Tautu (1973) [88], Mangel (1985) [114], Massy et al. (1970) [115], and Vidale and Wolfe (1957) [152].

*Example 1.3* (Controlled upwardly skip-free processes) The upwardly skip-free processes, also known as birth and death processes with catastrophes, belong to the category of *population processes* in Anderson (1991) [4] (Chap. 9, p. 292) with state space $S := \{0, 1, 2, \ldots\}$. Here we are interested in such processes with catastrophes of *two* sizes, and so the transition rates are as follows. For each $i \geq 2$,

$$q(j|i,a) := \begin{cases} \lambda i + a & \text{if } j = i+1, \\ -(\lambda i + \mu i + d(i) + a) & \text{if } j = i, \\ \mu i + d(i)\gamma_i^1 & \text{if } j = i-1, \\ d(i)\gamma_i^2 & \text{if } j = i-2, \\ 0 & \text{otherwise}, \end{cases} \tag{1.5}$$

where the constants $\lambda > 0$, $\mu > 0$, and $a \geq 0$ denote birth and death rates, and an *immigration* rate, respectively; $d(i)$ is a nonnegative number representing the rate at which the "catastrophes" occur when the process is in state $i \geq 2$. The numbers $\gamma_i^1$ and $\gamma_i^2$ are nonnegative and such that $\gamma_i^1 + \gamma_i^2 = 1$ for all $i \geq 2$, with $\gamma_i^k$ denoting the probability that the process makes a transition from $i$ to $i - k$ $(k = 1, 2)$, given that a catastrophe occurs when the process is in state $i \geq 2$. When the state $i$ is 0 or 1 (i.e., $i = 0$ or $i = 1$), it is natural to let

$$q(j|0,a) := \begin{cases} a & \text{if } j = 1, \\ -a & \text{if } j = 0, \\ 0 & \text{otherwise}, \end{cases}$$

$$q(j|1,a) := \begin{cases} \lambda + a & \text{if } j = 2, \\ -\lambda - \mu - a - d(1) & \text{if } j = 1, \\ \mu + d(1) & \text{if } j = 0, \\ 0 & \text{otherwise}, \end{cases}$$

where the constant $d(1)$ has a similar meaning as $d(i)$. On the other hand, we suppose that the immigration rate $a$ can be controlled and takes values in some interval $[0, \bar{a}]$ with $\bar{a} > 0$. Thus, we can interpret such an immigration rate $a$ as an action and then let the admissible action sets $A(i) := [0, \bar{a}]$ for $i \geq 0$. In addition, suppose that the benefit earned by each $a \in [0, \bar{a}]$ is represented by a real number $r(i, a)$ when the state of the process is $i$. This completes the specification of the continuous-time MDP in the sense of (1.4).

## 1.3 Summary of the Following Chapters

This book consists of twelve chapters and three appendices. In the present Chap. 1 we introduced some examples illustrating the class of problems we are interested in. The remainder of the book is organized as follows.

In Chap. 2, we formally introduce the concepts associated to a continuous-time MDP. Namely, the basic model of continuous-time MDPs and the concept of a Markov policy are stated in precise terms in Sect. 2.2. We also give, in Sect. 2.3, a precise definition of state and action processes in continuous-time MDPs, together with some fundamental properties of these two processes. Then, in Sect. 2.4, we introduce the basic optimality criteria that we are interested in.

Chapter 3 deals with *finite* models, that is, continuous-time MDPs with a finite number of states and actions. The long-run expected average reward (AR) criterion and the $n$-bias ($n = 0, 1, \ldots$) optimality criteria are introduced in Sect. 3.2. (Occasionally, we abbreviate expected average reward as EAR rather than expected AR.) For every $n = 0, 1, \ldots,$ formulas expressing the difference between the $n$-biases for any two stationary policies are provided in Sect. 3.3. These formulas are used in Sect. 3.4 to characterize $n$-bias optimal policies. The policy iteration and the linear programming algorithms for computing optimal policies for each of the $n$-bias criteria are given in Sects. 3.5 and 3.6, respectively.

Chapters 4 and 5 are on the denumerable state model with *nonnegative* costs. In Chap. 4, we study the expected $\alpha$-discounted cost criterion. After some technical preliminaries in Sects. 4.2 and 4.3, the corresponding discounted cost optimality equation and the existence of a discounted cost optimal stationary policy are established in Sects. 4.4 and 4.5, respectively. The convergence of value and policy iteration procedures is shown in Sects. 4.6 and 4.7, respectively. In Chap. 5, we turn to the expected average cost criterion. Using a so-called average cost minimum nonnegative solution approach provided in Sect. 5.3, we establish the average cost optimality inequality and the existence of average cost optimal policies in Sect. 5.4. We also obtain the average cost optimality equation in Sect. 5.5. Finally, in Sect. 5.6, we present an example showing that the existence of an average cost optimal policy does not imply the existence of a solution to the average cost optimality equation.

Chapter 6 concerns the expected $\alpha$-discounted reward criterion for continuous-time MDPs with *unbounded* transition rates and reward functions that may have neither upper nor lower bounds. We begin in Sect. 6.1 with introducing the "uniformization technique," widely used in continuous-time MDPs, and then, in Sect. 6.2, we establish the discounted reward optimality equation by using a value iteration technique. The existence of discounted reward optimal policies and a value iteration algorithm are given in Sects. 6.3 and 6.4, respectively. Furthermore, in Sect. 6.5, several examples are provided to illustrate the results of this chapter.

In Chap. 7, we study the EAR criterion for the same MDP model as in Chap. 6. After briefly introducing some basic facts in Sect. 7.2, we establish the average reward optimality equation and the existence of EAR optimal policies in Sect. 7.3. In Sect. 7.4, we provide a policy iteration algorithm for computing or at least approximating an EAR optimal policy. Finally, we illustrate the results in this chapter with several examples in Sect. 7.5.

Chapter 8 studies the pathwise average reward (PAR) criterion for the MDP model in Chaps. 6 and 7. First, in Sect. 8.1, we present an example showing the difference between the EAR and the PAR criteria. In Sects. 8.2 and 8.3, we introduce some basic facts that allow us to prove, in Sect. 8.4, the existence of PAR optimal policies. In Sect. 8.5, we provide policy and value iteration algorithms for computing a PAR optimal policy. We conclude with an example in Sect. 8.6.

Chapter 9 considers again the denumerable state continuous-time MDP model with unbounded transition rates and reward functions, but now with respect to the bias and other advanced criteria. Under suitable conditions, we obtain some interesting results on the existence and characterization of $n$-discount optimal policies in Sect. 9.2. These results are similar to those in Chap. 3, but of course in the context of this chapter. Finally, the existence of a *Blackwell optimal* policy is ensured in Sect. 9.3.

Chapter 10 deals with the minimization of the limiting average variance of a continuous-time MDP. Under some mild conditions, in Sect. 10.3, the limiting average variance of the reward process under any stationary policy is transformed into the usual long-run EAR problem with a new reward function. Then, the results developed in Chap. 7 are used in Sect. 10.4 to show the existence of a policy that minimizes the limiting average variance. An algorithm to compute a variance minimization optimal policy is also developed in Sect. 10.4. This chapter ends with an example in Sect. 10.5.

Chapters 11 and 12 are focused on continuous-time MDPs with *constraints*. In Chap. 11, we study the constrained model with discount criteria; that is, the optimality criterion is the discounted reward, and one constraint is imposed on a discounted cost. After some preliminaries in Sect. 11.2, in Sect. 11.3, we give conditions under which the existence of a discount constrained-optimal stationary policy is obtained by the Lagrange method. In Chap. 12, we turn to the expected average criteria. Using the same approach, we again establish the existence of an average constrained-optimal stationary policy. Examples are provided to illustrate the main results in these two chapters.

Except for Chaps. 1 and 2, each individual chapter is provided with a "Notes" section containing supplementary material and references.

Finally, some of the basic mathematical background is summarized in Appendices A, B, and C.

# Chapter 2
# Continuous-Time Markov Decision Processes

In this chapter we formally introduce the precise definitions of state and action processes in continuous-time MDPs, some fundamental properties of these two processes, and the basic optimality criteria that we are interested in.

## 2.1 Introduction

The main objective in this chapter is to formally introduce the continuous-time MDP that we are interested in. An informal discussion of the basic control model (1.4) and some application areas were presented in Sect. 1.2. In this chapter we begin, in Sect. 2.2, with a precise description of the control model (1.4) and the classes of policies—or more explicitly, "control policies"—that may be used. In Sect. 2.3, we show how to construct, for any given policy and any initial distribution, an underlying probability space on which one can define the state and the action processes of the corresponding continuous-time MDP. Finally, in Sect. 2.4, we introduce the basic optimality criteria we wish to analyze.

*Notation* Given a Borel space $X$ (i.e., a Borel subset of a complete separable metric space), its Borel $\sigma$-algebra is denoted by $\mathcal{B}(X)$. We denote by $P(X)$ the set of all probability measures on the measurable space $(X, \mathcal{B}(X))$, and by $B(X)$ the Banach space of real-valued bounded measurable functions on $X$. By convention, when referring to sets or functions, "measurable" means "Borel-measurable." In particular, when $X$ is a denumerable set (with the discrete topology), $\mathcal{B}(X)$ consists of all subsets of $X$, and thus any real-valued function on $X$ is measurable. Therefore, when dealing with denumerable spaces, the measurability of sets or functions will not be explicitly mentioned.

For any measurable function $w \geq 1$ on $X$, we define the $w$-weighted supremum norm $\|\cdot\|_w$ of a real-valued measurable function $u$ on $X$ by

$$\|u\|_w := \sup_{x \in X} \{ w(x)^{-1} |u(x)| \},$$

and the Banach space $B_w(X) := \{u : \|u\|_w < \infty\}$.

X. Guo, O. Hernández-Lerma, *Continuous-Time Markov Decision Processes*,
Stochastic Modelling and Applied Probability 62,
DOI 10.1007/978-3-642-02547-1_2, 

Obviously, with $w \equiv 1$, the spaces $B(X)$ and $B_1(X)$ are the same, but in general $B_w(X) \supseteq B(X)$ for any $w \geq 1$.

## 2.2 The Control Model

The control model associated with the continuous-time MDP that we are concerned with is the five-tuple

$$\big\{S, A, \big(A(i), i \in S\big), q(j|i,a), r(i,a)\big\} \tag{2.1}$$

with the following components:

(a) A denumerable set $S$, called the *state space*, which is the set of all states of the system under observation.
(b) A Borel space $A$, called the *action space*.
(c) A family $(A(i), i \in S)$ of nonempty measurable subsets $A(i)$ of $A$, where $A(i)$ denotes the set of actions or decisions available to the controller when the state of the system is $i \in S$. Let

$$K := \big\{(i,a) \mid i \in S, a \in A(i)\big\} \tag{2.2}$$

be the set of all feasible state–action pairs.
(d) The transition rates $q(j|i,a)$, which satisfy $q(j|i,a) \geq 0$ for all $(i,a) \in K$ and $j \neq i$. Moreover, we assume that the transition rates $q(j|i,a)$ are *conservative*, i.e.,

$$\sum_{j \in S} q(j|i,a) = 0 \quad \forall (i,a) \in K, \tag{2.3}$$

and *stable*, which means that

$$q^*(i) := \sup_{a \in A(i)} q_i(a) < \infty \quad \forall i \in S, \tag{2.4}$$

where $q_i(a) := -q(i|i,a) \geq 0$ for all $(i,a) \in K$. In addition, $q(j|i,a)$ is measurable in $a \in A(i)$ for each fixed $i, j \in S$.
(e) A measurable real-valued function $r(i,a)$ on $K$, called the *reward function*, which is assumed to be measurable in $a \in A(i)$ for each fixed $i \in S$. (As $r(i,a)$ is allowed to take positive and negative values, it can be interpreted as a *cost function* rather than a "reward" function.)

This completes the description of the control model (2.1). We will next explain how the *controller* or *decision-maker* chooses his/her actions. To this end, we first introduce the following concepts.

**Definition 2.1** (Randomized Markov policies) A *randomized Markov policy* is a real-valued function $\pi_t(C|i)$ that satisfies the following conditions:

(i) For all $i \in S$ and $C \in \mathcal{B}(A(i))$, the mapping $t \mapsto \pi_t(C|i)$ is measurable on $[0, \infty)$.
(ii) For all $i \in S$ and $t \geq 0$, $C \mapsto \pi_t(C|i)$ is a probability measure on $\mathcal{B}(A)$ and satisfies that $\pi_t(A(i)|i) = 1$, where $\pi_t(C|i)$ denotes the probability that an action in $C$ is taken when the system's state is $i$ at time $t$.

A randomized Markov policy $\pi_t(C|i)$ is said to be *randomized stationary* if $\pi_t(C|i) \equiv \pi(C|i)$ is independent of $t$.

A *deterministic stationary policy* is a function $f : S \to A$ such that $f(i)$ is in $A(i)$ for all $i \in S$. A deterministic stationary policy is simply referred to as a *stationary policy*.

Let $\Pi$ be the set of all randomized Markov policies, $\Pi^s$ the set of all randomized stationary policies, and $F$ the set of all stationary policies. Note that a stationary policy $f \in F$ can be viewed as a function $\pi_t(C|i) \in \Pi$ such that, for all $t \geq 0$ and $i \in S$, $\pi_t(\cdot|i)$ is the Dirac measure at $f(i)$. Thus, $F \subset \Pi^s \subset \Pi$.

Sometimes, a randomized Markov policy $\pi_t(C|i)$ will be simply referred to as a *Markov policy* and will be denoted as $(\pi_t)$ (or $\pi$ for simplicity). Similarly, a randomized stationary policy $\pi(C|i)$ in $\Pi^s$ will also be written as $\pi$. The subscript "$t$" in $\pi_t$ indicates the possible dependence of the policy on the time variable $t$. For notational ease, however, "$t$" will be omitted if there is no possibility of confusion.

*Evolution of the Control System* We now give a somewhat informal description of how the control system evolves. Suppose that the system is at state $i \in S$ at time $t \geq 0$ and that the decision-maker takes an action $a$ from the action set $A(i)$. Then the following "happen" on the time interval $[t, t + dt]$:

(i) The decision-maker receives an infinitesimal reward $r(i, a)dt$.
(ii) A transition from state $i$ to state $j$ (with $j \neq i$) occurs with probability $q(j|i, a)\,dt + o(dt)$, or the system remains at state $i$ with probability $1 + q(i|i, a)\,dt + o(dt)$. In the former case, the sojourn time at the state $i$ is exponentially distributed with parameter $q_i(a)$.

The aim of the decision-maker is to optimize (i.e., maximize or minimize) a given performance criterion, defined on a set of policies, which measures or evaluates in some sense the system's response to the different policies. We are particularly interested in the performance criteria defined by (2.19) and (2.21) below, but we will also consider other criteria.

For every given policy, we wish to guarantee the existence of an associated transition (probability) function on $S$ (equivalently, a Markov process with a unique transition function describing the random evolution of the system). To see how this is done, we first introduce some notation and terminology.

For each $(\pi_t) \in \Pi$, the associated transition rates are defined as

$$q(j|i, \pi_t) := \int_{A(i)} q(j|i, a)\pi_t(da|i) \quad \text{for } i, j \in S \text{ and } t \geq 0. \tag{2.5}$$

In the *stationary* case, that is, $\pi \in \Pi^s$ or $f \in F$, we write $q(j|i,\pi_t)$ as $q(j|i,\pi)$ or $q(j|i,f)$, respectively.

Let $Q(\pi_t) := [q(j|i,\pi_t)]$ be the associated matrix of transition rates with the $(i,j)$th element $q(j|i,\pi_t)$.

Note that, by (2.3) and (2.4), the integral (2.5) is well defined and bounded by $q^*(i)$, because

$$q(j|i,a) \le \sum_{j\neq i} q(j|i,a) = q_i(a) \le q^*(i) \quad \forall i,j \in S \text{ and } a \in A(i).$$

Moreover, $q(j|i,\pi_t)$ is measurable in $t \ge 0$ (for any fixed $i,j \in S$), and by (2.4) and (2.5), these transition rates are also stable and conservative, that is,

$$q_i(\pi_t) := -q(i|i,\pi_t) = \sum_{j\neq i} q(j|i,\pi_t) < +\infty \quad \forall i \in S \text{ and } t \ge 0.$$

Then, as a consequence of Proposition C.4 (in Appendix C), for each $\pi \in \Pi$, there exists a *minimal* transition function, denoted here by $p_\pi(s,i,t,j)$, with transition rates $q(j|i,\pi_t)$. More explicitly, for all $i,j \in S$ and $t > 0$, and a.e. $s \in (0,t)$, the Kolmogorov forward and backward equations

$$\begin{aligned}
\frac{\partial p_\pi(s,i,t,j)}{\partial t} &= \sum_{k\in S} p_\pi(s,i,t,k) q(j|k,\pi_t), \\
\frac{\partial p_\pi(s,i,t,j)}{\partial s} &= -\sum_{k\in S} q(k|i,\pi_s) p_\pi(s,k,t,j), \\
\frac{\partial p_\pi(0,i,t,j)}{\partial t} &= \sum_{k\in S} p_\pi(0,i,t,k) q(j|k,\pi) \\
&= \sum_{k\in S} q(k|i,\pi) p_\pi(0,k,t,j) \quad (\text{when } \pi \in \Pi_s)
\end{aligned} \tag{2.6}$$

hold, and, moreover,

$$\begin{gathered}
p_\pi(s,i,t,j) = \sum_{k\in S} p_\pi(s,i,v,k) p_\pi(v,k,t,j) \quad \forall v \in [s,t], \\
\lim_{\Delta t\downarrow 0} \frac{p_\pi(t,i,t+\Delta t,j) - \delta_{ij}}{\Delta t} = q(j|i,\pi_t), \qquad p_\pi(s,i,s,j) = \delta_{ij},
\end{gathered} \tag{2.7}$$

where $\delta_{ij}$ stands for Kronecker's delta, and "a.e." means "almost everywhere" with respect to the Lebesgue measure on $[0,\infty)$.

The transition function $p_\pi(s,i,t,j)$, however, might not be regular, where *regularity* is understood as

$$\sum_{j\in S} p_\pi(s,i,t,j) = 1 \quad \forall i \in S \text{ and } t \ge s \ge 0. \tag{2.8}$$

To guarantee the regularity condition (2.8) and thus the existence of an associated Markov process, we impose the following so-called *drift condition.*

**Assumption 2.2** There exist a function $w \geq 1$ on $S$ and constants $c_0 \neq 0,\ b_0 \geq 0$, and $L_0 > 0$ such that

(a) $\sum_{j\in S} w(j)q(j|i,a) \leq c_0 w(i) + b_0$ for all $(i,a) \in K$.
(b) $q^*(i) \leq L_0 w(i)$ for all $i \in S$, with $q^*(i)$ as in (2.4).

Since Assumption 2.2(b) is used *only* to guarantee the regularity of the minimum transition function, it is not required when either

- The transition rates are bounded (i.e., $\sup_{i\in S} q^*(i) < \infty$)

or the following conditions hold:

- The function $w$ is nondecreasing, and there exists a sequence $\{S_m, m \geq 1\}$ of subsets of $S$ such that (i) $S_m \uparrow S$ and $\sup_{i\in S_m} q^*(i) < \infty$ for each $m \geq 1$, and (ii) $\lim_{m\to\infty}[\inf_{j\notin S_m} w(j)] = +\infty$, where we suppose that $S = \{0, 1, 2, \ldots\}$, the set of nonnegative integers; see Çinlar (1975) [31], Chen (2004) [28], Guo and Hernández-Lerma (2003) [54], Hou and Guo (1988) [83], and Hou et al. (1994) [85].

In what follows, we will suppose that Assumption 2.2 holds and show by means of examples how it can be verified.

## 2.3 Continuous-Time Markov Decision Processes

*The material in this section is quite technical, and strictly speaking, it is not necessary to understand most of the ideas in this book. Hence the reader may skip this section on a first reading and come back to it as required.*

In this section we first construct a probability space associated with any given policy and any initial distribution, and then prove the existence of the associated state and action processes. This result is based on suitable versions of Kolmogorov's extension theorem, as, for instance, in Blumenthal and Getoor (1968) [15] and Rao (1995) [133].

Fix an arbitrary Markov policy $\pi = (\pi_t) \in \Pi$. By Assumption 2.2 and Proposition C.9, the corresponding transition function $p_\pi(s, i, t, j)$ is regular.

For any initial time $s \geq 0$, let

$$G^s := \big\{(t_0, t_1, \ldots, t_n) |\ s \leq t_0 < t_1 < \cdots < t_n < \infty \text{ for all } n \geq 0\big\},$$

directed by inclusion, i.e., $\tau := (t_0, t_1, \ldots, t_n) \ll \tau' \in G^s$ means that there exist some positive numbers $t_m > t_{m-1} > \cdots > t_n$ with $m \geq n$ such that $\tau' = (t_0, t_1, \ldots, t_n, \ldots, t_m)$.

Let $E := S \times A$, and let $\nu$ be any given probability distribution on $S$. Then, for each $\tau := (t_0, t_1, \ldots, t_n) \in G^s$, we can define a product probability measure $P_\tau^{\nu,\pi}$ on

$(E^{n+1}, \mathcal{B}(E^{n+1}))$ such that: for each $B = \{j_{t_0}\} \times C_{t_0} \times \{j_{t_1}\} \times C_{t_1} \times \cdots \times \{j_{t_n}\} \times C_{t_n}$ with $j_{t_k} \in S$ and $C_{t_k} \in \mathcal{B}(A)$ $(k = 0, 1, \ldots, n)$,

$$P_\tau^{\nu,\pi}(B) := \nu(j_{t_0})\pi_{t_0}(C_{t_0}|j_{t_0})p_\pi(t_0, j_{t_0}, t_1, j_{t_1})\pi_{t_1}(C_{t_1}|j_{t_1}) \cdots p_\pi(t_{n-1}, j_{t_{n-1}}, t_n, j_{t_n})\pi_{t_n}(C_{t_n}|j_{t_n}). \tag{2.9}$$

Let $T := [0, \infty)$, and let $(\Omega, \mathcal{B}(\Omega))$ be the (canonical) product measurable space with $\Omega := E^T$ being the set of all maps from $T$ to $E$. Let $\eta_\tau$ be the coordinate projection on $\Omega$, i.e.,

$$\eta_\tau(z) := z_\tau \quad \text{for each } z = \big(z(t), t \geq 0\big) \in \Omega, \tag{2.10}$$

where $z_\tau := (z(t_0), z(t_1), \ldots, z(t_n))$. We then have the following.

**Theorem 2.3** *Suppose that Assumption 2.2 holds. Then for each $\pi = (\pi_t) \in \Pi$ and each given (initial) distribution $\nu$ on $S$ at time $s \geq 0$, on the measurable space $(\Omega, \mathcal{B}(\Omega))$ above there exists a unique probability measure $P_{\nu,s}^\pi$ (depending on $s, \nu, \pi$) such that for each $t \geq s$ and $\tau := (t_0, t_1, \ldots, t_n)$,*

$$P_{\nu,s}^\pi\big(W_{s,i}(C)\big) = \nu(i)\pi_s(C|i), \tag{2.11}$$

$$P_{\nu,s}^\pi\big(W_{s,i}^{t,j}(C)\big) = \nu(i)p_\pi(s, i, t, j)\pi_t(C|j), \tag{2.12}$$

$$P_{\nu,s}^\pi(B_\tau) = \sum_{i \in S} \nu(i)p_\pi(s, i, t_0, j_{t_0})\pi_{t_0}(C_{t_0}|j_{t_0})p_\pi(t_0, j_{t_0}, t_1, j_{t_1})\pi_{t_1}(C_{t_1}|j_{t_1}) \cdots p_\pi(t_{n-1}, j_{t_{n-1}}, t_n, j_{t_n})\pi_{t_n}(C_{t_n}|j_{t_n}), \tag{2.13}$$

*where*

$$W_{s,i}(C) := \big\{z \in \Omega : z(s) \in \{i\} \times C\big\},$$

$$W_{s,i}^{t,j}(C) := \big\{z \in \Omega : \big(z(s), z(t)\big) \in \{i\} \times A \times \{j\} \times C\big\},$$

$$B_\tau := \big\{z \in \Omega : \eta_\tau(z) \in \{j_{t_0}\} \times C_{t_0} \times \cdots \times \{j_{t_n}\} \times C_{t_n}\big\},$$

*with $i, j, j_{t_k} \in S$, $C, C_{t_k} \in \mathcal{B}(A)$ $(k = 0, \ldots, n)$, and $s \leq t_0 < \cdots < t_n < \infty$.*

*Proof* Because $S$ and $A$ are Borel spaces (that is, Borel subsets of some complete separable metric spaces), the existence of a unique probability measure $P_\pi^{\nu,s}$ as in (2.11)–(2.13) follows from Rao (1995) [133, Theorem 4, p. 20] or Blumenthal and Getoor (1968) [15, Theorem 2.11, p. 17]. In fact, the results in (2.12) and (2.11) follow from (2.9), (2.10), and Rao (1995) [133, Proposition 1, p. 18]. □

We next use Theorem 2.3 to introduce the state and action processes.

**Definition 2.4** (State and action processes) For all $z = (z(t), t \geq 0) \in \Omega$ and $t \geq 0$, let $z(t) := (z^0(t), z^1(t)) \in E = S \times A$. In addition, define the random variables

$x(t)(z) := z^0(t)$ and $a(t)(z) := z^1(t)$ on the probability space $(\Omega, \mathcal{B}(\Omega), P^{\pi}_{\nu,s})$. The stochastic processes $\{x(t),\ t \geq 0\}$ and $\{a(t),\ t \geq 0\}$ are called the state and action processes, respectively, when using the policy $\pi \in \Pi$, and given the initial distribution $\nu$ at time $s \geq 0$.

The expectation operator with respect to $P^{\pi}_{\nu,s}$ is denoted by $E^{\pi}_{\nu,s}$. If $\nu$ is concentrated at the "initial state" $i$ (i.e., $\nu(\{i\}) = 1$, and $\nu(j) = 0$ for all $j \neq i$) at time $s$, then we write $E^{\pi}_{\nu,s}$ and $P^{\pi}_{\nu,s}$ as $E^{\pi}_{i,s}$ and $P^{\pi}_{i,s}$, respectively. In particular, when $s = 0$, we will further write $E^{\pi}_{i,0}$ and $P^{\pi}_{i,0}$ as $E^{\pi}_{i}$ and $P^{\pi}_{i}$, respectively.

More notation: for any real-valued measurable function $u(i, a)$ on $K$ and $(\pi_t) \in \Pi$, let

$$u(i, \pi_t) := \int_{A(i)} u(i, a)\pi_t(da|i) \quad \text{for all } i \in A(i) \text{ and } t \geq 0, \tag{2.14}$$

whenever the integral is well defined. If $(\pi_t) \in \Pi$ is replaced with $f \in F$, then (2.14) reduces to $u(i, f) := u(i, f(i))$.

With the notation in the last two paragraphs we can state the following facts.

**Theorem 2.5** *Suppose that Assumption 2.2 holds. Then for all $i, j \in S, \pi = (\pi_t) \in \Pi$, $C \in \mathcal{B}(A)$, and $t \geq s$:*

(a) $P^{\pi}_{i,s}(x(t) = j) = p_{\pi}(s, i, t, j)$, *and* $\sum_{k \in S} p_{\pi}(s, i, t, k) = 1$.

(b) $P^{\pi}_{i,s}(a(t) \in C | x(t) = j) = \pi_t(C|j)$.

(c) *Let $r(i, a)$ and $w(i)$ be as in* (2.1) *and Assumption* 2.2, *respectively, and suppose that, for some constant $M \geq 0$,*

$$\sup_{a \in A(i)} \left| r(i, a) \right| \leq M w(i) \quad \textit{for all } i \in S, \tag{2.15}$$

*then the expected reward $E^{\pi}_{i,s} r(x(t), a(t))$ is finite and measurable in $t \geq s$.*

*Proof* Parts (a) and (b) follow from (2.12) and Definition 2.4. On the other hand, by Definition 2.1 and (2.14), the reward $r(i, \pi_t)$ is measurable in $t \geq 0$ for every fixed $i \in S$. Moreover, (2.15) ensures that $E^{\pi}_{i,s} r(x(t), a(t))$ is finite-valued. In fact, Assumption 2.2 gives that

$$\sum_{j \in S} q(j|i, \pi_t) w(j) \leq |c_0| w(i) + b_0 \leq \big(|c_0| + b_0\big) w(i) \quad \forall i \in S,$$

where $b_0$ and $c_0$ are the constants in Assumption 2.2. Thus, by Proposition C.8 we have

$$\left| E^{\pi}_{i,s} r\big(x(t), a(t)\big) \right| \leq M w(i) e^{(|c_0| + b_0)(t - s)} < \infty.$$

Finally, by (2.12) and (2.14) we have

$$E^{\pi}_{i,s} r\big(x(t), a(t)\big) = \sum_{j \in S} r(j, \pi_t) p_{\pi}(s, i, t, j),$$

which is Borel measurable in $t$, because $p_\pi(s,i,t,j)$ is continuous in $t \geq s$; see, for instance, Proposition B.2(e) or Feller (1940) [42]. Hence part (c) follows. □

As a consequence of Theorem 2.5, we have the following important special case.

**Theorem 2.6** *Suppose that Assumption* 2.2 *holds. Then, for all* $\nu \in P(S), \pi = (\pi_t) \in \Pi$, *and any measurable function* $u$ *on* $K$ *that satisfies* (2.15) *in lieu of* $r$,

(a) $E^\pi_{\nu,0} u(x(t),a(t)) = \sum_{i,j\in S} u(j,\pi_t) p_\pi(0,i,t,j)\nu(i)$ *for all* $t \geq 0$.
(b) $E^\pi_{\nu,0} u(x(t),a(t)) = E^\pi_{\nu,0} u(x(t),\pi_t)$ *is measurable in* $t \geq 0$.

Let us now consider an arbitrary initial distribution $\nu$ on $S$. By Theorem 2.3 and Definition 2.4, for all $t \geq s$, $j \in S$, and $C \in \mathcal{B}(A)$, we have

$$P^\pi_{\nu,0}\big(x(0)=i\big) = \nu(i), \tag{2.16}$$

$$P^\pi_{\nu,0}\big(x(t)=j|x(s)=i\big) = p_\pi(s,i,t,j), \tag{2.17}$$

$$P^\pi_{\nu,0}\big(a(t)\in C|x(t)=i\big) = \pi_t(C|i). \tag{2.18}$$

Moreover, the state process $\{x(t)\}$ is a Markov process with transition (probability) function $p_\pi(s,i,t,j)$, that is, for all $0 \leq t_1 < \cdots < t_n < s < t$, $i, j, i_k \in S$ $(k = 1,\ldots,n)$, and $n \geq 1$,

$$\begin{aligned} &P^\pi_{\nu,0}\big(x(t)=j|x(t_1)=i_1,\ldots,x(t_n)=i_n,x(s)=i\big) \\ &\quad = P^\pi_{\nu,0}\big(x(t)=j|x(s)=i\big) = p_\pi(s,i,t,j). \end{aligned}$$

Furthermore, it is well known (e.g., Anderson (1991) [4], Chung (1970) [30], Chung (1967) [29], Gihman and Skorohod (1979) [47], Williams (1979) [157]) that for a given transition function $p_\pi(s,i,t,j)$, there exists a right process that has $p_\pi(s,i,t,j)$ as its transition function. (As noted in the Appendix B.1, a stochastic process $\{x(t),\ t\geq 0\}$ is called a *right process* if each sample path is right continuous and has a finite left limit at every $t$.) Thus, without loss of generality we can suppose that the state process $\{x(t)\}$ is a right process. Of course, $\{x(t)\}$ depends on the particular policy $\pi$ being used and also on the given initial distribution $\nu$. Hence, strictly speaking, we should write, for instance, $x(t)^{\pi,\nu}$ rather than just $x(t)$. However, we shall keep the simpler notation $x(t)$ since it will always be clear from the context what particular $\pi$ and $\nu$ are being used.

## 2.4 Basic Optimality Criteria

After the preliminaries above, we now define the two main optimality criteria dealt with in this book. These are the expected *discounted reward* criterion for infinite-horizon models, and the long-run expected *average reward* criterion also for infinite-horizon problems. Other optimality criteria are variations of these two criteria.

**Definition 2.7** (The expected discounted reward criterion) The expected discounted reward of a policy $\pi = (\pi_t) \in \Pi$, when the initial state is $i \in S$, is defined as

$$V_\alpha(i,\pi) := E_i^\pi\left[\int_0^\infty e^{-\alpha t} r\big(x(t),\pi_t\big)\,dt\right], \tag{2.19}$$

where $r(i,\pi_t)$ is given by (2.14) with $u(i,a) = r(i,a)$, i.e.,

$$r(i,\pi_t) := \int_{A(i)} r(i,a)\pi_t(da|i). \tag{2.20}$$

The corresponding *optimal discounted reward function* (or optimal value function of the discount optimality problem) is

$$V_\alpha^*(i) := \sup_{\pi\in\Pi} V_\alpha(i,\pi)$$

for every initial state $i \in S$.

Given $\varepsilon \geq 0$, *discounted reward $\varepsilon$-optimality* of a Markov policy $\pi^* \in \Pi$ means that

$$V_\alpha\big(i,\pi^*\big) \geq V_\alpha^*(i) - \varepsilon \quad \forall i \in S,$$

whereas $\pi^* \in \Pi$ is said to be *discounted reward optimal* (or $\alpha$-discounted reward optimal) if the latter inequality holds for $\varepsilon = 0$, i.e.,

$$V_\alpha\big(i,\pi^*\big) = V_\alpha^*(i) \quad \forall i \in S.$$

**Definition 2.8** (The expected average reward criterion) The long-run expected average reward (AR), also referred to as the *gain* of a policy $\pi \in \Pi$, is given by

$$\bar{V}(i,\pi) := \liminf_{T\to\infty} \frac{1}{T} V_T(i,\pi), \tag{2.21}$$

where

$$V_T(i,\pi) := E_i^\pi\left[\int_0^T r\big(x(t),a(t)\big)\,dt\right] \tag{2.22}$$

denotes the total expected reward over the time horizon $[0,T]$ of the policy $\pi \in \Pi$ when the initial state of the system is $i \in S$.

The corresponding optimal AR function (i.e., the optimal value function of the AR optimality problem) is

$$\bar{V}^*(i) := \sup_{\pi\in\Pi} \bar{V}(i,\pi) \quad \text{for } i \in S. \tag{2.23}$$

Of course, AR $\varepsilon$-optimality and AR optimality are given the obvious definitions, that is, $\pi^* \in \Pi$ is AR $\varepsilon$-optimal or AR optimal (also referred to as *gain optimal*) if

$$\bar{V}\big(i, \pi^*\big) \geq \bar{V}^*(i) - \varepsilon \quad \forall i \in S$$

or

$$\bar{V}\big(i, \pi^*\big) = \bar{V}^*(i) \quad \forall i \in S,$$

respectively.

In this book, we do not analyze finite-horizon continuous-time MDPs (the interested reader is nonetheless referred to Hernández-Lerma (1994) [70] or Prieto-Rumeau and Hernández-Lerma (2006b) [126]).

*The Dynamic Programming Approach* The solution of a continuous-time MDP typically consists of two main steps: First, we must determine the optimal value function (for the corresponding optimality criterion), and then we must find optimal (or $\varepsilon$-optimal) policies. In this book, we mainly follow the dynamic programming approach (introduced by Bellman (1957) [10]), which consists in characterizing the MDP's optimal value function, say $v^*$, as a solution of a certain equation called the *optimality equation* (also known as *Bellman's equation, Hamilton–Jacobi–Bellman equation*, or the *dynamic programming equation*). This equation usually takes the form

$$v^*(i) = \sup_{a \subset A(i)} \big(\mathbf{H}v^*\big)(i, a) \quad \forall i \in S \tag{2.24}$$

for a suitably defined operator $\mathbf{H}$. The next step consists in proving that a deterministic stationary policy $f^* \in \mathbf{F}$ attaining the supremum in (2.24), i.e.,

$$v^*(i) = \big(\mathbf{H}v^*\big)\big(i, f^*(i)\big) \quad \forall i \in S,$$

is optimal.

This is the approach that we will follow in the subsequent chapters, with the corresponding particularities to account for each specific model.

Among other possible approaches, a popular one is to use linear programming techniques, in which the MDP is transformed into an equivalent linear programming problem; see, for instance, Hernández-Lerma and Lasserre (1996) [74], and Puterman (1994) [129].

# Chapter 3
# Average Optimality for Finite Models

In this chapter we will focus on a *finite* control model, by which we mean the control model (2.1) with finite state and finite action spaces. Our main goal here is to study the expected average reward criterion. Results for the discounted reward case will be presented in Sect. 4.8 of Chap. 4.

## 3.1 Introduction

The long-run expected average reward is a common optimality criterion for MDPs, in particular, for finite control models. However, since the expected average reward concentrates on the asymptotic or limiting behavior of a system, it ignores transient performance. Hence, we may have, for instance, two policies that yield the same long-run expected average reward but quite different finite-horizon rewards. Thus, the expected average reward is *underselective* because it does not distinguish between two such policies. To address this feature or, say, "deficiency" of the long-run average reward, one should propose and investigate more selective optimality criteria. These criteria include, for instance, bias optimality, $n$-discount optimality, and Blackwell optimality, some of which are studied here and also in later chapters.

In this chapter we analyze an interesting *self-contained* approach recently proposed by Guo, Song, and Zhang (2008) [66] and Zhang and Cao (2009) [166] for finite models, based on $n$-bias ($n = -1, 0, 1, \ldots$). We proceed as follows:

(i) We first provide difference formulas for the $n$-biases of two stationary policies ($n \geq -1$); see Theorem 3.6 below.
(ii) Assuming the existence of a stationary policy $f^*$ that satisfies the bias optimality conditions, we prove that such a policy $f^*$ is $n$-bias optimal for all $n \geq -1$. To this end, we use the difference formulas for $n$-biases and an interesting observation on the *canonical* form of the transition function for a continuous-time Markov chain; see (3.17) and (3.18).

X. Guo, O. Hernández-Lerma, *Continuous-Time Markov Decision Processes*,
Stochastic Modelling and Applied Probability 62,
DOI 10.1007/978-3-642-02547-1_3, 

(iii) We show the existence of a policy $f^*$ as in (ii) by using the so-called $n$-bias policy iteration algorithms. Such a policy can be obtained in a *finite* number of policy iterations. The convergence of the $n$-bias policy iteration algorithms follows from the difference formulas for $n$-biases and (3.18) again.

The remainder of the chapter is organized as follows. In Sect. 3.2, we define the $n$-bias optimality criteria that we are concerned with. In Sect. 3.3, we provide difference formulas for the $n$-biases of two stationary policies. Some optimality characterizations of an $n$-bias optimal policy are presented in Sect. 3.4. In Sect. 3.5, we introduce the $n$-bias policy iteration algorithms to prove the existence of an $n$-bias optimal policy and to compute an $n$-bias optimal policy. The chapter concludes with Sect. 3.6, where we present a linear programming approach to ergodic and to multichain models.

## 3.2 $n$-bias Optimality Criteria

In this section we introduce the $n$-bias optimality criteria.

In what follows, we suppose that all operations on matrices or vectors, such as limits of sequences, are component-wise. Without risk of confusion, we denote by "0" both the matrix and the vector with all zero components, by $e$ the vector with all components 1, and by $I$ the identity matrix. Also, for each vector $u$, we denote by $u(i)$ the $i$th component of $u$. For each stationary policy $f \in F$, Proposition C.5 implies that the corresponding transition function $p_f(s,i,t,j)$ is *stationary*, that is, $p_f(s,i,t,j) = p_f(0,i,t-s,j)$ for all $i, j \in S$ and $t \geq s \geq 0$. We denote by $P(t,f)$ the stationary *transition (probability) matrix* with $(i,j)$th element $p_f(i,t,j) := p_f(0,i,t,j)$, and by $Q(f)$ the *transition rate matrix* $[q(j|i,f)]$.

Using the above matrix notation, the Kolmogorov equations (2.6) can be rewritten as

$$\frac{d}{dt}P(t,f) = Q(f)P(t,f) = P(t,f)Q(f) \quad \forall t \geq 0, \text{ and } P(0,f) = I. \tag{3.1}$$

Moreover (by Proposition B.10), the limits

$$P^*(f) := \lim_{t\to\infty} P(t,f) \tag{3.2}$$

exist. For each $f \in F$, let $r(f)$ be the vector with $i$th component $r(i,f)$ for each $i \in S$. Then, by Definition 2.8 (with $\pi = f \in F$), the long-run expected average reward of policy $f$ can be seen as the vector

$$\bar{V}(f) := \liminf_{T\to\infty} \frac{\int_0^T P(t,f)r(f)\,dt}{T} \tag{3.3}$$

with $i$th component $\bar{V}(i,f)$. We also have the following lemma.

**Lemma 3.1** *For all $f \in F$ and $t \geq 0$, the following assertions hold*:

(a) $P(t, f)P^*(f) = P^*(f)P(t, f) = P^*(f)P^*(f) = P^*(f)$, *and* $P^*(f)e = e$.
(b) $Q(f)P^*(f) = P^*(f)Q(f) = 0$, *and* $\bar{V}(f) = P^*(f)r(f)$.
(c) $(P(t, f) - P^*(f))^n = P(nt, f) - P^*(f)$ *for all integers* $n \geq 1$.
(d) $\int_0^\infty \|P(t, f) - P^*(f)\| \, dt < \infty$, *where* $\|D\| := \sup_{i\in S} \sum_{j\in S} |d_{ij}|$ *for any matrix* $D = [d_{ij}]_{|S|\times|S|}$.

*Proof* (a) By the Chapman–Kolmogorov equation (Proposition B.2), $P(t + s, f) = P(t, f)P(s, f) = P(s, f)P(t, f)$. Hence, letting $s \to \infty$, from (3.2) we obtain (a).

(b) The first statement follows from (3.1), (3.2), and Proposition B.10(f). The second statement follows from (3.2) and (3.3).

(c) By the Chapman–Kolmogorov equation, $P(t, f)^n = P(nt, f)$. This fact and (a), together with a straightforward induction argument, give (c).

(d) By (3.2) and the finiteness of the state space $S$, we can pick $t_0 > 0$ such that $\|P(t_0, f) - P^*(f)\| < \frac{1}{2}$. Noting that $\|P(s, f) - P^*(f)\| \leq 2$ for all $s \geq 0$, by (a) and (c) we have

$$
\begin{aligned}
\int_0^\infty \big\|P(s, f) - P^*(f)\big\| \, ds &= \sum_{m=0}^\infty \int_{mt_0}^{(m+1)t_0} \big\|P(s, f) - P^*(f)\big\| \, ds \\
&= \sum_{m=0}^\infty \int_0^{t_0} \big\|P(mt_0 + s, f) - P^*(f)\big\| \, ds \\
&\leq 2\, t_0 + \sum_{m=1}^\infty \int_0^{t_0} \big\|\big(P(t_0, f) - P^*(f)\big)^m P(s, f)\big\| \, ds \\
&\leq 2\, t_0 + \int_0^{t_0} \big\|P(s, f)\big\| \, ds \sum_{m=1}^\infty \frac{1}{2^m}
\end{aligned}
$$

(because $\|MN\| \leq \|M\| \|N\|$). The latter inequality yields (d). □

Choose an arbitrary policy $f \in F$. We define a bias operator $H_f$ from $B(S)$ into itself (see the index of notation for the definition of $B(S)$) by

$$
H_f u := \int_0^\infty \big[P(t, f) - P^*(f)\big] u \, dt \quad \forall u \in B(S). \tag{3.4}
$$

By Lemma 3.1(d) and the finiteness of $S$, we see that $H_f u$ is indeed in $B(S)$ for every $u \in B(S)$.

For each $f \in F$, the *bias* of $f$ is defined by

$$
\begin{aligned}
g_0(f) &:= \int_0^\infty \big[P(t, f) - P^*(f)\big] r(f) \, dt \\
&= \int_0^\infty \big[P(t, f)r(f) - \bar{V}(f)\big] \, dt,
\end{aligned} \tag{3.5}
$$

where the second equality follows from Lemma 3.1(b). Note that we can interpret the bias of $f$ as the total difference between the immediate reward $P(t, f)r(f)$ and the long-run EAR $\bar{V}(f)$.

*Remark 3.2* The reader should be warned that the terminology in this chapter is not standard. Indeed, some authors (for instance, Cao (1998) [18], Cao (2000) [19], Cao (2003) [20], Cao and Chen (1997) [22], Cao and Guo (2004) [23]) use the term "potential", whereas other authors (for example, Prieto-Rumeau and Hernández-Lerma (2006) [122] and (2006a) [125]) use the term "bias." Here, we will us the bias term.

For each $f \in F$, by (3.4) and (3.5), we have $g_0(f) = H_f r(f)$. In general, for each $n \geq 0$, by (3.4) we define inductively

$$g_n(f) := (-1)^n \big(H_f^{n+1} r(f)\big). \tag{3.6}$$

By Lemma 3.1(d), $g_n(f)$ is finite. Moreover, for convenience in the arguments to follow, for $n = -1$, we define

$$g_{-1}(f) := \bar{V}(f) \quad \forall f \in F.$$

**Definition 3.3** For each $f \in F$ and each integer $n \geq -1$, $g_n(f)$ is called the $n$-bias of $f$.

As was already mentioned, by (3.6) and (3.5), the 0-bias of $f$ (i.e., $g_0(f)$) is the same as the bias of $f$. Similarly, $g_1(f)$ is the bias of $f$ when the reward function $r(f)$ is replaced with $-g_0(f)$. In general, $g_n(f)$ is the bias of $f$ when the reward function is $-g_{n-1}(f)$. Thus, by (3.5), (3.6), and the results in Taylor (1976) [148] (see the proof of Theorem 9.7 below) we have

$$\frac{g_{-1}(f)}{\alpha} + \sum_{n=0}^{\infty} \alpha^n g_n(f) = V_\alpha(f) \quad \forall \alpha > 0, \tag{3.7}$$

where $V_\alpha(f)$ is the $\alpha$-discounted reward in (2.19). The left-hand side of (3.7) is the so-called *Laurent series* of the $\alpha$-discounted reward $V_\alpha(f)$.

From (3.7) we see that the long-run expected average reward criterion concerns the optimality of the first term $g_{-1}(f) = \bar{V}(f)$ of the Laurent series, whereas optimizing the bias $g_0(f)$ amounts to optimizing the second term of the Laurent series given that the first term has been optimized. This remark obviously suggests to introduce the following concept of $n$-bias optimality.

**Definition 3.4** A policy $f^*$ in $F$ is said to be $(-1)$-bias optimal (that is, AR optimal) if $g_{-1}(f^*) \geq g_{-1}(f)$ for all $f \in F$. For every integer $n \geq -1$, let $F_n^*$ be the set of all $n$-bias optimal policies. Then a policy $f^* \in F_n^*$ is called $(n+1)$-bias optimal if $g_{n+1}(f^*) \geq g_{n+1}(f)$ for all $f \in F_n^*$, in which case we write $f^* \in F_{n+1}^*$.

From Definition 3.4 and (3.7) we see that our $n$-bias optimality is equivalent to what some authors call *$n$-discount optimality* (see, for example, Guo, Hernández-Lerma, and Prieto-Rumeau (2006) [65]). Moreover, it is evident that an $(n+1)$-bias optimal policy is also $n$-bias optimal, and so we have $F_{n+1}^* \subset F_n^*$ for all $n \geq -1$.

The main goal of the following Sects. 3.3–3.5 is to provide a *self-contained* approach to show the existence of $n$-bias optimal policies and also to prove the convergence of policy iteration algorithms for computing such policies.

For ease of presentation, we will assume that the reward function is such that $r \geq 1$ on $K$. This entails no loss of generality because, since $r$ is bounded, from (3.4) and Definition 3.4 we see that $n$-bias optimality is not altered if we replace $r$ with $r + L$ for any fixed constant $L > 0$.

## 3.3 Difference Formulas of $n$-biases

In this section we first provide formulas for the difference of the $n$-biases of two stationary policies. These formulas are then used to prove the existence of a policy that is $n$-bias optimal for all $n \geq -1$.

Note that, for all $f \in F$ and $n \geq 0$, by (3.4) and (3.6) we have

$$g_{n+1}(f) = -\int_0^\infty [P(t, f) - P^*(f)] g_n(f)\, dt. \tag{3.8}$$

The following lemma shows how to obtain $g_n(f)$.

**Lemma 3.5** *For each $f \in F$, the following hold:*

(a) *The so-called Poisson equation $g_{-1}(f) = r(f) + Q(f)g_0(f)$ holds, and $g_{-1}(f)$ and $g_0(f)$ are a unique solution to the equations*

$$g_{-1}(f) = r(f) + Q(f)g_0(f), \qquad Q(f)g_{-1}(f) = 0, \qquad P^*(f)g_0(f) = 0.$$

(b) $P^*(f)g_n(f) = 0$ *for all $n \geq 0$.*
(c) *For each $n \geq 0$, $g_{n+1}(f)$ is the unique solution to the equations*

$$P^*(f)g_{n+1}(f) = 0, \tag{3.9}$$

$$Q(f)g_{n+1}(f) = g_n(f). \tag{3.10}$$

*Proof* (a) By Lemma 3.1(b) and (3.5) we have

$$\begin{aligned} Q(f)g_0(f) &= Q(f)\int_0^\infty [P(t, f) - P^*(f)] r(f)\, dt \\ &= \int_0^\infty Q(f)P(t, f)r(f)\, dt. \end{aligned}$$

Therefore, by (3.1)–(3.2) and Lemma 3.1(b),

$$\begin{aligned}Q(f)g_0(f) &= \lim_{T\to\infty}\int_0^T \frac{d}{dt}P(t,f)r(f)\,dt\\ &= \lim_{T\to\infty}\big(P(T,f)-I\big)r(f)\\ &= P^*(f)r(f)-r(f)\\ &= g_{-1}(f)-r(f),\end{aligned}$$

and so the Poisson equation follows. Moreover, by Lemma 3.1 and (3.5) we see that $Q(f)g_{-1}(f)=0$, $P^*(f)g_0(f)=0$. This means that $g_{-1}(f)$ and $g_0(f)$ are a solution to the equations in (a). To prove the uniqueness of the solution, suppose that $x$ and $y$ satisfy

$$x=r(f)+Q(f)y, \qquad Q(f)x=0, \qquad P^*(f)y=0.$$

Thus, $P(t,f)Q(f)x=0$ for all $t\geq 0$, and so (by (3.1)) we have $[P(T,f)-I]x=0$ for all $T\geq 0$. Letting $T\to\infty$, from (3.2) we get $P^*(f)x=x$. Moreover, by Lemma 3.1(b) and the hypothesis we have

$$x=P^*(f)x=P^*(f)r(f)+P^*(f)Q(f)y=g_{-1}(f).$$

Thus, it remains to show that $y=g_0(f)$. In fact, by Lemma 3.1(b) and $x=g_{-1}(f)=P^*(f)r(f)$, from $x=r(f)+Q(f)y$ we have

$$P(t,f)Q(f)y=P(t,f)x-P(t,f)r(f)=-\big[P(t,f)r(f)-g_{-1}(f)\big],$$

which, together with (3.1) and a straightforward calculation, gives

$$P(T,f)y-y=-\int_0^T\big[P(t,f)r(f)-g_{-1}(f)\big]dt \quad \forall T>0. \tag{3.11}$$

Letting $T\to\infty$ in (3.11), by $P^*(f)y=0$ and (3.5) we have $y=g_0(f)$.

(b) By Lemma 3.1(a) and (3.5), part (b) holds for $n=0$. In general, by Lemma 3.1(a) and (3.8) we have that, for each $n\geq 0$,

$$\begin{aligned}P^*(f)g_{n+1}(f) &= -P^*(f)\int_0^\infty\big[P(t,f)-P^*(f)\big]g_n(f)\,dt\\ &= -\int_0^\infty\big[P^*(f)P(t,f)-P^*(f)P^*(f)\big]g_n(f)\,dt\\ &= 0.\end{aligned} \tag{3.12}$$

This gives (b).

(c) Of course, (3.9) follows from part (b); see (3.12). Moreover, premultiplying (3.8) by $Q(f)$ and then using (3.1)–(3.2) and Lemma 3.1(b), we have

$$\begin{aligned}Q(f)g_{n+1}(f) &= -Q(f)\int_0^\infty \big[P(t,f)-P^*(f)\big]g_n(f)\,dt \\ &= -\int_0^\infty \big[Q(f)P(t,f)\big]g_n(f)\,dt \\ &= -\big[P^*(f)-I\big]g_n(f) = g_n(f).\end{aligned}$$

This gives (3.10).

To prove the uniqueness, suppose that

$$P^*(f)x = 0 \quad \text{and} \quad Q(f)x = g_n(f).$$

Then, by (b) we have

$$\begin{aligned}P(t,f)Q(f)x = P(t,f)g_n(f) &= P(t,f)g_n(f) - P^*(f)g_n(f) \\ &= \big[P(t,f)-P^*(f)\big]g_n(f),\end{aligned}$$

which, together with (3.1) and a straightforward calculation, gives

$$\big[P(T,f)-I\big]x = \int_0^T \big[P(t,f)-P^*(f)\big]g_n(f)\,dt.$$

Now let $T \to \infty$. Then, since $P^*(f)x = 0$, by (3.2) and (3.8)we get

$$x = -\int_0^\infty \big[P(t,f)-P^*(f)\big]g_n(f)\,dt = g_{n+1}(f),$$

and so the uniqueness follows. □

Now we give some results on the difference of the $n$-biases of two policies.

**Theorem 3.6** *Suppose that $f$ and $h$ are both in $F$. Then*

(a) $g_{-1}(h) - g_{-1}(f) = P^*(h)[r(h) + Q(h)g_0(f) - g_{-1}(f)] + [P^*(h) - I]g_{-1}(f)$.

(b) *If $g_{-1}(h) = g_{-1}(f)$, then*

$$\begin{aligned}g_0(h) - g_0(f) &= \int_0^\infty P(t,h)\big[r(h) + Q(h)g_0(f) - g_{-1}(f)\big]\,dt \\ &\quad + P^*(h)\big[Q(h) - Q(f)\big]g_1(f) \\ &= \int_0^\infty P(t,h)\big[r(h) + Q(h)g_0(f) - g_{-1}(f)\big]\,dt \\ &\quad + P^*(h)\big[g_0(h) - g_0(f)\big].\end{aligned}$$

(c) *For some $n \geq 0$, if $g_n(h) = g_n(f)$, then*

$$g_{n+1}(h) - g_{n+1}(f) = \int_0^\infty P(t,h)\big[Q(h) - Q(f)\big]g_{n+1}(f)\,dt + P^*(h)\big[Q(h) - Q(f)\big]g_{n+2}(f).$$

*Proof* (a) Since $P^*(h)Q(h) = 0$ (by Lemma 3.1(b)), from Lemma 3.5(a) we have

$$\begin{aligned} g_{-1}(h) - g_{-1}(f) &= P^*(h)r(h) - g_{-1}(f) \\ &= P^*(h)\big[r(h) + Q(h)g_0(f) - g_{-1}(f)\big] \\ &\quad + P^*(h)g_{-1}(f) - g_{-1}(f) \\ &= P^*(h)\big[r(h) + Q(h)g_0(f) - g_{-1}(f)\big] + \big[P^*(h) - I\big]g_{-1}(f). \end{aligned}$$

This is (a).

(b) Since $\bar{V}(f) = g_{-1}(f)) = g_{-1}(h) = \bar{V}(h)$, by (3.5) we have

$$\begin{aligned} g_0(h) - g_0(f) &= \int_0^\infty \big[P(t,h)r(h) - P(t,f)r(f)\big]\,dt \\ &= \int_0^\infty P(t,h)\big[r(h) + Q(h)g_0(f) - g_{-1}(f)\big]\,dt + \Delta, \end{aligned} \tag{3.13}$$

where $\Delta := -\int_0^\infty P(t,h)Q(h)g_0(f)\,dt + \int_0^\infty [P(t,h)g_{-1}(f) - P(t,f)r(f)]\,dt$.

Then, by (3.1) and (3.2), a straightforward calculation gives

$$\Delta = g_0(f) - P^*(h)g_0(f) + \int_0^\infty \big[P(t,h)g_{-1}(f) - P(t,f)r(f)\big]\,dt.$$

Thus, using Lemma 3.1 and that $P^*(f)r(f) = g_{-1}(f) = g_{-1}(h) = P^*(h)r(h)$, we obtain

$$\begin{aligned} \Delta &= g_0(f) - P^*(h)g_0(f) + \int_0^\infty \big[P(t,h)g_{-1}(h) - P(t,f)r(f)\big]\,dt \\ &= g_0(f) - P^*(h)g_0(f) + \int_0^\infty \big[P^*(h)r(h) - P(t,f)r(f)\big]\,dt \\ &= g_0(f) - P^*(h)g_0(f) + \int_0^\infty \big[P^*(f)r(f) - P(t,f)r(f)\big]\,dt \\ &= g_0(f) - P^*(h)g_0(f) - \int_0^\infty \big(P(t,f) - P^*(f)\big)r(f)\,dt \\ &= -P^*(h)g_0(f). \end{aligned}$$

The latter fact, together with (3.13), Lemma 3.5(b), and (3.10), gives

$$\begin{aligned}
g_0(h) - g_0(f) &= \int_0^\infty P(t,h)\big[r(h) + Q(h)g_0(f) - g_{-1}(f)\big]\,dt \\
&\quad + P^*(h)Q(h)g_1(f) - P^*(h)Q(f)g_1(f) \\
&= \int_0^\infty P(t,h)\big[r(h) + Q(h)g_0(f) - g_{-1}(f)\big]\,dt \\
&\quad + P^*(h)\big[Q(h) - Q(f)\big]g_1(f) \\
&= \int_0^\infty P(t,h)\big[r(h) + Q(h)g_0(f) - g_{-1}(f)\big]\,dt \\
&\quad + P^*(h)\big[g_0(h) - g_0(f)\big].
\end{aligned}$$

This implies (b).

(c) Since $g_n(f) = g_n(h)$, from Lemma 3.5(b) and from (3.8) and (3.10) we have

$$\begin{aligned}
&g_{n+1}(h) - g_{n+1}(f) \\
&\quad = \int_0^\infty \big[P(t,f) - P(t,h)\big]g_n(f)\,dt \\
&\quad = \int_0^\infty \big[P(t,f) - P(t,h)\big]Q(f)g_{n+1}(f)\,dt \\
&\quad = \int_0^\infty P(t,h)\big[Q(h) - Q(f)\big]g_{n+1}(f)\,dt + \int_0^\infty P(t,f)Q(f)g_{n+1}(f)\,dt \\
&\qquad - \int_0^\infty P(t,h)Q(h)g_{n+1}(f)\,dt \\
&\quad = \int_0^\infty P(t,h)\big[Q(h) - Q(f)\big]g_{n+1}(f)\,dt + P^*(f)g_{n+1}(f) - P^*(h)g_{n+1}(f) \\
&\quad = \int_0^\infty P(t,h)\big[Q(h) - Q(f)\big]g_{n+1}(f)\,dt - P^*(h)g_{n+1}(f). \qquad (3.14)
\end{aligned}$$

Therefore, by (3.10) and Lemma 3.1(b) it follows that

$$\begin{aligned}
g_{n+1}(h) - g_{n+1}(f) &= \int_0^\infty P(t,h)\big[Q(h) - Q(f)\big]g_{n+1}(f)\,dt \\
&\quad - P^*(h)Q(f)g_{n+2}(f) \\
&= \int_0^\infty P(t,h)\big[Q(h) - Q(f)\big]g_{n+1}(f)\,dt \\
&\quad + P^*(h)\big[Q(h) - Q(f)\big]g_{n+2}(f),
\end{aligned}$$

which gives (c). □

Theorem 3.6 gives formulas for the difference of the $n$-biases of two stationary policies. These formulas will be used below to prove some interesting characterization of an $n$-bias optimal policy.

In the following proof we use the sets $F_n^*$ in Definition 3.4.

**Theorem 3.7** *Let $|S|$ be the cardinality of $S$, that is, the number of states in $S$. If $f \in F$ is $|S|$-bias optimal, then $f$ is also $n$-bias optimal for all $n \geq -1$.*

*Proof* By Definition 3.4 and the uniqueness of $P(t, f)$ determined by $Q(f)$, it suffices to show that $g_n(h) = g_n(f)$ for all $n \geq |S|$ and $h \in F_{|S|}^*$ with $Q(h) \neq Q(f)$. In fact, for each $h \in F_{|S|}^*$, as $f \in F$ is $|S|$-bias optimal, we have $g_n(h) = g_n(f)$ for all $-1 \leq n \leq |S|$. Since this theorem is obviously true for the case $|S| = 1$, we next consider the case $|S| \geq 2$. By Lemma 3.1 and (3.10) we have

$$\big(Q(h) - Q(f)\big)g_{-1}(f) = Q(h)g_{-1}(h) - Q(f)g_{-1}(f) = 0, \quad \text{and}$$
$$\big(Q(h) - Q(f)\big)g_n(f) = Q(h)g_n(h) - Q(f)g_n(f) = g_{n-1}(h) - g_{n-1}(f) = 0$$

for all $1 \leq n \leq |S|$. These equalities and (3.6) give

$$\big(Q(h) - Q(f)\big)g_{-1}(f) = 0, \quad \text{and} \quad \big(Q(h) - Q(f)\big)H_f^n r(f) = 0 \tag{3.15}$$

for all $2 \leq n \leq |S| + 1$. Therefore, the vectors $g_{-1}(f)$ and $H_f^n r(f)$ (for $2 \leq n \leq |S| + 1$) belong to the null space of $Q(h) - Q(f)$ (that is, the space $\{u : (Q(h) - Q(f))u = 0\}$). Since $Q(h) - Q(f) \neq 0$, we see that the rank of $Q(h) - Q(f)$ is at least 1, and so the dimension of the null space of $Q(h) - Q(f)$ is at most $|S| - 1$. Hence, $g_{-1}(f)$ and $H_f^n r(f)$ (for $2 \leq n \leq |S| + 1$) are linearly dependent. Thus, because $g_{-1}(f) = \bar{V}(f) \geq 1$ (by (3.3) and $r(f) \geq 1$), there exists an integer $2 \leq k \leq |S|$ such that $H_f^{k+1} r(f)$ is a linear combination of $g_{-1}(f)$ and $H_f^n r(f)$ for all $2 \leq n \leq k$.

We now show by induction that, for each $m \geq 2$, $H_f^m r(f)$ is a linear combination of $g_{-1}(f)$ and $H_f^n r(f)$ for all $2 \leq n \leq k$. To see this, suppose that this conclusion holds for some $m$ $(\geq k + 1)$. More explicitly, there exist $k$ numbers $\lambda_l$ such that

$$H_f^m r(f) = \lambda_1 g_{-1}(f) + \sum_{l=2}^{k} \lambda_l \big(H_f^l r(f)\big),$$

which, together with (3.4) and $H_f g_{-1}(f) = 0$, gives

$$H_f^{m+1} r(f) = \sum_{l=2}^{k-1} \lambda_l \big(H_f^{l+1} r(f)\big) + \lambda_k \big(H_f^{k+1} r(f)\big). \tag{3.16}$$

Since $H_f^{k+1} r(f)$ is a linear combination of $g_{-1}(f)$ and $H_f^n r(f)$ for all $2 \leq n \leq k$, it follows from (3.16) that $H_f^{m+1} r(f)$ is also a linear combination of $g_{-1}(f)$ and

$H^n_f r(f)$ for all $2 \le n \le k$, and so the desired conclusion is proved. Therefore, by (3.15) and (3.6) we have

$$\big(Q(h) - Q(f)\big)g_{-1}(f) = 0 \quad \text{and} \quad \big(Q(h) - Q(f)\big)g_n(f) = 0 \quad \forall n \ge |S|.$$

This fact and Theorem 3.6(c) give that, for each $n \ge |S|$,

$$\begin{aligned} & g_n(h) - g_n(f) \\ &= \int_0^\infty P(t,h)\big[Q(h) - Q(f)\big]g_n(f)\,dt - P^*(h)\big[Q(h) - Q(f)\big]g_{n+1}(f) \\ &= 0. \end{aligned}$$

This completes the proof of Theorem 3.7. □

Theorem 3.7 shows that to obtain a policy that is $n$-bias optimal for *all* $n \ge -1$, it suffices to find an $|S|$-bias optimal policy.

## 3.4 Characterization of $n$-bias Policies

To obtain an $n$-bias optimal policy for *all* $n \ge -1$, by Theorem 3.7 we only need to focus on the existence and calculation of an $|S|$-bias optimal policy. We begin with some remarks and simple but useful lemmas.

*Remark 3.8*

(a) If a state is recurrent (or transient) with respect to the transition matrix $P(t, f)$, it will also be said to be recurrent (or transient) under the corresponding transition rate matrix $Q(f)$. For the definition of recurrent and transient states, see Appendix B.2, in particular Definition B.12.
(b) Consider a discrete- or continuous-time Markov chain with a finite state space $S$. Then $S$ can be written as a disjoint union $C_1 \cup \cdots \cup C_m \cup C_{m+1}$, where $C_1, \dots, C_m$ are closed irreducible sets of recurrent states, and $C_{m+1}$ is a set of transient states. This is a well-known fact from the elementary theory of Markov chains, and it is used in the proof of Lemma 3.9 below.

By Remark 3.8(b) and reordering the states if necessary, for each $f \in F$, we can write $Q(f)$ in the *canonical form* (depending on $f$)

$$Q(f) = \begin{bmatrix} Q_1(f) & 0 & \cdots & 0 & 0 \\ 0 & Q_2(f) & \cdots & 0 & 0 \\ \vdots & \vdots & \ddots & \vdots & \vdots \\ 0 & \cdots & \cdots & Q_m(f) & 0 \\ W_1(f) & W_2(f) & \cdots & W_m(f) & W_{m+1}(f) \end{bmatrix}, \tag{3.17}$$

in which $Q_k(f)$ corresponds to the transitions among recurrent states in $C_k^f$ for $k = 1, 2, \ldots, m$ (depending on $f$), and $W_k(f)$, for each $k = 1, 2, \ldots, m$, corresponds to the transitions from the transient states in $C_{m+1}^f$ to the recurrent states in $C_k^f$. Finally, $W_{m+1}(f)$ corresponds to the transitions among the transient states in $C_{m+1}^f$.

Moreover, by (3.17) and Proposition C.12 we see that $P^*(f)$ can also be expressed in the *canonical form*

$$P^*(f) = \begin{bmatrix} P_1^*(f) & 0 & \cdots & 0 & 0 \\ 0 & P_2^*(f) & \cdots & 0 & 0 \\ \vdots & \vdots & \ddots & \vdots & \vdots \\ 0 & \cdots & \cdots & P_m^*(f) & 0 \\ W_1^*(f) & W_2^*(f) & \cdots & W_m^*(f) & 0 \end{bmatrix}, \tag{3.18}$$

in which $P_k^*(f) = e_k \pi_k(f)$, where $\pi_k(f)$ is the invariant probability measure (i.p.m.) of $Q_k(f)$ obtained by solving

$$\pi_k(f) Q_k(f) = 0 \quad \text{subject to} \quad \pi_k(f) e_k = 1,$$

where $e_k$ is the column vector with all components 1, and $W_k(f) P_k^*(f) + W_{m+1}(f) W_k^*(f) = 0$ for $k = 1, \ldots, m$.

From the canonical forms (3.18) we have the following simple fact, which will be repeatedly used below.

**Lemma 3.9** *Fix $f \in F$.*

(a) *If $P^*(f)u = 0$ and $u(i) \le 0$ (or $u(i) \ge 0$) for all recurrent states $i$ under $Q(f)$, then $u(i) = 0$ for all recurrent states $i$ under $Q(f)$.*
(b) *If $u(i) \ge 0$ for all recurrent states $i$ under $Q(f)$, then $P^*(f)u \ge 0$.*

*Proof* Pick an arbitrary $f \in F$.

(a) By (3.18), the columns in $P^*(f)$ corresponding to transient states are all zeros, and thus all $u(i)$ with $i$ being transient states contribute nothing to $P^*(f)u$. Therefore, part (a) follows.

(b) This part follows from the canonical form (3.18) of $P^*(f)$. □

**Lemma 3.10** *For all $f, h \in F$, the following assertions hold:*

(a) *If $g_{-1}(h) = g_{-1}(f)$, then $Q(f) g_{-1}(h) = 0$ and $P^*(h)[r(h) + Q(h) g_0(f) - g_{-1}(f)] = 0$.*
(b) *If $g_n(h) = g_n(f)$ for some $n \ge 0$, then $P^*(h)[Q(h) - Q(f)] g_{n+1}(f) = 0$.*

*Proof* (a) Since $g_{-1}(f) = g_{-1}(h)$, by Lemma 3.1(b) we have

$$Q(f) g_{-1}(h) = Q(f) g_{-1}(f) = Q(f) P^*(f) r(f) = 0,$$

and so the first part of (a) follows. Now, since $g_{-1}(f) = g_{-1}(h) = P^*(h)r(h)$, from Lemma 3.1(a) we obtain

$$\big(P^*(h) - I\big)g_{-1}(f) = \big(P^*(h) - I\big)g_{-1}(h) = 0,$$

which, together with Theorem 3.6(a), gives the second part of (a).

(b) By Lemma 3.5 and the hypothesis that $g_n(f) = g_n(h)$, we have

$$0 = P^*(h)g_n(h) = P^*(h)g_n(f) = P^*(h)Q(f)g_{n+1}(f).$$

Hence, Lemma 3.1(b) yields (b). □

We next give another characterization of an $n$-bias optimal policy. To this end, we introduce the following notation.

For all $f \in F$, $n \geq 2$, and $i \in S$, define

$$A_0^f(i) := \Big\{a \in A(i) : \sum_{j \in S} q(j|i,a)g_{-1}(f)(j) = 0\Big\}, \tag{3.19}$$

$$A_1^f(i) := \Big\{a \in A_0^f(i) : \ r(i,a) + \sum_{j \in S} q(j|i,a)g_0(f)(j) = g_{-1}(f)(i)\Big\},$$

$$\vdots \qquad \vdots \qquad \vdots$$

$$A_n^f(i) := \Big\{a \in A_{n-1}^f(i) : \ \sum_{j \in S} q(j|i,a)g_{n-1}(f)(j) = g_{n-2}(f)(i)\Big\}. \tag{3.20}$$

By Lemma 3.1(a) and Lemma 3.5, $f(i)$ is in $A_n^f(i)$ for all $i \in S$ and $n \geq 0$.

**Theorem 3.11** *Suppose that there exists a policy $f^* \in F$ satisfying the following $n+3$ ($n \geq 0$) bias optimality conditions*:

$$\max_{a \in A(i)} \Big\{\sum_{j \in S} q(j|i,a)g_{-1}\big(f^*\big)(j)\Big\} = 0, \tag{3.21}$$

$$\max_{a \in A_0^{f^*}(i)} \Big\{r(i,a) + \sum_{j \in S} q(j|i,a)g_0\big(f^*\big)(j)\Big\} = g_{-1}\big(f^*\big)(i), \tag{3.22}$$

$$\max_{a \in A_{k+1}^{f^*}(i)} \Big\{\sum_{j \in S} q(j|i,a)g_{k+1}\big(f^*\big)(j)\Big\} = g_k\big(f^*\big)(i) \quad \forall 0 \leq k \leq n \tag{3.23}$$

*for all $i \in S$. Then,*

(a) *$f^*$ is $n$-bias optimal.*
(b) *If in addition $n \geq |S|$, then $f^*$ is $k$-bias optimal for all $k \geq 0$.*

*Remark 3.12* The bias optimality conditions (3.21)–(3.23) are sometimes referred to as the *bias optimality equations* or as the *average reward optimality conditions*.

*Proof* (a) For each fixed $f \in F$, let

$$u := Q(f)g_{-1}(f^*), \quad \text{and} \quad v := r(f) + Q(f)g_0(f^*) - g_{-1}(f^*).$$

Since $f^*$ satisfies (3.21), we have that $u = Q(f)g_{-1}(f^*) \le 0$. Therefore, by Lemma 3.9(a) and Lemma 3.1(b) we have $u(i) = 0$ for all recurrent states $i$ under $Q(f)$. This implies, by (3.22) and (3.19), that

$$v(i) = [r(f) + Q(f)g_0(f^*) - g_{-1}(f^*)](i) \le 0$$

for all recurrent states $i$ under $Q(f)$. Thus, by Lemma 3.9(b) we conclude that $P^*(f)v \le 0$. On the other hand, since $Q(f)g_{-1}(f^*) \le 0$, we have

$$P(t, f)Q(f)g_{-1}(f^*) \le 0 \quad \text{for all } t \ge 0.$$

Hence, by (3.1) and a straightforward calculation, we obtain $[P(T, f) - I]g_{-1}(f^*) \le 0$ for all $T \ge 0$, which, together with Lemma 3.1(a) and letting $T \to \infty$, gives $(P^*(f) - I)g_{-1}(f^*) \le 0$. Finally, by Theorem 3.6(a) we have

$$g_{-1}(f) - g_{-1}(f^*) = P^*(f)v + (P^*(f) - I)g_{-1}(f^*) \le 0.$$

This means that $f^*$ is in $F^*_{-1}$.

We next show that $f^*$ is in $F^*_0$. Since for each $f \in F^*_{-1}$ we have $g_{-1}(f^*) = g_{-1}(f)$, we obtain $Q(f)g_{-1}(f^*) = Q(f)g_{-1}(f) = 0$ (by Lemma 3.1(b)). The latter fact and (3.19) imply that $f(i)$ is in $A_0^{f^*}(i)$ for all $i \in S$. Thus, by (3.22) we have $g_{-1}(f^*) - r(f) - Q(f)g_0(f^*) \ge 0$ and $P^*(f)[g_{-1}(f^*) - r(f) - Q(f)g_0(f^*)] = P^*(f)[g_{-1}(f) - r(f)] = 0$ (by Lemma 3.1(b)). Therefore, by Lemma 3.9(a) we have

$$r(i, f(i)) + \sum_{j \in S} q(j|i, f(i))g_0(f^*)(j) = g_{-1}(f^*)(i)$$

for all recurrent states $i$ under $Q(f)$, and thus $f(i)$ is in $A_1^{f^*}(i)$ for all recurrent states $i$ under $Q(f)$. Therefore, by (3.23) (for $k = 0$) and Lemma 3.5(c), we obtain

$$\sum_{j \in S} q(j|i, f(i))g_1(f^*)(j) \le g_0(f^*)(i) = \sum_{j \in S} q(j|i, f^*(i))g_1(f^*)(j)$$

for all recurrent states $i$ under $Q(f)$, and so $P^*(f)[Q(f) - Q(f^*)]g_1(f^*) \le 0$ (by Lemma 3.9(b)). Thus, since $r(f) + Q(f)g_0(f^*) - g_{-1}(f^*) \le 0$ (just proved), we have

$$\int_0^\infty P(t, f)[r(f) + Q(f)g_0(f^*) - g_{-1}(f^*)]\,dt$$
$$+ P^*(f)[Q(f) - Q(f^*)]g_1(f^*) \le 0.$$

Hence Theorem 3.6(b) gives

$$g_0(f) - g_0(f^*) \leq 0,$$

and so $f^*$ is in $F_0^*$. This means that (a) holds for $n = 0$.

We now consider the case $n \geq 1$. By induction, let us suppose that $f^*$ is in $F_m^*$ for some $0 \leq m \leq n-1$. Then, by Definition 3.4, to show that $f^*$ is in $F_{m+1}^*$, we need to prove that $g_{m+1}(f^*) \geq g_{m+1}(f)$ for all $f \in F_m^*$. To this end, first note that, for each $f \in F_m^*$, the definition of $F_m^*$ and $f^* \in F_m^*$ (the induction hypothesis) give that $g_l(f^*) = g_l(f)$ for all $-1 \leq l \leq m$. Hence, by Lemma 3.1(b) and Lemma 3.5(b),

$$P^*(f)\big[Q(f)g_{m+1}(f^*) - g_m(f^*)\big] = 0. \tag{3.24}$$

Moreover, again using $g_l(f^*) = g_l(f)$ for all $-1 \leq l \leq m$, by Lemma 3.5 and (3.20) we see that $f(i)$ is in $A_{m+1}^{f^*}(i)$ for all $i \in S$, and then, from (3.23) and (3.10) it follows that

$$Q(f)g_{m+1}(f^*) \leq g_m(f^*) = Q(f^*)g_{m+1}(f^*). \tag{3.25}$$

Hence, using (3.24) and (3.25), by Lemma 3.9(a) we have

$$\sum_{j \in S} q\big(j|i, f(i)\big)g_{m+1}(f^*)(j) = g_m(f^*)(i)$$

for all recurrent states $i$ under $Q(f)$, and so $f(i)$ is in $A_{m+2}^{f^*}(i)$ for all such recurrent states $i$ under $Q(f)$. Thus, by (3.23) and (3.10), we have

$$\sum_{j \in S} q\big(j|i, f(i)\big)g_{m+2}(f^*)(j) \leq g_{m+1}(f^*)(i) = \sum_{j \in S} q\big(j|i, f^*(i)\big)g_{m+2}(f^*)(j)$$

for all recurrent states $i$ under $Q(f)$, and so $P^*(f)[Q(f) - Q(f^*)]g_{m+2}(f^*) \leq 0$ (by Lemma 3.9(b)). Thus, using (3.25) and that $P(t, f) \geq 0$, we obtain

$$\int_0^\infty P(t, f)\big[Q(f) - Q(f^*)\big]g_{m+1}(f^*)\,dt + P^*(f)\big[Q(f) - Q(f^*)\big]g_{m+2}(f^*) \leq 0.$$

This inequality and Theorem 3.6(c) together give

$$g_{m+1}(f) - g_{m+1}(f^*) \leq 0,$$

and so $f^*$ is in $F_{m+1}^*$. Hence (a) is proved.

(b) This part obviously follows from (a) and Theorem 3.7. □

To conclude this section, we will provide a further characterization of a $(-1)$-bias optimal policy (that is, an AR optimal policy) based on the first two bias optimality equations (3.21) and (3.22).

**Theorem 3.13** *Suppose that there exists a policy $f^* \in F$ satisfying the following two* average reward optimality equations: *for every $i \in S$,*

$$\max_{a\in A(i)}\left\{\sum_{j\in S} q(j|i,a)g_{-1}(f^*)(j)\right\} = 0 \quad \textit{and} \tag{3.26}$$

$$\max_{a\in A_0^{f^*}(i)}\left\{r(i,a) + \sum_{j\in S} q(j|i,a)g_0(f^*)(j)\right\} = g_{-1}(f^*)(i) \tag{3.27}$$

*with (by* (3.19))

$$A_0^{f^*}(i) := \left\{a \in A(i) : \sum_{j\in S} q(j|i,a)g_{-1}(f^*)(j) = 0\right\}. \tag{3.28}$$

*Then,*

(a) *There exists a vector $h^*$ such that $(g_{-1}(f^*), h^*)$ is a solution to the modified average reward optimality equations: for every $i \in S$,*

$$\max_{a\in A(i)}\left\{\sum_{j\in S} q(j|i,a)g_{-1}(f^*)(j)\right\} = 0, \tag{3.29}$$

$$g_{-1}(f^*)(i) = \max_{a\in A(i)}\left\{r(i,a) + \sum_{j\in S} q(j|i,a)h^*(j)\right\}. \tag{3.30}$$

(b) $\bar{V}(f^*) = \bar{V}^*$, *and so $f^*$ is AR optimal.*
(c) *Each $f \in F$ that attains the maximum in* (3.26) *and* (3.27) *is also AR optimal.*

*Proof* (a) The idea of the proof is to show that (3.30) holds with $h^* := g_0(f^*) + \hat{\delta}g_{-1}(f^*)$ for some $\hat{\delta} \geq 0$.

For each fixed $i \in S$, since $\sum_{j\in S} q(j|i,a)g_{-1}(f^*)(j) = 0$ for all $a \in A_0^{f^*}(i)$, by (3.27) and the finiteness of $A(i)$, there exists $\hat{a} \in A_0^{f^*}(i)$ such that

$$r(i,\hat{a}) + \sum_{j\in S} q(j|i,\hat{a})\left[g_0(f^*)(j) + \delta g_{-1}(f^*)(j)\right] - g_{-1}(f^*)(i) = 0 \quad \text{for all } \delta \geq 0.$$

Thus, to prove (3.30), it suffices to find some constant $\hat{\delta} \geq 0$ such that for all $i \in S$ and $a \in A(i)$,

$$r(i,a) + \sum_{j\in S} q(j|i,a)\left[g_0(f^*)(j) + \hat{\delta}g_{-1}(f^*)(j)\right] - g_{-1}(f^*)(i) \leq 0. \tag{3.31}$$

Indeed, if $r(i,a) + \sum_{j\in S} q(j|i,a)g_0(f^*)(j) - g_{-1}(f^*)(i) \le 0$ for all $i \in S$ and $a \in A(i)$, then (3.31) holds with $\hat{\delta} = 0$. On the other hand, suppose that the set

$$\hat{K} := \left\{(i',a') \in K \mid r(i',a') + \sum_{j\in S} q(j|i',a')g_0\big(f^*\big)(j) - g_{-1}\big(f^*\big)(i') > 0\right\}$$

is nonempty. Then, for any $(i',a') \in \hat{K}$, by (3.27) we see that $a'$ is not in $A_0^{f^*}(i')$. Therefore, it follows from (3.26) and (3.28) that $\sum_{j\in S} q(j|i',a')g_{-1}(f^*)(j) < 0$. Thus, we can choose $\delta(i',a') > 0$ (depending on $i'$ and $a'$) such that (3.31) holds for such $i'$ and $a'$. Furthermore, by the finiteness of $\hat{K}$, we can choose $\hat{\delta} := \max_{(i'a')\in\hat{K}} \delta(i',a') > 0$ for which (3.31) holds for all $(i',a') \in \hat{K}$, and so (3.31) holds for all $i \in S$ and $a \in A(i)$ since (3.31) is obviously true for any $(i,a) \notin \hat{K}$ and any $\delta > 0$.

Hence, $(g_{-1}(f^*), h^*)$ is a solution to (3.29) and (3.30), where $h^* := g_0(f^*) + \hat{\delta} g_{-1}(f^*)$ with $\hat{\delta}$ as defined above. So (a) follows.

(b) For each $\pi = (\pi_t) \in \Pi$, let $P_\pi(s,t) := [p_\pi(s,i,t,j)]$ be the transition matrix with $(i,j)$th element $p_\pi(s,i,t,j)$. Then, by (3.29) and (2.5) we have

$$Q(\pi_t)g_{-1}\big(f^*\big) \le 0 \quad \forall t \ge 0,$$

which, together with (2.6) and a straightforward calculation, gives

$$\big[P_\pi(0,t) - I\big]g_{-1}\big(f^*\big) \le 0 \quad \forall t \ge 0. \tag{3.32}$$

On the other hand, by (3.30), (2.5), and (2.20) we have

$$g_{-1}\big(f^*\big) \ge r(\pi_t) + Q(\pi_t)h^* \quad \forall t \ge 0,$$

and so

$$P_\pi(0,t)g_{-1}\big(f^*\big) \ge P_\pi(0,t)r(\pi_t) + P_\pi(0,t)Q(\pi_t)h^* \quad \forall t \ge 0.$$

This inequality and (3.32) give

$$g_{-1}\big(f^*\big) \ge P_\pi(0,t)r(\pi_t) + P_\pi(0,t)Q(\pi_t)h^*.$$

Therefore,

$$Tg_{-1}\big(f^*\big) \ge \int_0^T P_\pi(0,t)r(\pi_t)\,dt + \big[P_\pi(0,T) - I\big]h^*,$$

which, together with (2.21), yields $g_{-1}(f^*) \ge \bar{V}(\pi)$, and so (b) follows.

(c) Under the condition in (c), all the inequalities in the proof of (b) (with $\pi := f$) become equalities, and thus we have $\bar{V}(f) = g_{-1}(f^*) = \bar{V}^*$. □

**Corollary 3.14** *Suppose that there exist vectors $g$ and $h$ satisfying the modified average optimality equations*

$$\max_{a\in A(i)}\left\{\sum_{j\in S} q(j|i,a)g(j)\right\}=0 \quad \forall i\in S, \tag{3.33}$$

$$g(i)=\max_{a\in A(i)}\left\{r(i,a)+\sum_{j\in S} q(j|i,a)h(j)\right\} \quad \forall i\in S. \tag{3.34}$$

*Then $g\geq \bar{V}^*$.*

*Proof* Comparing (3.33)–(3.34) with (3.29)–(3.30), we see that the desired result follows from the proof of Theorem 3.13(b). □

In the following section, we will show the existence of $f^*$ satisfying (3.21)–(3.23) and how it can be calculated.

## 3.5 Computation of $n$-bias Optimal Policies

In this section, we introduce policy iteration algorithms for computing $n$-bias optimal policies for $n\geq -1$. First, we consider the case $n=-1$ (that is, average optimality); then $n=0$, and finally $n\geq 1$.

*Notation* For two vectors $u$ and $v$, we write $u\succeq v$ if $u\geq v$ and $u(i)>v(i)$ for at least one $i\in S$.

### 3.5.1 The Policy Iteration Algorithm for Average Optimality

In this subsection, we provide a *policy iteration algorithm for average optimality*, which computes a $(-1)$-bias optimal policy, that is, an AR optimal policy. The basic problem, then, is to solve (3.26), (3.27).

To state the policy iteration algorithm, we need to introduce some notation and also a preliminary result, Lemma 3.15.

For any given $f\in F$, $i\in S$, and $a\in A(i)$, let

$$w_f(i,a):=r(i,a)+\sum_{j\in S} q(j|i,a)g_0(f)(j) \tag{3.35}$$

and

$$B_0^f(i):=\left\{a\in A(i):\ \begin{array}{c}\sum_{j\in S} q(j|i,a)g_{-1}(f)(j)>0, \quad \text{or}\\ w_f(i,a)>g_{-1}(f)(i)\\ \text{when } \sum_{j\in S} q(j|i,a)g_{-1}(f)(j)=0\end{array}\right\}. \tag{3.36}$$

We then define an *improvement policy* $h \in F$ (depending on $f$) as follows:

$$h(i) \in B_0^f(i) \quad \text{if } B_0^f(i) \neq \emptyset, \quad \text{and} \quad h(i) := f(i) \quad \text{if } B_0^f(i) = \emptyset. \tag{3.37}$$

The policy $h$ is indeed an "improvement" of $f$ because, as shown in the following lemma, we either have

$$g_{-1}(h) \succeq g_{-1}(f)$$

or if $g_{-1}(h) = g_{-1}(f)$ and $h \neq f$, then

$$g_0(h) \succeq g_0(f).$$

Note that such a policy $h$ may not be unique since there may be more than one action in $B_0^f(i)$ for some state $i \in S$. Let

$$u_f^h := Q(h)g_{-1}(f), \qquad v_f^h := r(h) + Q(h)g_0(f) - g_{-1}(f). \tag{3.38}$$

**Lemma 3.15** *For any given $f \in F$, let $h \in F$ be defined as in* (3.37). *Then*:

(a) $g_{-1}(h) \geq g_{-1}(f)$, *and $v_f^h(i) \geq 0$ for all recurrent states $i$ under $Q(h)$.*
(b) *If $v_f^h(i) > 0$ for some recurrent state $i$ under $Q(h)$, then $g_{-1}(h) \succeq g_{-1}(f)$.*
(c) *If $Q(h)g_{-1}(f) \neq 0$, then $g_{-1}(h) \succeq g_{-1}(f)$.*
(d) *If $g_{-1}(h) = g_{-1}(f)$ and $h \neq f$, then $g_0(h) \succeq g_0(f)$.*

*Proof* (a) By (3.36) and (3.37) we see that $u_f^h = Q(h)g_{-1}(f) \geq 0$. Then, by Lemma 3.1(b) and Lemma 3.9(a), we have $u_f^h(i) = 0$ for all recurrent states $i$ under $Q(h)$. Hence, it follows from (3.36) and (3.38) that $v_f^h(i) \geq 0$ for all recurrent states $i$ under $Q(h)$, which implies the second part of (a). Moreover, since $Q(h)g_{-1}(f) \geq 0$, by (3.1) and a straightforward calculation we have

$$\big[P(T,h) - I\big]g_{-1}(f) \geq 0 \quad \text{for all } T \geq 0,$$

which, together with (3.2) gives $(P^*(h) - I)g_{-1}(f) \geq 0$. Thus, by Theorem 3.6(a) and Lemma 3.9(b) we have

$$g_{-1}(h) - g_{-1}(f) = P^*(h)v_f^h + \big(P^*(h) - I\big)g_{-1}(f) \geq 0,$$

and so the first part of (a) follows.

(b) From the proof of (a) we have $(P^*(h) - I)g_{-1}(f) \geq 0$ and $v_f^h(i) \geq 0$ for all the recurrent states $i$ under $Q(h)$. On the other hand, by (a) and the condition in (b) we have $P^*(h)v_f^h \succeq 0$, and so

$$g_{-1}(h) - g_{-1}(f) = P^*(h)v_f^h + \big[P^*(h) - I\big]g_{-1}(f) \succeq 0.$$

This inequality gives (b).

(c) By (a), it suffices to prove that $g_{-1}(h) \neq g_{-1}(f)$. Suppose that $g_{-1}(h) = g_{-1}(f)$. Then, by Lemma 3.1(b) we have

$$Q(h)g_{-1}(f) = Q(h)g_{-1}(h) = Q(h)P^*(h)r(h) = 0, \tag{3.39}$$

which contradicts the condition in (c).

(d) We first prove that $g_0(h) \geq g_0(f)$. To do so, let $\Delta g := g_0(h) - g_0(f)$. Then, since $g_{-1}(h) = g_{-1}(f)$, as in the proof of (3.39), we can see that $Q(h)g_{-1}(f) = 0$. Moreover, by (a) and (b) and (3.36)–(3.38) we have $v_f^h \geq 0$, and, moreover, $v_f^h(i) = 0$ for all recurrent states $i$ under $Q(h)$, and so $h(i)$ is not in $B_0^f(i)$ for any recurrent $i$ under $Q(h)$. Thus, by Theorem 3.6(b) and (3.36) we have that

$$\bigl(I - P^*(h)\bigr)\Delta g = \int_0^\infty P(t,h)v_f^h\,dt \geq 0.$$

Thus, by Lemma 3.9(a) and (3.37) we conclude that $[(I - P^*(h))\Delta g](i) = 0$ and $h(i) = f(i)$ for all recurrent states $i$ under $Q(h)$. Since each recurrent class under $Q(h)$ (denoted as $C_k^h$ depending on $h$, with $k$ as the index of a recurrent class) is closed and irreducible, from $[(I - P^*(h))\Delta g](i) = 0$ for all $i \in C_k^h$ we see that $\Delta g$ is a constant on $C_k^h$, denoted as $\rho_k$. Since $h(i) = f(i)$ for all $i \in C_k^h$, it follows that $C_k^h$ is also a closed irreducible recurrent class under $Q(f)$. Thus, $P^*(h)$ and $P^*(f)$ have the same elements on $C_k^h$. Therefore, by Lemma 3.5(b) (for $n = 0$) we have

$$\bigl[P^*(h)\Delta g\bigr](i) = P^*(h)\bigl[g_0(h) - g_0(f)\bigr](i) = 0 \quad \text{for all } i \in C_k^h,$$

which, together with $(I - P^*(h))\Delta g(i) = 0$ and the fact that $\Delta g$ is a constant $\rho_k$ on $C_k^h$, gives $\rho_k = 0$, and thus $g_0(h)(i) = g_0(f)(i)$ for all $i \in C_k^h$.

Similarly, we have $g_0(h)(i) = g_0(f)(i)$ for all recurrent states $i$ under $Q(h)$, which gives $P^*(h)[g_0(h) - g_0(f)] = 0$. Thus, by Theorem 3.6(b) and $v_f^h \geq 0$, we have

$$g_0(h) - g_0(f) = \int_0^\infty P(t,h)v_f^h\,dt + P^*(h)\bigl[g_0(h) - g_0(f)\bigr] \geq 0.$$

Summarizing, we have $g_0(h) \geq g_0(f)$. Hence, it remains to show that $g_0(h) \neq g_0(f)$.

Arguing by contradiction, suppose that $g_0(h) = g_0(f)$. By Lemma 3.5(a) and $g_{-1}(h) = g_{-1}(f)$, we have

$$\begin{aligned} r(h) + Q(h)g_0(f) &= r(h) + Q(h)g_0(h) \\ &= r(f) + Q(f)g_0(f). \end{aligned} \tag{3.40}$$

On the other hand, since $h \neq f$ and $Q(h)g_{-1}(f) = 0$, from (3.35), (3.36), and (3.37) we have

$$r(h) + Q(h)g_0(f) \succeq r(f) + Q(f)g_0(f), \tag{3.41}$$

which contradicts (3.40). Hence, we must have $g_0(h) \succeq g_0(f)$. □

Having Lemma 3.15, we can easily state the following policy iteration algorithm.

**Policy iteration algorithm for average optimality** (also referred to as the $(-1)$-*bias policy iteration algorithm*)

1. Let $k = 0$ and pick an arbitrary policy $f_k \in F$.
2. (Policy evaluation) Obtain $g_0(f_k)$ and $g_{-1}(f_k)$ (by means of Lemma 3.1(b) and Lemma 3.5(b)).
3. (Policy improvement) Obtain an improvement policy $f_{k+1}$ from (3.37) with $f$ and $h$ replaced by $f_k$ and $f_{k+1}$, respectively.
4. If $f_{k+1} = f_k$, then stop because $f_{k+1}$ is $(-1)$-bias optimal (by Theorem 3.16 below). Otherwise, increment $k$ by 1 and return to Step 2.

We will now prove the existence of $f^*$ satisfying (3.26) and (3.27) by using the $(-1)$-bias policy iteration algorithm.

**Theorem 3.16** *In a finite number of iterations, the* $(-1)$*-bias policy iteration algorithm yields a* $(-1)$*-bias optimal policy, denoted by* $f^*_{-1}$*, which satisfies* (3.26) *and* (3.27).

*Proof* Let $\{f_k\}$ $(k \geq 0)$ be the sequence of policies in the $(-1)$-bias policy iteration algorithm above. Then, by Lemma 3.15(a), we have $g_{-1}(f_{k+1}) \geq g_{-1}(f_k)$. That is, as $k$ increases, $g_{-1}(f_k)$ either increases or stays the same. Furthermore, by Lemma 3.15(d), when $g_{-1}(f_k)$ stays the same, $g_0(f_k)$ increases. Thus, any two policies in the sequence of $\{f_k,\ k = 0, 1, \ldots\}$ either have different long-run average rewards or have different 0-biases. Thus, every policy in the iteration sequence is different. Since the number of policies in $F$ is finite, the iterations must stop after a finite number because, otherwise, we can find the next improved policy in the policy iteration. Suppose that the algorithm stops at a policy denoted by $f^*_{-1}$. Then $f^*_{-1}$ must satisfy (3.26) and (3.27). Thus, by Theorem 3.13, $f^*_{-1}$ is $(-1)$-bias optimal. □

### *3.5.2 The 0-bias Policy Iteration Algorithm*

In this subsection we provide a 0-*bias policy iteration algorithm* for computing a 0-bias optimal policy. More explicitly, this is an algorithm for solving (3.21)–(3.23) with $n = 0$.

**Lemma 3.17** *Let* $F^*_{-1}$ *be as in Definition* 3.4.

(a) *For all* $f^* \in F^*_{-1}$ *and* $f \in F$*, if the following two conditions hold*
   (i) $Q(f)g_{-1}(f^*) = 0$
   (ii) $r(f) + Q(f)g_0(f^*) \geq g_{-1}(f^*)$
   *then* $g_{-1}(f) = g_{-1}(f^*)$.
(b) *Under the two conditions in (a), if, in addition,* $Q(f)g_1(f^*)(i) \geq g_0(f^*)(i)$ *for all states* $i$ *such that* $[r(f) + Q(f)g_0(f^*)](i) = g_{-1}(f^*)(i)$*, then*

$$g_{-1}(f) = g_{-1}\big(f^*\big) \quad \textit{and} \quad g_0(f) \geq g_0\big(f^*\big).$$

*Proof* (a) Let $u := [Q(f) - Q(f^*)]g_1(f^*) = Q(f)g_1(f^*) - g_0(f^*)$ (by (3.10)), and $v := r(f) + Q(f)g_0(f^*) - g_{-1}(f^*) \geq 0$ (by condition (ii)). Then, by (i), we have $P(t, f)Q(f)g_{-1}(f^*) = 0$ for all $t \geq 0$, and so it follows from (3.1) and a straightforward calculation that $[P^*(f) - I]g_{-1}(f^*) = 0$. Thus, by Theorem 3.6(a) we have

$$g_{-1}(f) - g_{-1}\big(f^*\big) = P^*(f)v + \big[P^*(f) - I\big]g_{-1}\big(f^*\big) = P^*(f)v \geq 0, \qquad (3.42)$$

which together with the AR optimality of $f^*$ gives

$$g_{-1}(f) = g_{-1}\big(f^*\big), \qquad (3.43)$$

and so (a) follows.

(b) From (3.42) and (3.43) we see that $P^*(f)v = 0$. Since $v \geq 0$, by Lemma 3.9(a) we further have that $v(i) = 0$ for all the recurrent states $i$ under $Q(f)$. Hence, by the conditions in (b) we have $u(i) \geq 0$ for all recurrent states $i$ under $Q(f)$. It follows from Lemma 3.9(b) that $P^*(f)u \geq 0$, and so by Theorem 3.6(b) we have

$$g_0(f) - g_0\big(f^*\big) = \int_0^\infty P(t, f)v\,dt + P^*(f)u \geq 0.$$

This inequality and (3.43) give (b). □

From Theorem 3.11 we see that a policy $f^* \in F$ satisfying (3.21)–(3.23) (with $n = 0$) is 0-bias optimal. We now wish to present a policy iteration algorithm for computing one such a policy. To state this algorithm we will use the following notation.

For a given $f \in F^*_{-1}$ (such as $f^*_{-1}$ in Theorem 3.16) and $i \in S$, recall the definitions of $A_0^f(i)$ in (3.19) and $w_f(i, a)$ in (3.35), i.e.,

$$\begin{aligned} A_0^f(i) &= \Big\{a \in A(i) \Big| \sum_{j \in S} q(j|i, a)g_{-1}(f)(j) = 0\Big\}, \\ w_f(i, a) &:= r(i, a) + \sum_{j \in S} q(j|i, a)g_0(f)(j). \end{aligned} \qquad (3.44)$$

Let

$$B_1^f(i) := \left\{ a \in A_0^f(i) : \begin{array}{c} w_f(i, a) > g_{-1}(f)(i); \quad \text{or} \\ \sum_{j \in S} q(j|i, a)g_1(f)(j) > g_0(f)(i) \\ \text{when } w_f(i, a) = g_{-1}(f)(i) \end{array} \right\}. \qquad (3.45)$$

We now define an *improvement policy* $h \in F$ (depending on $f$) as follows:

$$h(i) \in B_1^f(i) \quad \text{if } B_1^f(i) \neq \emptyset \quad \text{and} \quad h(i) := f(i) \quad \text{if } B_1^f(i) = \emptyset. \qquad (3.46)$$

The comments after (3.37) concerning $h$ and $f$ remain valid in the present case, with obvious notational changes. Moreover, from (3.19), Lemma 3.1(b), and Lemma 3.5 we see that $f(i)$ is in $A_0^f(i)$ for each $i \in S$, but $f(i)$ is not in $B_1^f(i)$ for any $i \in S$.

**Lemma 3.18** *For any given $f \in F_{-1}^*$, let $h$ be as in* (3.46). *Then*

(a) $g_{-1}(h) = g_{-1}(f)$, *and* $g_0(h) \geq g_0(f)$.
(b) *If* $g_0(h) = g_0(f)$ *and* $h \neq f$, *then* $g_1(h) \succeq g_1(f)$.

*Proof* (a) In Lemma 3.17, take $f^*$ and $f$ as $f$ and $h$ here, respectively. Hence, by (3.19) and (3.46) the conditions in Lemma 3.17(b) hold. Thus (a) follows from Lemma 3.17(b).

(b) Since $g_0(h) = g_0(f)$, by (a) and Lemma 3.5(a) we have

$$\begin{aligned} r(h) + Q(h)g_0(f) &= r(h) + Q(h)g_0(h) \\ &= g_{-1}(h) \\ &= g_{-1}(f) \\ &= r(f) + Q(f)g_0(f). \end{aligned} \tag{3.47}$$

Hence, by Theorem 3.6(b), using again $g_0(h) = g_0(f)$, we obtain

$$P^*(h)\big[Q(h) - Q(f)\big]g_1(f) = 0. \tag{3.48}$$

By (3.47) and (3.44) we see that $w_f(i, h(i)) = w_f(i, f(i)) = g_{-1}(f)(i)$ for all $i \in S$. Hence, by (3.45) and (3.46) we have

$$\big[Q(h) - Q(f)\big]g_1(f) = Q(h)g_1(f) - g_0(f) \geq 0. \tag{3.49}$$

This implies (by Lemma 3.9(a), (3.48) and (3.46)) that $Q(h)g_1(f)(i) = g_0(f)(i)$ for all the recurrent states $i$ under $Q(h)$, and so we have

$$h(i) = f(i) \quad \forall \text{ recurrent state } i \text{ under } Q(h).$$

Therefore (from (3.17) and (3.18))

$$P^*(h)\big[Q(h) - Q(f)\big] = 0, \tag{3.50}$$

which, together with (3.49) and Theorem 3.6(c), gives

$$\begin{aligned} g_1(h) - g_1(f) &= \int_0^\infty P(t,h)\big[Q(h) - Q(f)\big]g_1(f)\,dt \\ &\quad + P^*(h)\big[Q(h) - Q(f)\big]g_2(f) \\ &= \int_0^\infty P(t,h)\big[Q(h) - Q(f)\big]g_1(f)\,dt \geq 0, \end{aligned}$$

that is, $g_1(h) \geq g_1(f)$. It remains to show that $g_1(h) \neq g_1(f)$.

By contradiction, suppose that $g_1(h) = g_1(f)$. Then, since by assumption $g_0(h) = g_0(f)$, from (3.10) we have

$$Q(f)g_1(f) = g_0(f) = g_0(h) = Q(h)g_1(h) = Q(h)g_1(f). \tag{3.51}$$

On the other hand, since $h \neq f$, by (3.45)–(3.47) we obtain

$$Q(h)g_1(f) \succeq Q(f)g_1(f),$$

which contradicts (3.51). □

Having Lemma 3.18, we can state the following *policy iteration algorithm* for computing a 0-bias optimal policy.

**0-bias policy iteration algorithm:**

1. Let $k = 0$, and take $f_k := f_{-1}^*$ as in Theorem 3.16.
2. (Policy evaluation) Use Lemma 3.5 and (3.18) to obtain $g_{-1}(f_k)$, $g_0(f_k)$, and $g_1(f_k)$.
3. (Policy improvement) Obtain an improvement policy $f_{k+1}$ from (3.46) (with $f_k$ and $f_{k+1}$ in lieu of $f$ and $h$, respectively).
4. If $f_{k+1} = f_k$, then stop because $f_{k+1}$ is 0-bias optimal (by Theorem 3.19 below). Otherwise, increment $k$ by 1 and return to Step 2.

The 0-bias policy iteration algorithm ensures the existence of a policy satisfying (3.21)–(3.23) (with $n = 0$). The precise statement is as follows.

**Theorem 3.19** *Starting from a* $(-1)$*-bias optimal policy* $f_{-1}^*(\in F_{-1}^*)$ *in Theorem* 3.16, *the* 0*-bias policy iteration algorithm stops at a* 0*-bias optimal policy (denoted as* $f_0^*$*) satisfying* (3.21)–(3.23) *(with* $n = 0$*) in a finite number of iterations.*

*Proof* Let $\{f_k\}$ be the sequence of policies obtained by the 0-bias policy iteration algorithm. Then, by the definition of $\{f_k\}$ and Lemma 3.18(a) we see that the $f_k$ are all in $F_{-1}^*$ and $g_0(f_{k+1}) \geq g_0(f_k)$. Hence, as $k$ increases, $g_0(f_k)$ either increases or stays the same. Furthermore, by Lemma 3.18(b), when $g_0(f_k)$ remains the same, $g_1(f_k)$ increases. Thus, any two policies in the sequence $\{f_k\}$ either have different 0-biases or have different 1-biases. Thus, every policy in the iteration sequence is different. Since the number of policies in $F_{-1}^*$ is finite, the algorithm must stop after a finite number of iterations. Suppose that it stops at a policy denoted as $f_0^*$. Then $f_0^*$ must satisfy the optimality conditions (3.22) and (3.23) (with $n = 0$) because, otherwise, we can find the next improved policy in the policy iteration algorithm. On the other hand, since $f_0^*$ is also in $F_{-1}^*$, we have $g_{-1}(f_0^*) = g_{-1}(f_{-1}^*)$. Thus, by Theorem 3.16 we see that $f_0^*$ also satisfies (3.21), and so $f_0^*$ satisfies (3.21)–(3.23) (with $n = 0$). Hence, by Theorem 3.11, $f_0^*$ is 0-bias optimal. □

### 3.5.3 n-bias Policy Iteration Algorithms

As we have seen above, starting from an arbitrary policy $f$ in $F$, we can obtain a $(-1)$-bias optimal policy satisfying (3.21) and (3.22) by using the $(-1)$-bias policy iteration algorithm. Then, starting from such a $(-1)$-bias optimal policy, we can further obtain a 0-bias optimal policy satisfying (3.21)–(3.23) with $n=0$, by using the 0-bias policy iteration algorithm. Following a similar procedure, we will now prove the existence of a policy satisfying (3.21)–(3.23) for every $n \geq 0$. To do so, we will use *induction* on $n$ in (3.23) and the $n$-bias policy iteration algorithm introduced below. We will proceed as in the previous subsection.

Suppose that there exists $f$ satisfying (3.21)–(3.23) for some $n \geq 0$. We then need to show the existence of a policy $h$ satisfying (3.21)–(3.23) for $n+1$. Let $f$ be as in the induction hypothesis, that is, (3.21)–(3.23) hold for all $0 \leq k \leq n$ when $f^*$ there is replaced with $f$ above. Hence, by Theorem 3.11(a), $f$ is $n$-bias optimal. Now recall the sets $A_k^f(i)$ in (3.19)–(3.20) and the definitions of $B_0^f(i)$ and $B_1^f(i)$ in (3.36) and (3.45) respectively. Furthermore, for all $i \in S$ and $n \geq 0$, let

$$B_{n+2}^f(i) := \left\{ a \in A_{n+1}^f(i) : \begin{array}{r} \sum_{j \in S} q(j|i,a) g_{n+1}(f)(j) > g_n(f)(i), \quad \text{or} \\ \sum_{j \in S} q(j|i,a) g_{n+2}(f)(j) > g_{n+1}(f)(i) \\ \text{when } \sum_{j \in S} q(j|i,a) g_{n+1}(f)(j) = g_n(f)(i) \end{array} \right\}. \tag{3.52}$$

We then define an *improvement policy* $h \in F$ (depending on $f$) as follows:

$$h(i) \in B_{n+2}^f(i) \quad \text{if } B_{n+2}^f(i) \neq \emptyset, \quad \text{and} \quad h(i) := f(i) \quad \text{if } B_{n+2}^f(i) = \emptyset. \tag{3.53}$$

By Lemma 3.5, $f(i)$ is in $A_k^f(i)$ for all $i \in S$ and $0 \leq k \leq n+1$, but it follows from (3.52) and (3.53) that $f(i)$ is not in $B_{n+2}^f(i)$ for any $i$.

**Lemma 3.20** *Suppose that $f$ satisfies* (3.21)–(3.23) *for some $n \geq 0$. Let $h$ be defined as in* (3.53). *Then,*

(a) *$g_k(h) = g_k(f)$ for all $-1 \leq k \leq n$, and $g_{n+1}(h) \geq g_{n+1}(f)$.*
(b) *If $g_{n+1}(h) = g_{n+1}(f)$ and $h \neq f$, then $g_{n+2}(h) \succeq g_{n+2}(f)$.*

*Proof* (a) As we have already remarked, $f(i)$ is in $A_k^f(i)$ for all $0 \leq k \leq n+1$ and $i \in S$. Hence, by (3.53) and (3.20) we see that $h(i)$ is also in $A_k^f(i)$ for $0 \leq k \leq n+1$ and $i \in S$, and so we have

$$Q(h) g_{-1}(f) = 0,$$

$$r(h) + Q(h) g_0(f) = g_{-1}(f), \tag{3.54}$$

$$Q(h) g_{k+1}(f) = g_k(f) = Q(f) g_{k+1}(f) \quad \forall 0 \leq k \leq n-1, \tag{3.55}$$

$$Q(h) g_{n+1}(f) \geq g_n(f) = Q(f) g_{n+1}(f). \tag{3.56}$$

(When $n=0$, (3.55) is replaced by (3.56) with $n=0$.)

By (3.54)–(3.56), Lemma 3.17(a), and Theorem 3.6, we obtain

$$g_k(h) = g_k(f) \quad \forall -1 \le k \le n-1, \text{ and } g_n(h) \ge g_n(f),$$

which, together with $g_n(f) \ge g_n(h)$ (by Theorem 3.11 and the $n$-bias optimality of $f$), gives the first part of (a). Since $g_k(h) = g_k(f)$ $(k = n-1, n)$, by Theorem 3.6(c) and (3.55) we have $P^*(h)[Q(h) - Q(f)]g_{n+1}(f) = 0$, which, together with (3.56) and Lemma 3.9, implies that $Q(h)g_{n+1}(f)(i) = Q(f)]g_{n+1}(f)(i)$ for recurrent states $i$ under $Q(h)$. Thus, from (3.52)–(3.53) and Lemma 3.5(c) we see

$$Q(h)g_{n+2}(f)(i) \ge g_{n+1}(f)(i) = Q(f)g_{n+2}(f)(i)$$

for recurrent states $i$ under $Q(h)$, and therefore (by Lemma 3.9)

$$P^*(h)\big[Q(h) - Q(f)\big]g_{n+2}(f) \ge 0. \tag{3.57}$$

Using $g_n(h) = g_n(f)$ in the first part of (a), by Theorem 3.6(c) and (3.56)–(3.57) we obtain $g_{n+1}(h) \ge g_{n+1}(f)$, and so (a) follows.

(b) Since $g_{n+1}(h) = g_{n+1}(f)$, by (a) and Lemma 3.5(c) we have

$$Q(h)g_{n+1}(f) = Q(h)g_{n+1}(h) = g_n(h) = g_n(f) = Q(f)g_{n+1}(f),$$

which, together with Theorem 3.6(c) and $g_{n+1}(h) = g_{n+1}(f)$, implies that

$$P^*(h)\big[Q(h) - Q(f)\big]g_{n+2}(f) = 0, \quad \text{and} \quad Q(h)g_{n+2} \ge g_{n+1}(f). \tag{3.58}$$

Thus, by (3.57), (3.58), and Lemma 3.9(a), we obtain $[Q(h) - Q(f)]g_{n+2}(f)(i) = 0$ for all recurrent states $i$ under $Q(h)$. Hence, it follows from (3.52) and (3.53) that

$$h(i) = f(i) \quad \forall \text{ recurrent state } i \text{ under } Q(h),$$

and so

$$P^*(h)\big[Q(h) - Q(f)\big] = 0.$$

This fact, together with Theorem 3.6(c) and (3.58), gives

$$g_{n+2}(h) - g_{n+2}(f) = \int_0^\infty P(t,h)\big[Q(h) - Q(f)\big]g_{n+2}(f)\,dt \ge 0.$$

Thus, to complete the proof of part (b), it only remains to show that $g_{n+2}(h) \ne g_{n+2}(f)$.

Suppose that $g_{n+2}(h) = g_{n+2}(f)$. Then, it follows from (3.10) that

$$Q(f)g_{n+2}(f) = g_{n+1}(f) = g_{n+1}(h) = Q(h)g_{n+2}(h) = Q(h)g_{n+2}(f). \tag{3.59}$$

However, since $h \ne f$, by (3.52)–(3.53) we have

$$Q(h)g_{n+2}(f) \succeq g_{n+1}(f) = Q(f)g_{n+2}(f),$$

which contradicts (3.59). □

Finally, with Lemma 3.20 we can state a *policy iteration algorithm* for computing an $n$-bias optimal policy as follows.

**$n$-bias policy iteration algorithm:** Fix $n \geq 1$.

1. Let $k = 0$ and pick a policy $f_k \in F_{n-1}^*$.
2. (Policy evaluation) Obtain (by Lemma 3.5) $g_l(f_k)$ for all $0 \leq l \leq n+1$.
3. (Policy improvement) Obtain a policy $f_{k+1}$ from (3.52)–(3.53).
4. If $f_{k+1} = f_k$, then stop because $f_{k+1}$ is $n$-bias optimal (by Theorem 3.21 below). Otherwise, increment $k$ by 1 and return to Step 2.

By analogy with Theorem 3.19, we now obtain the following.

**Theorem 3.21** *Fix $n \geq 1$. Starting from an $(n-1)$-bias optimal policy satisfying the first $n+2$ bias optimality conditions* (3.21)–(3.23), *the $n$-bias policy iteration algorithm stops, in a finite number of iterations, at an $n$-bias optimal policy satisfying the $n+3$ bias optimality conditions* (3.21)–(3.23).

*Proof* Let $\{f_k\}$ be a sequence of policies obtained by the $n$-bias policy iteration algorithm, with $f_0$ being $(n-1)$-bias optimal and satisfying the first $n+1$ bias optimality conditions (3.21)–(3.23). Hence, as $k$ increases, from the construction of $\{f_k\}$ and Lemma 3.20 we see that $g_n(f_k)$ either increases or remains the same. Furthermore, by Lemma 3.20(b), when $g_n(f_k)$ remains the same, $g_{n+1}(f_k)$ increases in $k$. Thus, any two policies in the sequence $\{f_k\}$ either have different $n$-biases or have different $(n+1)$-biases. Thus, every policy in the iteration sequence is different. Since the number of policies in $F_{n-1}^*$ is finite, the iteration must stop after a finite number of iterations; otherwise, we can find the next improved policy in the policy iteration. Suppose that the algorithm stops at a policy denoted $f^*$. Thus, by (3.20) and (3.52) we see that $f^*$ must satisfy

$$\max_{a \in A_{n+1}^{f^*}(i)} \left\{ \sum_{j \in S} q(j|i,a) g_{n+1}(f^*)(j) \right\} = g_n(f^*)(i) \quad \forall i \in S, \tag{3.60}$$

$$\max_{a \in A_{n+2}^{f^*}(i)} \left\{ \sum_{j \in S} q(j|i,a) g_{n+2}(f^*)(j) \right\} = g_{n+1}(f^*)(i) \quad \forall i \in S. \tag{3.61}$$

Moreover, from Lemma 3.20 we also have

$$g_l(f^*) = g_l(f_0) \quad \forall -1 \leq l \leq n,$$

which, together with (3.20), gives $A_l^{f^*}(i) = A_l^{f_0}(i)$ for all $0 \leq l \leq n+1$ and $i \in S$. Therefore, $f^*$ also satisfies the first $n+1$ bias optimality conditions from (3.21) to (3.23) because $f_0$ does. Thus, by the construction of $f^*$ and (3.60)–(3.61), we see that $f^*$ satisfies the $n+3$ bias optimality conditions (3.21)–(3.23). Thus, by Theorem 3.11, $f^*$ is $n$-bias optimal. □

Inductively, by Theorems 3.7, 3.16, 3.19, and 3.21 we conclude that we can use policy iteration algorithms to obtain a policy that is $n$-bias optimal. In particular, in a finite number of iterations, we can obtain a policy that is $n$-bias optimal for all $n \geq -1$ by using the $|S|$-bias policy iteration algorithm.

## 3.6 The Linear Programming Approach

We cannot close this chapter on finite MDPs without mentioning the linear programming formulation, which was one of the earliest solution approaches. See, for instance, Kallenberg (1983) [96] or Chap. 6 in Hernández-Lerma and Lasserre (1996) [73] for references going back to the early 1960s. We will consider ergodic and multichain models in Sects. 3.6.1 and 3.6.2, respectively.

### *3.6.1 Linear Programming for Ergodic Models*

This subsection deals with the important special case of *ergodic models*, that is, we suppose that our continuous-time MDP under each (randomized or deterministic) stationary policy becomes an *ergodic* Markov chain. (For the definition of ergodicity, see the paragraph after Proposition B.15 in Appendix B.2.)

Thus, as a consequence of Lemma 3.1, (3.3), and (3.18), $\bar{V}(f) = g_{-1}(f)$ is a constant vector for each $f \in F$, and so $\bar{V}^*(i) \equiv \bar{V}^*$ is independent of $i$ and will be still denoted by $\bar{V}^*$. It follows from Corollary 3.14 (at the end of Sect. 3.4) that $g \geq \bar{V}^*$ whenever there exist a function $h$ on $S$ and a constant $g$ such that

$$g \geq r(i,a) + \sum_{j \in S} q(j|i,a)h(j) \quad \forall i \in S \text{ and } a \in A(i). \tag{3.62}$$

Moreover, by Theorems 3.13 and 3.16 we further see that $\bar{V}^*$ is the smallest $g$ for which there exists an $h$ on $S$ satisfying (3.62). This yields the following linear programming model.

**Primal linear program (P-LP).**

Minimize $g$

subject to

$$g - \sum_{j \in S} q(j|i,a)h(j) \geq r(i,a) \quad \forall i \in S,\ a \in A(i),$$

with constants $g$ and functions $h$ unconstrained. (Observe that the latter inequality is just a way of rewriting (3.62).)

We will see, however, that it is more informative to analyze this P-LP problem through its dual linear program below.

**Dual linear program (D-LP).**

$$\text{Maximize} \sum_{i\in S}\sum_{a\in A(i)} r(i,a)x(i,a)$$

subject to

$$\sum_{i\in S}\sum_{a\in A(i)} q(j|i,a)x(i,a) = 0 \quad \forall j \in S, \tag{3.63}$$

$$\sum_{i\in S}\sum_{a\in A(i)} x(i,a) = 1, \tag{3.64}$$

and $x(i,a) \geq 0$ for all $a \in A(i)$ and $i \in S$.

By definition, $x(i,a)$ is a *feasible solution* to this D-LP if $x(i,a)$ is nonnegative and satisfies (3.63) and (3.64). A feasible solution $x^*(i,a)$ to the D-LP is said to be an *optimal solution* if

$$\sum_{i\in S}\sum_{a\in A(i)} r(i,a)x^*(i,a) \geq \sum_{i\in S}\sum_{a\in A(i)} r(i,a)x(i,a)$$

for all feasible solutions $x(i,a)$ to the D-LP.

Observe that (3.63)–(3.64) can be jointly written in the form $Bx = b$, with $B$ an $(|S|+1) \times |K|$-matrix, with $K$ as in (2.2), and $b$ a column vector in $\mathbf{R}^{|S|+1}$, and $B$ and $b$ are uniquely given with components 0, 1, and $q(j|i,a)$.

A *basis* of a matrix $B$ is a *maximal* set of linearly independent columns of $B$, and refer to the columns in the basis as *basic columns*. (Of course, such a basis of $B$ may not be unique, and so neither the basic columns.) A feasible solution to the D-LP in the form $Bx = b$ is called a *basic feasible solution* if its components not corresponding to the basic columns of some basis of $B$ equal 0. Note that one of the equations in (3.63) is *redundant* because of the conservativeness (2.3) of the MDP. Moreover, in the above D-LP, a basic feasible solution has at most $|S|$ nonzero components.

For each $\pi \in \Pi^s$, we recall from Definition 2.1 that $\pi(a|i)$ denotes the probability that $\pi$ chooses action $a$ at state $i$ and from (2.5) that $q(j|i,\pi) := \sum_{a\in A(i)} q(j|i,a)\pi(a|i)$. Moreover, it follows from Proposition C.5 that $p_\pi(s,i,t,j)$ is homogeneous and is denoted by $p_\pi(i,t,j) := p_\pi(0,i,t,j)$. In this case, if there exists an i.p.m. of $p_\pi(i,t,j)$, then it is also said that it is an i.p.m. of $Q(\pi) := [q(j|i,\pi)]$.

By (2.4) and Proposition C.12 we have the following fact.

**Lemma 3.22** *Fix an arbitrary policy $\pi \in \Pi^s$. Then, for the ergodic model, the equations*

$$\sum_{i\in S} q(j|i,\pi)y_i = 0 \quad \forall j \in S \tag{3.65}$$

*subject to*

$$\sum_{i\in S} y_i = 1, \quad \textit{and} \quad y_i > 0 \quad \forall i \in S \tag{3.66}$$

*have a unique solution* $\mu_\pi$, *which is the unique i.p.m. of* $[q(j|i,\pi)]$.

The following result establishes a relationship between feasible solutions to the D-LP and randomized stationary policies.

**Theorem 3.23**

(a) *For all* $\pi \in \Pi^s$, $i \in S$, *and* $a \in A(i)$, *let*

$$x_\pi(i,a) := \pi(a|i)\mu_\pi(i), \tag{3.67}$$

*where* $\mu_\pi$ *is the unique solution of* (3.65)–(3.66). *Then* $x_\pi := \{x_\pi(i,a), a \in A(i), i \in S\}$ *is a feasible solution to the D-LP problem.*

(b) *Let* $x := \{x(i,a), a \in A(i), i \in S\}$ *be a feasible solution to the D-LP problem. Then* $\sum_{a\in A(i)} x(i,a) > 0$ *for all* $i \in S$. *Define a randomized stationary policy* $\pi^x$ *by*

$$\pi^x(a|i) := \frac{x(i,a)}{\sum_{b\in A(i)} x(i,b)} \quad \forall a \in A(i) \textit{ and } i \in S. \tag{3.68}$$

*Then* $\pi^x$ *is in* $\Pi^s$, *and* $x_{\pi^x}(i,a) = x(i,a)$ *for all* $a \in A(i)$ *and* $i \in S$.

*Proof* (a) Since $\sum_{a\in A(i)} \pi(a|i) = 1 = \sum_{i\in S} \mu_\pi(i)$, by (3.67) and (3.66) we have

$$\sum_{i\in S}\sum_{a\in A(i)} x_\pi(x,a) = 1.$$

Moreover, since $q(j|i,\pi) = \sum_{a\in A(i)} q(j|i,a)\pi(a|i)$ for all $i, j \in S$, by (3.67) and (3.65) we have

$$\sum_{i\in S}\sum_{a\in A(i)} q(j|i,a)x_\pi(i,a) = \sum_{i\in S} q(j|i,\pi)\mu_\pi(i) = 0 \quad \forall j \in S.$$

Hence, $x_\pi$ is a feasible solution to D-LP.

(b) Let $\hat{u}(i) := \sum_{a\in A(i)} x(i,a)$ for each $i \in S$, and $S' := \{i \in S, \hat{u}(i) > 0\}$. Then, by (3.64) $S'$ is not empty. Thus, for each $i \in S'$, (3.68) implies that

$$x(i,a) = \pi^x(a|i)\hat{u}(i). \tag{3.69}$$

Since $\sum_{a\in A(i)} \pi^x(a|i) = 1$ and $q(j|i,\pi^x) = \sum_{a\in A(i)} q(j|i,a)\pi^x(a|i)$, substituting (3.69) into (3.63) and (3.64) shows that

$$\sum_{i\in S} q(j|i,\pi^x)\hat{u}(i) = 0, \quad \text{and} \quad \sum_{i\in S} \hat{u}(i) = 1.$$

Thus, by Proposition C.12, $\hat{u}(i) = \mu_{\pi^x}(i) > 0$ for all $i \in S$, and so $S' = S$. From (3.69) together with $\hat{u}(i) = \mu_{\pi^x}(i)$ and (a), we obtain (b). □

Theorem 3.23 establishes a one-to-one relationship between randomized stationary policies and feasible solutions to the D-LP problem through (3.67) and (3.68). The next result further establishes a one-to-one relationship between deterministic stationary policies and basic feasible solutions to the D-LP problem.

**Theorem 3.24**

(a) *If $f \in F$, then the i.p.m. $\mu_f$ of $[q(j|i, f)]$ is a basic feasible solution to the D-LP problem.*
(b) *Let $x := \{x(i,a), a \in A(i), i \in S\}$ be a basic feasible solution to the D-LP problem, and let $\pi^x$ be as in (3.68). Then $\pi^x$ is in $F$ and $\pi^x(i) = a$ if $x(i,a) > 0$.*
(c) *There exists a finite optimal basic feasible solution $x^*$ to the D-LP, and the policy $f^{x^*}$ for which $f^{x^*}(i) = a$ if $x^*(i,a) > 0$ is an AR optimal deterministic stationary policy.*

*Proof* Obviously, (a) follows from Theorem 3.23(a) and (3.67).

To prove (b), we note that a basic feasible solution has at most $|S|$ nonzero components. Since Theorem 3.23(b) establishes that $\sum_{a \in A(i)} x(i,a) > 0$ for all $i \in S$, there must be one nonzero entry $x(i,a)$ for each $i$, and so (b) follows.

(c) Since $S$ and $A(i)$ are both finite, and $r(i,a)$ is a real-valued function, the existence of an optimal basic feasible solution $x^*$ to the D-LP is a well-known fact from elementary linear programming. Hence, the desired result follows from (b). □

### *3.6.2 Linear Programming for Multichain Models*

The results in Sect. 3.6.1 are for ergodic models. Now we will extend those results to general multichain models. The term "multichain" refers to Markov chains as in Remark 3.8(b) (see also (3.17)), with several closed irreducible sets and perhaps a set of transient states. Our arguments here follow the approach by Puterman (1994) [129], which requires to use the *uniformization technique* to transform the multichain continuous-time MDP into a *discrete-time* MDP.

By the properties of solutions of the modified average reward optimality equations (3.33)–(3.34), it follows from Corollary 3.14 (at the end of Sect. 3.4) that if there exist two functions $g$ and $h$ on $S$ satisfying that

$$\sum_{j \in S} q(j|i,a) g(j) \le 0 \quad \forall a \in A(i),\ i \in S, \tag{3.70}$$

$$g(i) \ge r(i,a) + \sum_{j \in S} q(j|i,a) h(j) \quad \forall a \in A(i),\ i \in S, \tag{3.71}$$

then $g(i) \geq \bar{V}^*(i)$ for all $i \in S$. Consequently, the "minimal" $g$ that satisfies (3.70) and for which there exists $h$ satisfying (3.71) equals the optimal average reward $\bar{V}^*$. Furthermore, in the multichain model, $g(i)$ may vary with the state $i$. Therefore, now we will minimize positive linear combinations of components of $g$. Thus, the above discussion suggests the following primal linear programming problem to characterize $\bar{V}^*$.

**Primal linear program for multichain MDPs (P-LP-M).**

$$\text{Minimize} \sum_{j \in S} \hat{\alpha}_j g(j)$$

subject to

$$\sum_{j \in S} q(j|i,a) g(j) \leq 0 \quad \forall a \in A(i),\ i \in S, \tag{3.72}$$

$$g(i) \geq r(i,a) + \sum_{j \in S} q(j|i,a) h(j) \quad \forall a \in A(i),\ i \in S, \tag{3.73}$$

with $g$ and $h$ unconstrained, $\hat{\alpha}_j > 0$ for all $j \in S$, and $\sum_{j \in S} \hat{\alpha}_j = 1$.

As in the case for ergodic models, we will analyze this problem by means of its dual linear program.

**Dual linear program for multichain MDPs (D-LP-M).**

$$\text{Maximize} \sum_{i \in S} \sum_{a \in A(i)} r(i,a) x(i,a)$$

subject to

$$\sum_{i \in S} \sum_{a \in A(i)} q(j|i,a) x(i,a) = 0, \quad j \in S, \tag{3.74}$$

$$\sum_{a \in A(i)} x(i,a) - \sum_{i \in S} \sum_{a \in A(i)} q(j|i,a) y(i,a) = \hat{\alpha}_j, \quad j \in S, \tag{3.75}$$

and $x(i,a) \geq 0$, $y(i,a) \geq 0$ for all $a \in A(i)$ and $i \in S$.

This is the generalization of the ergodic case in Sect. 3.6.1; the multichain problem requires the inclusion of (3.72) in the primal linear program and the corresponding additional set of variables $y(i,a)$ in the dual problem. In the ergodic case, $g(\cdot) \equiv g$ is a *constant*, and so (3.72) trivially holds. Furthermore, at least one of the equations in (3.74) is redundant, and so (3.75) generalizes the constraint

$$\sum_{i \in S} \sum_{a \in A(i)} x(i,a) = 1$$

in the ergodic model. To see this, it suffices to sum (3.75) over $j \in S$.

We will next establish the relationship between optimal solutions to the D-LP-M and average reward optimal policies. (In fact, the relationship between feasible solutions and randomized stationary policies can also be obtained with arguments as for the above ergodic models and those by Puterman (1994) [129, pp. 462–468]. However, we omit the details.)

By the finiteness of $S$ and $A(i)$, we see that $\|q\| := \sup_{a\in A(i),i\in S} |q(i|i,a)| < \infty$. Now let

$$\beta := \frac{\|q\|}{1+\|q\|}, \qquad \bar{r}(i,a) := \frac{r(i,a)}{1+\|q\|}, \quad \text{and}$$
$$\bar{p}(j|i,a) := \frac{q(j|i,a)}{1+\|q\|} + \delta_{ij} \tag{3.76}$$

for all $i, j \in S$ and $a \in A(i)$. Then, by our hypotheses on the model (2.1) it follows that $\bar{p}(j|i,a) \geq 0$ and $\sum_{j\in S} \bar{p}(j|i,a) = 1$ for all $a \in A(i)$ and $i \in S$. Hence, $\{S, A(i), \bar{p}(j|i,a), \bar{r}(i,a)\}$ is the model of a *discrete-time* MDP; see Puterman (1994) [129] or Sennott (1999) [141], for instance. This approach of transforming a continuous-time MDP into a discrete-time MDP is called the *uniformization technique*. Observe from (3.76) that this technique requires $\|q\| < \infty$.

Let $\bar{P}(f) := [\bar{p}(j|i, f(i))]$ (for any fixed $f \in F$), which of course is a transition probability matrix. It generates a deviation matrix denoted by $H_p^f$ (see Puterman (1994) [129, Theorem A.7(b)], for instance), which by Proposition C.12(b) is given by

$$\begin{aligned} H_p^f &:= \left[I - \bar{P}(f) + P^*(f)\right]^{-1}\left(I - P^*(f)\right) \\ &= \left[P^*(f) - \left(1+\|q\|\right)^{-1} Q(f)\right]^{-1}\left(I - P^*(f)\right). \end{aligned}$$

The $(i, j)$-element of $H_p^f$ is denoted by $H_p^f(j|i)$. The $(i, j)$-element of $P^*(f)$ is denoted by $p_f^*(j|i)$.

**Theorem 3.25**

(a) *Suppose that $f \in F$ is AR optimal, and let*

$$\begin{aligned} x_f\left(i, f(i)\right) &:= \sum_{j\in S} \hat{\alpha}_j p_f^*(j|i), \\ y_f\left(i, f(i)\right) &:= \sum_{j\in S} \left[\hat{\alpha}_j H_p^f(j|i) + \gamma_j p_f^*(j|i)\right], \\ x_f(i,a) &= y_f(i,a) := 0 \quad \text{for } a \neq f(i), \end{aligned}$$

*where*

$$\gamma_j := \begin{cases} \max_{l\in C_k^f}\{-\sum_{i\in S} \hat{\alpha}_i H_p^f(l|i)) / \sum_{i\in C_k^f} p_f^*(l|i)\}, & j \in C_k^f \\ 0 & j \in T^f \end{cases}$$

*with $C_k^f$ being the kth recurrent class of $Q(f)$ and $T^f$ the set of transient states of $Q(f)$. Then $(x_f, y_f)$ is a basic feasible solution to the D-LP-M problem.*

(b) *Suppose that $(x^*, y^*)$ is an optimal solution to the D-LP-M problem. Then the following policy $\pi^{(x^*,y^*)}$ is AR optimal:*

$$\pi^{(x^*,y^*)}(a|i) := \begin{cases} x(i,a)/\sum_{a'\in A(i)} x(i,a'), & i \in S_x, \\ y(i,a)/\sum_{a'\in A(i)} y(i,a'), & i \notin S_x, \end{cases} \tag{3.77}$$

*where $S_x := \{i \in S : \sum_{a\in A(i)} x(i,a) > 0\}$.*

*Proof* A direct calculation together with (3.76) shows that (3.74) and (3.75) can be rewritten as

$$\text{Maximize} \sum_{i\in S}\sum_{a\in A(i)} r(i,a)x(i,a)$$

subject to

$$\sum_{a\in A(j)} x(j,a) - \sum_{i\in S}\sum_{a\in A(i)} \bar{p}(j|i,a)x(i,a) = 0 \quad \forall j \in S, \tag{3.78}$$

$$\sum_{a\in A(j)} x(j,a) + \sum_{a\in A(j)} z(j,a) - \sum_{i\in S}\sum_{a\in A(i)} \bar{p}(j|i,a)z(i,a) = \hat{\alpha}_j \quad \forall j \in S, \tag{3.79}$$

with $z(i,a) := (1 + \|q\|)y(i,a)$. Now we see that (3.78) and (3.79) are the same as the dual linear program (9.3.5) and (9.3.6) in Puterman (1994) [129]. Thus, the desired results follow from Theorems 9.3.5 and 9.3.8 by Puterman (1994) [129]. □

*Remark 3.26* In contrast to Theorem 3.24 for ergodic models, the optimal policy $\pi^{(x^*,y^*)}$ in Theorem 3.25 may *not* be deterministic even if the optimal solution $(x^*, y^*)$ is basic feasible. In fact, there are examples for which a basic feasible solution to the D-LP-M does not generate a deterministic stationary policy through (3.77). However, Theorem 3.25, together with Theorem 3.16, shows that if an optimal solution of D-LP-M does not generate a deterministic stationary policy, then we can find a different optimal solution which does by using the uniformization technique; for instance, see Examples 9.3.2 and 9.3.1 by Puterman (1994) [129, p. 471].

## 3.7 Notes

When the state and the action spaces are both finite, the so-called average reward optimality equations and the methods to determine optimal stationary policies have been investigated by Howard (1960) [86], Lembersky (1974) [106], Miller (1968) [117], and many other authors. Here, in Sects. 3.2–3.5, we follow the approach by Guo, Song, and Zhang (2009) [66] and Zhang and Cao (2009) [166], whose proofs are direct and simple. Moreover, the approach is self-contained in the sense that it does not require results from other problems, for instance, from discount discrete-time or continuous-time MDPs.

The approach of transforming a continuous-time MDP into an equivalent discrete-time MDP, as in (3.76), is called the *uniformization technique*; see Bertsekas (2001) [12], Howard (1960) [86], Lewis and Puterman (2001) [107], Puterman (1994) [129], Serfozo (1979) [142], Sennott (1999) [141], and Veinott (1969) [151], for instance. When the latter approach is applicable, many results about continuous-time MDPs (such as the existence of solutions to the AROE, convergence of the value and the policy iteration algorithms, etc.) can be obtained from those for the discrete-time case. One should keep in mind, however, that the uniformization technique requires $\|q\|$ to be finite! Moreover, the uniformization technique cannot show the fact that EAR optimality over the class of all randomized Markov policies is the same as that over the class of all stationary policies.

For discrete-time MDPs, the reader may consult, for instance, Bertsekas (1995) [11], Bertsekas (2001) [12], Borkar (1991) [17], Dynkin and Yushkevich (1979) [38], Feinberg and Shwartz (2002) [41], Guo and Shi (2001) [60], Hernández-Lerma and Lasserre (1999) [74], Hou and Guo (1998) [84], Howard (1960) [86], Puterman (1994) [129], Sennott (1999) [141], Ross (1968) [137], Ross (1983) [138], and Yushkevich (1973) [161]. For controlled diffusion processes, see Borkar (1989) [16], and Fleming and Soner (1993) [44]. For controlled piecewise deterministic Markov processes, see Davis (1993) [32].

# Chapter 4
# Discount Optimality for Nonnegative Costs

This chapter deals with the general continuous-time MDP in (2.1) with a *denumerable* space $S$, in contrast to Chap. 3 in which we concentrated on finite models. The main concern now is to minimize the expected discounted *cost* and find conditions that ensure the existence of discounted-cost optimal policies.

## 4.1 Introduction

As mentioned above, in this chapter we go back to the general continuous-time MDP in (2.1) with a *denumerable* (finite or infinitely countable) state space $S$, in contrast to Chap. 3 in which we concentrated on finite control models. Our main concern now is to minimize the expected discounted *cost*, so that the reward $r(i,a)$ in (2.19)–(2.20) should be replaced with a *cost* $c(i,a)$. In addition, we will introduce two simplifying assumptions. The first one is that $c(i,a)$ is supposed to be *nonnegative*, and the second is that Assumption 2.2 (in Sect. 2.2) is supposed to hold throughout this chapter.

The rest of the chapter is organized as follows. In Sect. 4.2, we introduce the nonnegative model that we are concerned with. The discounted-cost optimality equation is established in Sect. 4.4, after some preliminary results given in Sect. 4.3. Under suitable hypotheses on the MDP model, in Sect. 4.5, we show the existence of optimal policies. In Sect. 4.6, we use a "value iteration" scheme to obtain some approximation results. We then specialize the MDP to a *finite* model in Sect. 4.7 and introduce a policy iteration algorithm to find a discounted-cost optimal stationary policy. Section 4.8 presents some examples that illustrate our results, and the chapter concludes in Sect. 4.9 with notes on the literature related to this chapter.

## 4.2 The Nonnegative Model

In this chapter, we consider the nonnegative cost MDP model

$$\{S, (A(i), i \in S), q(j|i,a), c(i,a)\}, \tag{4.1}$$

X. Guo, O. Hernández-Lerma, *Continuous-Time Markov Decision Processes*, Stochastic Modelling and Applied Probability 62,
DOI 10.1007/978-3-642-02547-1_4, 

which is the same as (2.1), except that the reward function $r(i, a)$ in (2.1) is replaced with a *nonnegative* cost function $c(i, a)$ in (4.1), that is, $c(i, a) \geq 0$ on $K$.

To introduce the optimality criterion, we recall some notation from Sects. 2.2 and 2.3: For all $i \in S$, $\pi = (\pi_t) \in \Pi$, $t \geq 0$, and $p_\pi(s, i, t, j)$,

$$\begin{aligned} c(i, \pi_t) &= \int_{A(i)} c(i, a)\pi_t(da|i), \\ E_i^\pi c\big(x(t), \pi_t\big) &= \sum_{j \in S} c(j, \pi_t) p_\pi(0, i, t, j). \end{aligned} \tag{4.2}$$

In particular, when $\pi = f \in F$, we write $c(i, \pi_t)$ as $c(i, f) := c(i, f(i))$.

Then, for every (fixed) discount factor $\alpha > 0$, we define the expected discounted-cost criterion

$$J_\alpha(i, \pi) := \int_0^\infty e^{-\alpha t} E_i^\pi c\big(x(t), \pi_t\big)\, dt, \tag{4.3}$$

and the corresponding optimal discounted cost or value function

$$J_\alpha^*(i) := \inf_{\pi \in \Pi} J_\alpha(i, \pi)$$

for all $i \in S$.

**Definition 4.1** For each $\varepsilon \geq 0$, a policy $\pi^* \in \Pi$ is called discounted-cost $\varepsilon$-optimal if

$$J_\alpha\big(i, \pi^*\big) \leq J_\alpha^*(i) + \varepsilon \quad \forall i \in S.$$

A discounted-cost 0-optimal policy is simply called discounted-cost optimal.

*Remark 4.2* Under Assumption 2.2, the transition probability function $p_\pi(s, i, t, j)$ is regular; see Theorem 2.5(a). Then, without loss of generality, in (4.3) we may replace the cost $c$ with $c + L$ for any constant $L$. Therefore, under Assumption 2.2, the condition "$c \geq 0$" in our model (4.1) can be weakened to "$c$ is bounded below."

## 4.3 Preliminaries

This section presents some technical results required to obtain the optimality equation in Theorem 4.6 below.

The next result establishes a relationship between the discounted cost and the (pointwise) minimal nonnegative solution to some equations.

**Lemma 4.3** *For all $f \in F$ and $i \in S$, let $q_i(f) := q_i(f(i))$. Then the following statements hold*:

(a) *$J_\alpha(f)$ is the minimum nonnegative solution to the equation*

$$u(i) = \frac{c(i, f)}{\alpha + q_i(f)} + \frac{1}{\alpha + q_i(f)} \sum_{j \neq i} u(j) q(j|i, f), \quad i \in S. \tag{4.4}$$

(b) *If a nonnegative function $u$ on $S$ satisfies*

$$u(i) \geq \frac{c(i, f)}{\alpha + q_i(f)} + \frac{1}{\alpha + q_i(f)} \sum_{j \neq i} u(j) q(j|i, f) \quad \forall i \in S,$$

*then $u \geq J_\alpha(f)$.*

*Proof* (a) We will first show that $J_\alpha(i, f)$ can be expressed as in (4.6) below. Choose an arbitrary $f \in F$. For all $i, j \in S$ and $n \geq 1$, let

$$\varphi_{ij}^{(n)}(f) := \begin{cases} \frac{\delta_{ij}}{\alpha + q_i(f)} & \text{for } n = 1, \\ \frac{1}{\alpha + q_i(f)} [\delta_{ij} + \sum_{k \neq i} q(k|i, f) \varphi_{kj}^{(n-1)}(f)] & \text{for } n \geq 2. \end{cases} \tag{4.5}$$

Noting that $p_f(s, i, t, j)$ is homogeneous (since $f \in F$ is stationary) and that $\varphi_{ij}^{(n)}(f)$ is nondecreasing in $n \geq 1$, by Proposition C.6 we have

$$\int_0^\infty e^{-\alpha t} p_f(0, i, t, j)\, dt = \lim_{n \to \infty} \varphi_{ij}^{(n)}(f).$$

Therefore, since $c(i, a) \geq 0$, the monotone convergence theorem gives

$$\begin{aligned} J_\alpha(i, f) &= \sum_{j \in S} \int_0^\infty e^{-\alpha t} c(j, f) p_f(0, i, t, j)\, dt \\ &= \sum_{j \in S} c(j, f) \Big[ \lim_{n \to \infty} \varphi_{ij}^{(n)}(f) \Big] \\ &= \lim_{n \to \infty} \sum_{j \in S} c(j, f) \varphi_{ij}^{(n)}(f). \end{aligned} \tag{4.6}$$

It is now easily seen that $J_\alpha(f)$ satisfies (4.4). Indeed, for any $n \geq 1$, from (4.5) we see that

$$\begin{aligned} &\sum_{j \in S} c(j, f) \varphi_{ij}^{(n+1)}(f) \\ &= \sum_{j \in S} c(j, f) \left[ \frac{\delta_{ij}}{\alpha + q_i(f)} + \frac{1}{\alpha + q_i(f)} \sum_{k \neq i} \varphi_{kj}^{(n)}(f) q(k|i, f) \right] \\ &= \frac{c(i, f)}{\alpha + q_i(f)} + \frac{1}{\alpha + q_i(f)} \sum_{k \neq i} \left[ \sum_{j \in S} c(j, f) \varphi_{kj}^{(n)}(f) \right] q(k|i, f), \end{aligned} \tag{4.7}$$

and then letting $n \to \infty$, from the monotone convergence theorem and (4.6)–(4.7) we obtain

$$J_\alpha(i, f) = \frac{c(i, f)}{\alpha + q_i(f)} + \frac{1}{\alpha + q_i(f)} \sum_{k \neq i} J_\alpha(k, f) q(k|i, f).$$

Hence, $J_\alpha(f)$ satisfies (4.4).

Now, to prove that $J_\alpha(f)$ is the minimum nonnegative solution to (4.4), let $u$ be an arbitrary nonnegative solution of (4.4). To prove that $J_\alpha(f) \leq u$, by (4.6) it is sufficient to show that

$$\sum_{j \in S} c(j, f) \varphi_{ij}^{(n)}(f) \leq u(i) \quad \forall i \in S \text{ and } n \geq 1. \tag{4.8}$$

This is obviously true for $n = 1$. In fact, since $u(i) \geq 0$ and $q(j|i, f) \geq 0$ for $i \neq j$, we have

$$\begin{aligned} u(i) &= \frac{c(i, f)}{\alpha + q_i(f)} + \frac{1}{\alpha + q_i(f)} \sum_{j \neq i} u(j) q(j|i, f) \\ &\geq \frac{c(i, f)}{\alpha + q_i(f)} = \sum_{j \in S} c(j, f) \varphi_{ij}^{(1)}(f). \end{aligned}$$

To prove (4.8) by induction, suppose that (4.8) holds for some $n \geq 1$. Then, by (4.7) and the induction hypothesis, we have

$$\begin{aligned} \sum_{j \in S} c(j, f) \varphi_{ij}^{(n+1)}(f) &= \frac{c(i, f)}{\alpha + q_i(f)} + \frac{1}{\alpha + q_i(f)} \sum_{k \neq i} \left[ \sum_{j \in S} c(j, f) \varphi_{kj}^{(n)}(f) \right] q(k|i, f) \\ &\leq \frac{c(i, f)}{\alpha + q_i(f)} + \frac{1}{\alpha + q_i(f)} \sum_{k \neq i} u(k) q(k|i, f) \\ &= u(i). \end{aligned}$$

Hence, (4.8) is valid for $n + 1$. This completes the proof of (a).

(b) Suppose that $u$ satisfies the inequality in (b). Then there exists a nonnegative function $v$ on $S$ such that

$$u(i) = \frac{c(i, f) + v(i)}{\alpha + q_i(f)} + \frac{1}{\alpha + q_i(f)} \sum_{j \neq i} u(j) q(j|i, f) \quad \forall i \in S.$$

It follows from (a) and (4.6) (with $c(j, f) + v(j)$ in lieu of $c(j, f)$) that

$$u(i) \geq \sum_{j \in S} \lim_{n \to \infty} [c(j, f) + v(j)] \varphi_{ij}^{(n)}(f) \geq \sum_{j \in S} \lim_{n \to \infty} c(j, f) \varphi_{ij}^{(n)}(f) = J_\alpha(i, f),$$

which yields (b). □

We next present a useful property about the discounted-cost optimality equation.

**Lemma 4.4** *Suppose that the optimal discounted cost $J_\alpha^*(i) < \infty$ for all $i \in S$. Then the following two optimality equations are equivalent, in the sense that if $J_\alpha^*$ satisfies one of them, then it satisfies the other. For every $i \in S$, we have*

$$J_\alpha^*(i) = \inf_{a \in A(i)} \left\{ \frac{c(i,a)}{\alpha + q_i(a)} + \frac{1}{\alpha + q_i(a)} \sum_{j \neq i} J_\alpha^*(j) q(j|i,a) \right\}, \tag{4.9}$$

$$\alpha J_\alpha^*(i) = \inf_{a \in A(i)} \left\{ c(i,a) + \sum_{j \in S} J_\alpha^*(j) q(j|i,a) \right\}. \tag{4.10}$$

*Proof* Let $J_\alpha^*$ satisfy (4.9), and choose an arbitrary $i \in S$ and $\varepsilon > 0$. Then there exists $f \in F$ (depending on $\varepsilon$) such that

$$J_\alpha^*(i) \geq \frac{c(i,f)}{\alpha + q_i(f)} + \frac{1}{\alpha + q_i(f)} \sum_{j \neq i} J_\alpha^*(j) q(j|i,f) - \frac{\varepsilon}{\alpha + q_i(f)} \quad \forall i \in S,$$

which, together with $|J_\alpha^*(i) q_i(f)| \leq |J_\alpha^*(i)| q^*(i) < \infty$, implies

$$\begin{aligned} \alpha J_\alpha^*(i) &\geq c(i,f) + \sum_{j \in S} J_\alpha^*(j) q(j|i,f) - \varepsilon \\ &\geq \inf_{a \in A(i)} \left\{ c(i,a) + \sum_{j \in S} J_\alpha^*(j) q(j|i,a) \right\} - \varepsilon. \end{aligned}$$

Thus, letting $\varepsilon \to 0$, we obtain

$$\alpha J_\alpha^*(i) \geq \inf_{a \in A(i)} \left\{ c(i,a) + \sum_{j \in S} J_\alpha^*(j) q(j|i,a) \right\}. \tag{4.11}$$

On the other hand, from (4.9) we have

$$J_\alpha^*(i) \leq \frac{c(i,a)}{\alpha + q_i(a)} + \frac{1}{\alpha + q_i(a)} \sum_{j \neq i} J_\alpha^*(j) q(j|i,a) \quad \forall a \in A(i),$$

and so (using $|J_\alpha^*(i) q_i(a)| < \infty$ again)

$$\alpha J_\alpha^*(i) \leq c(i,a) + \sum_{j \in S} J_\alpha^*(j) q(j|i,a) \quad \forall a \in A(i).$$

Hence,

$$\alpha J_\alpha^*(i) \leq \inf_{a \in A(i)} \left\{ c(i,a) + \sum_{j \in S} J_\alpha^*(j) q(j|i,a) \right\}.$$

This inequality and (4.11) give (4.10). The converse, that (4.10) implies (4.9), can be proved similarly. □

For any given $\pi \in \Pi$, let

$$\tau_1 := \inf\{t > 0 : x(t) \neq x(0)\} \tag{4.12}$$

be the holding time of the Markov process $\{x(t)\}$ with transition function $p_\pi(s, i, t, j)$ at initial state $x(0)$ (i.e., the first jump time of the process from the initial state $x(0)$).

**Lemma 4.5** *For each $\pi \in \Pi$, each initial state $x(0) = i \in S$, and each nonnegative function $u$ on $S$, we have*

(a) $P_i^\pi(\tau_1 > t) = e^{\int_0^t q(i|i,\pi_v)\,dv}$ *for all $t \geq 0$.*
(b) $E_i^\pi[e^{-\alpha\tau_1} u(x(\tau_1))] = \int_0^\infty e^{-\alpha t} e^{\int_0^t q(i|i,\pi_v)\,dv} \sum_{j\neq i} u(j) q(j|i,\pi_t)\,dt.$

*Proof* These results are well known; see Gihman and Skorohod (1975) [46, p. 89], Kitaev and Rykov (1995) [98, p. 150], or Yin and Zhang (1998) [160, p. 19], for instance. (Also, by Proposition B.8 we see that this lemma is true.) □

## 4.4 The Discounted Cost Optimality Equation

In Lemma 4.4, it is tacitly *assumed* that the optimal discounted cost $J_\alpha^*$ satisfies (4.9)—or equivalently (4.10). In this section we show that $J_\alpha^*$ indeed satisfies the so-called discounted-cost optimality equation (4.9), and we also show the existence of discounted-cost $\varepsilon(> 0)$-optimal policies. The precise statement is as follows.

**Theorem 4.6**

(a) *The function $J_\alpha^*$ satisfies the discounted-cost optimality equation* (4.9), *i.e.*,

$$J_\alpha^*(i) = \inf_{a\in A(i)} \left\{ \frac{c(i,a)}{\alpha + q_i(a)} + \frac{1}{\alpha + q_i(a)} \sum_{j\neq i} J_\alpha^*(j) q(j|i,a) \right\} \quad \forall i \in S.$$

(b) *For each $\varepsilon > 0$, there exists a discounted-cost $\varepsilon$-optimal stationary policy.*

*Proof* We prove (a) and (b) together. Take an arbitrary policy $\pi = (\pi_t) \in \Pi$ and an arbitrary initial state $i \in S$. By (4.3) and (4.12),

$$J_\alpha(i,\pi) = E_i^\pi\left[\int_0^{\tau_1} e^{-\alpha t} c(x(t),\pi_t)\,dt + \int_{\tau_1}^\infty e^{-\alpha t} c(x(t),\pi_t)\,dt\right]. \tag{4.13}$$

Since $\tau_1$ is the holding time at state $i$ under $\pi$, by Lemma 4.5 we have

$$E_i^\pi\left[\int_0^{\tau_1} e^{-\alpha t} c(x(t),\pi_t)\,dt\right] = \int_0^\infty e^{-\alpha t + \int_0^t q(i|i,\pi_v)\,dv} c(i,\pi_t)\,dt. \tag{4.14}$$

For each $t_0 \geq 0$, we now define the so-called "$t_0$-shift policy" $\pi^{t_0}$ by $\pi_t^{t_0}(\cdot|i) := \pi_{t_0+t}(\cdot|i)$ for all $i \in S$. Then, by the strong Markov property (B.2) and Lemma 4.5, we have

$$
\begin{aligned}
&E_i^\pi\left[\int_{\tau_1}^\infty e^{-\alpha t} c\big(x(t), \pi_t\big)\, dt\right] \\
&= E_i^\pi\left[\int_0^\infty e^{-\alpha\tau_1} e^{-\alpha t} c\big(x(\tau_1+t), \pi_{\tau_1+t}\big)\, dt\right] \\
&= E_i^\pi\left[E_i^\pi\left(\int_0^\infty e^{-\alpha\tau_1} e^{-\alpha t} c\big(x(\tau_1+t), \pi_{\tau_1+t}\big)\, dt | \mathcal{F}_{\tau_1}\right)\right] \\
&= E_i^\pi\left[e^{-\alpha\tau_1} J_\alpha\big(x(\tau_1), \pi^{\tau_1}\big)\right] \\
&\geq E_i^\pi\left[e^{-\alpha\tau_1} J_\alpha^*\big(x(\tau_1)\big)\right] \\
&= \int_0^\infty e^{-\alpha t} e^{\int_0^t q(i|i,\pi_v)\, dv} \sum_{j\neq i} J_\alpha^*(j) q(j|i,\pi_t)\, dt.
\end{aligned} \tag{4.15}
$$

Thus, by (4.13)–(4.15) we have

$$
\begin{aligned}
J_\alpha(i,\pi) &\geq \int_0^\infty e^{-\alpha t + \int_0^t q(i|i,\pi_v)\, dv}\left[c(i,\pi_t) + \sum_{j\neq i} J_\alpha^*(j) q(j|i,\pi_t)\right] dt \\
&= \int_0^\infty \int_{A(i)} e^{-\alpha t + \int_0^t q(i|i,\pi_v)\, dv}\left[c(i,a) + \sum_{j\neq i} J_\alpha^*(j) q(j|i,a)\right] \pi_t(da|i)\, dt.
\end{aligned} \tag{4.16}
$$

Noting that

$$
\begin{aligned}
&\int_0^\infty \int_{A(i)} e^{-\alpha t + \int_0^t q(i|i,\pi_v)\, dv}\big(\alpha + q_i(a)\big)\pi_t(da|i)\, dt \\
&= \int_0^\infty e^{-\alpha t + \int_0^t q(i|i,\pi_v)\, dv}\big(\alpha - q(i|i,\pi_t)\big)\, dt = 1,
\end{aligned}
$$

by (4.16) we have

$$
J_\alpha(i,\pi) \geq \inf_{a\in A(i)}\left\{\frac{c(i,a)}{\alpha + q_i(a)} + \frac{1}{\alpha + q_i(a)}\sum_{j\neq i} J_\alpha^*(j) q(j|i,a)\right\}.
$$

This implies (since $\pi$ was arbitrary)

$$
J_\alpha^*(i) \geq \inf_{a\in A(i)}\left\{\frac{c(i,a)}{\alpha + q_i(a)} + \frac{1}{\alpha + q_i(a)}\sum_{j\neq i} J_\alpha^*(j) q(j|i,a)\right\}. \tag{4.17}
$$

On the other hand, by Remark 4.2 we may assume that $c(i,a) \geq L$ for some $L > 0$. Now fix an arbitrary $\varepsilon > 0$, and for each $i \in S$, let $\varepsilon_i$ be such that $0 < \alpha\varepsilon_i \leq \min\{\alpha\varepsilon, L\}$. Then, (4.17) gives the existence of $f_\varepsilon \in F$ (depending on $\varepsilon$) such that

$$\begin{aligned} J_\alpha^*(i) &\geq \inf_{a\in A(i)} \left\{ \frac{c(i,a)}{\alpha + q_i(a)} + \frac{1}{\alpha + q_i(a)} \sum_{j\neq i} J_\alpha^*(j) q(j|i,a) \right\} \\ &\geq \frac{c(i, f_\varepsilon) - \alpha\varepsilon_i}{\alpha + q_i(f_\varepsilon)} + \frac{1}{\alpha + q_i(f_\varepsilon)} \sum_{j\neq i} J_\alpha^*(j) q(j|i, f_\varepsilon). \end{aligned}$$

Since $c(i, f_\varepsilon(i)) - \alpha\varepsilon_i \geq 0$, by Lemma 4.3(b) we see that the expected discounted cost with respect to the cost "$c(i,a) - \alpha\varepsilon_i$" is less than $J_\alpha^*$, that is,

$$\int_0^\infty e^{-\alpha t} E_i^{f_\varepsilon} \big(c\big(x(t), f_\varepsilon\big) - \alpha\varepsilon_{x(t)}\big)\, dt \leq J_\alpha^*(i) \quad \forall i \in S.$$

Moreover, because $\varepsilon_i \leq \varepsilon$ for all $i \in S$, we have $\alpha \int_0^\infty e^{-\alpha t} E_i^{f_\varepsilon}[\varepsilon_{x(t)}]\, dt \leq \varepsilon$. It follows that

$$J_\alpha^*(i) \geq J_\alpha(i, f_\varepsilon) - \varepsilon \geq J_\alpha^*(i) - \varepsilon.$$

Therefore,

$$\begin{aligned} J_\alpha^*(i) &\geq \inf_{a\in A(i)} \left\{ \frac{c(i,a)}{\alpha + q_i(a)} + \frac{1}{\alpha + q_i(a)} \sum_{j\neq i} J_\alpha^*(j) q(j|i,a) \right\} \\ &\geq \frac{c(i, f_\varepsilon) - \alpha\varepsilon_i}{\alpha + q_i(f_\varepsilon)} + \frac{1}{\alpha + q_i(f_\varepsilon)} \sum_{j\neq i} J_\alpha^*(j) q(j|i, f_\varepsilon) \\ &\geq \frac{c(i, f_\varepsilon) - \alpha\varepsilon}{\alpha + q_i(f_\varepsilon)} + \frac{1}{\alpha + q_i(f_\varepsilon)} \sum_{j\neq i} \big[J_\alpha(f_\varepsilon, j) - \varepsilon\big] q(j|i, f_\varepsilon) \\ &= J_\alpha(f_\varepsilon, i) - \varepsilon \\ &\geq J_\alpha^*(i) - \varepsilon, \end{aligned}$$

and so (a) and (b) follow. □

*Remark 4.7* Theorem 4.6 establishes the *discounted-cost optimality equation* and the existence of a discounted-cost $\varepsilon$-optimal stationary policy in the class $\Pi$ of all randomized Markov policies. The conditions in Theorem 4.6 are *very weak*; in particular, it is allowed that $J_\alpha^*(i) = \infty$ for some $i \in S$.

In the following section, we show how to obtain discounted-cost optimal (as apposed to $\varepsilon$-optimal) policies.

## 4.5 Existence of Optimal Policies

In this section, we establish the existence of discounted-cost optimal policies, assuming that there are policies in $F$ that attain the minimum in the optimality equation in Theorem 4.6(a). Hence, we suppose the following.

**Assumption 4.8** There exists $f_\alpha \in F$ (depending on the discount factor $\alpha$) that attains the minimum in the right-hand side of the optimality equation (4.9); that is, by Theorem 4.6(a),

$$\begin{aligned} J_\alpha^*(i) &= \frac{c(i, f_\alpha)}{\alpha + q_i(f_\alpha)} + \frac{1}{\alpha + q_i(f_\alpha)} \sum_{j \neq i} J_\alpha^*(j) q(j|i, f_\alpha) \\ &= \min_{a \in A(i)} \left\{ \frac{c(i,a)}{\alpha + q_i(a)} + \frac{1}{\alpha + q_i(a)} \sum_{j \neq i} J_\alpha^*(j) q(j|i,a) \right\} \quad \forall i \in S. \end{aligned} \tag{4.18}$$

*Remark 4.9* The obvious question is when does Assumption 4.8 hold? Clearly, it holds when the sets $A(i)$ are all finite. A sufficient condition that can be used to verify Assumption 4.8 is given in Assumption 4.12 below. Other conditions implying Assumption 4.8 have been given by Guo and Cao (2005) [52], Guo and Hernández-Lerma (2003) [53, 54, 56], Guo and Liu (2001) [58], Guo and Zhu (2002) [62, 63], Kakumanu (1971) [92], Kitaev and Rykov (1995) [98], Miller (1968) [117], Puterman (1994) [129], and Sennott (1999) [141], for instance.

**Theorem 4.10** *Let $f_\alpha$ be as in Assumption* 4.8. *Then $f_\alpha$ is discounted-cost optimal.*

*Proof* Since $J_\alpha^*$ is the optimal discounted cost, $J_\alpha^* \le J_\alpha(f_\alpha)$. On the other hand, if $f_\alpha \in F$ satisfies (4.18), then Lemma 4.3(a) gives that $J_\alpha^* \ge J_\alpha(f_\alpha)$. Hence, $J_\alpha(f_\alpha) = J_\alpha^*$, and Theorem 4.10 follows. □

## 4.6 Approximation Results

We now turn to the approximation of the optimal discounted-cost function $J_\alpha^*$ and of a discounted-cost optimal stationary policy.

Define the operators $T_f$ (for each fixed $f \in F$) and $T$ as follows: For any nonnegative function $v$ on $S$, let

$$T_f v(i) := \frac{c(i,f)}{\alpha + q_i(f)} + \frac{1}{\alpha + q_i(f)} \sum_{j \neq i} v(j) q(j|i,f), \tag{4.19}$$

$$Tv(i) := \inf_{a \in A(i)} \left\{ \frac{c(i,a)}{\alpha + q_i(a)} + \frac{1}{\alpha + q_i(a)} \sum_{j \neq i} v(j) q(j|i,a) \right\} \tag{4.20}$$

for all $i \in S$. Observe that these operators are *monotone*, that is, $v \geq v'$ implies $T_f v \geq T_f v'$, and similarly for $T$. (Compare the right-hand sides of (4.20) and (4.9). We can express (4.9) as $J_\alpha^* = T J_\alpha^*$, which states that $J_\alpha^*$ is a "fixed point" of $T$.)

**Lemma 4.11**

(a) *For each* $f \in F$, *let*

$$V_0^f := 0, \quad \text{and} \quad V_{n+1}^f := T_f V_n^f \quad \text{for } n \geq 0. \tag{4.21}$$

*Then* $\{V_n^f\}$ *is nondecreasing in* $n \geq 0$, *and* $\lim_{n\to\infty} V_n^f = J_\alpha(f)$.

(b) *Let*

$$V_0^* := 0, \quad \text{and} \quad V_{n+1}^* := T V_n^* \quad \text{for } n \geq 0. \tag{4.22}$$

*Then* $\{V_n^*\}$ *is nondecreasing in* $n \geq 0$, *and* $\lim_{n\to\infty} V_n^* \leq J_\alpha(f)$ *for all* $f \in F$.

*Proof* (a) This part obviously follows from the proof of Lemma 4.3(a).

(b) Since $c(i,a) \geq 0$ and the operator $T$ is monotone, from (4.20) and (4.22) we see that $V_{n+1}^* \geq V_n^*$ for all $n \geq 0$, and so the first statement in (b) follows. We will prove the second statement in (b) by induction. Take an arbitrary $f \in F$. It follows from (4.21) and (4.22) that $V_0^* \leq V_0^f$. Suppose now that $V_n^* \leq V_n^f$ for some $n \geq 0$. Then, (4.21) and (4.22) yield

$$V_{n+1}^* = T V_n^* \leq T V_n^f \leq T_f V_n^f = V_{n+1}^f,$$

and letting $n \to \infty$, then by (a) we have $\lim_{n\to\infty} V_n^* \leq J_\alpha(f)$. Hence, the proof of (b) is completed. □

To strengthen the result in Lemma 4.11(b) so that $V_n^* \to J_\alpha^*$, we need the following continuity–compactness condition, which clearly is stronger than Assumption 4.8.

**Assumption 4.12**

(a) $A(i)$ is compact for each $i \in S$.
(b) For all $i, j \in S$, the functions $c(i,a)$ and $q(j|i,a)$ are continuous on $A(i)$.
(c) There exists $\hat{f} \in F$ such that $J_\alpha(\hat{f}) < \infty$, and $\sum_{j\in S} J_\alpha(j,\hat{f}) q(j|i,a)$ is continuous on $A(i)$ for each $i \in S$.

*Remark 4.13* Assumption 4.12 is a continuity–compactness condition widely used for MDPs; see Haviv and Puterman (1998) [74] and Puterman (1994) [129], for instance. Furthermore, *Assumption* 4.12 *implies Assumption* 4.8. Many conditions and examples that verify Assumption 4.12 are given by Guo and Cao (2005) [52], Guo and Hernández-Lerma (2003) [53–55], Guo and Liu (2001) [58], Guo and Zhu (2002) [62, 63], Kakumanu (1971) [92], Kitaev and Rykov (1995) [98], Miller (1968) [117], Puterman (1994) [129], and Sennott (1999) [141], for instance. Moreover, when the cost $c(i,a)$ is *bounded*, Assumption 4.12(c) is not required.

We next state the approximation result of this section, in which we use the following concept: a policy $f \in F$ is said to be a *limit point* of a sequence $\{f_n\}$ in $F$ if there exists a subsequence $\{m\}$ of $\{n\}$ such that $f_m(i) \to f(i)$ as $m \to \infty$ for each $i \in S$.

**Theorem 4.14** *Suppose that Assumption* 4.12 *holds, and let* $V_n^*$ *be as in* (4.22). *Then*:

(a) $\lim_{n\to\infty} V_n^* = J_\alpha^*$.
(b) *Let* $f_n \in F$ $(n \geq 0)$ *be such that*

$$V_{n+1}^*(i) = \min_{a\in A(i)} \left\{ \frac{c(i,a)}{\alpha + q_i(a)} + \frac{1}{\alpha + q_i(a)} \sum_{j\neq i} V_n^*(j) q(j|i,a) \right\}$$
$$= \frac{c(i, f_n(i))}{\alpha + q_i(f_n(i))} + \frac{1}{\alpha + q_i(f_n(i))} \sum_{j\neq i} V_n^*(j) q\big(j|i, f_n(i)\big) \tag{4.23}$$

*for all* $i \in S$. *Then any limit point of* $\{f_n\}$ *is a discounted-cost optimal stationary policy.*

*Proof* We will prove (a) and (b) together. First, note that we can write $F$ as the product space $\prod_{i\in S} A(i)$. Therefore, under Assumption 4.12(a), Tychonoff's theorem (see Proposition A.6) yields that $F$ is compact. Moreover, by Assumption 4.12 and Proposition A.4, the function $\sum_{j\neq i} u(j)q(j|i,a)$ is continuous in $a \in A(i)$ for any $0 \leq u \leq J_\alpha(\hat{f})$. Thus, the existence of a limit point, say $f^*$, of $\{f_n\}$ satisfying (4.23) is indeed guaranteed. Therefore, there exists a subsequence $\{f_m\}$ of $\{f_n\}$ such that $\lim_{m\to\infty} f_m(i) = f^*(i)$ for all $i \in S$. Let $u^* := \lim_{n\to\infty} V_n^* \geq 0$. Observe that $u^* \geq 0$ (by Lemma 4.11(b)). Then, by (4.23) and Fatou's lemma (see Proposition A.3), we have

$$u^*(i) \geq \frac{c(i, f^*(i))}{\alpha + q_i(f^*)} + \frac{1}{\alpha + q_i(f^*)} \sum_{j\neq i} u^*(j) q\big(j|i, f^*(i)\big) \quad \forall i \in S.$$

Thus, by Lemma 4.3(b), we have $u^* \geq J_\alpha(f^*)$, which, together with Lemma 4.11(b), gives

$$J_\alpha\big(f^*\big) = u^* = \inf_{f\in F} J_\alpha(f). \tag{4.24}$$

Moreover, under Assumption 4.8, there exists $f_\alpha \in F$ such that, for every $i \in S$,

$$J_\alpha^*(i) = \min_{a\in A(i)} \left\{ \frac{c(i,a)}{\alpha + q_i(a)} + \frac{1}{\alpha + q_i(a)} \sum_{j\neq i} J_\alpha^*(j) q(j|i,a) \right\}$$
$$= \frac{c(i, f_\alpha)}{\alpha + q_i(f_\alpha)} + \frac{1}{\alpha + q_i(f_\alpha)} \sum_{j\neq i} J_\alpha^*(j) q(j|i, f_\alpha). \tag{4.25}$$

By Lemma 4.3(a) and (4.25) we have $J_\alpha^* \geq J_\alpha(f_\alpha)$, and so $J_\alpha^* = \inf_{f\in F} J_\alpha(f)$. The latter fact and (4.24) imply that $J_\alpha(f^*) = u^* = J_\alpha^*$, and so the proof is completed. □

*Remark 4.15* Theorem 4.14(a) provides a "value iteration" algorithm (which is so named because it is based on iterations of the value operator $T$) to approximate the optimal discounted-cost function $J_\alpha^*$. Theorem 4.14(b) further shows that a discounted-cost optimal stationary policy can be obtained as a limit point of $\{f_n\}$ satisfying the value iteration steps given in (4.23). It should be mentioned that Assumptions 4.12(a)–(b) in Theorem 4.14 are used only as a *sufficient condition* for the existence of $\{f_n\}$ and $f_\alpha$ satisfying (4.23) and (4.25), respectively, and for the existence of a limit point $f^*$ of $\{f_n\}$. Therefore, in particular, Theorem 4.14 is true when the $A(i)$ are all finite. On the other hand, Assumption 4.12(c) is a technical condition that allows us to use "dominated convergence" to interchange limits and summations.

## 4.7 The Policy Iteration Approach

In this section, we provide a policy iteration algorithm for solving the discounted-cost problem for the important special case of a *finite* MDP, in which the state and action spaces are both finite. For such a finite model, the associated cost is obviously bounded. Thus, using Remark 4.2, without loss of generality we may assume that the cost function $c(i,a) \geq 0$ for all $(i,a) \in K$.

Then, from Lemma 4.3 we have the following fact.

**Lemma 4.16** *For the finite MDP model, $J_\alpha(f)$ is a unique bounded solution to the equation*

$$\alpha u(i) = c(i,f) + \sum_{j\in S} u(j) q(j|i,f) \quad \forall i \in S \tag{4.26}$$

*for every $f \in F$.*

*Proof* Fix an arbitrary $f \in F$. A direct calculation shows that (4.4) and (4.26) are equivalent; hence, by Remark 4.2 and Lemma 4.3 we see that $J_\alpha(f)$ satisfies (4.26). To prove the uniqueness, suppose that a function $V'$ on $S$ also satisfies (4.26), and let $\rho := \max_{i\in S} \frac{q_i(f)}{\alpha + q_i(f)} \in (0,1)$ (by the finiteness of $S$ and $A(i)$). Then, using the equivalent form (4.4) of (4.26), we have

$$\begin{aligned} \|J_\alpha(f) - V'\| &\leq \rho \|J_\alpha(f) - V'\| \\ &\leq \rho^n \|J_\alpha(f) - V'\| \quad \text{for all } n \geq 1. \end{aligned}$$

Thus, letting $n \to \infty$, we obtain that $J_\alpha(f) = V'$. □

We now wish to state a policy iteration algorithm to find $J_\alpha^*$ and $f_\alpha$ that satisfy (4.18). To this end, we need to introduce some notation.

For every given $f \in F$, $i \in S$, and $a \in A(i)$, let

$$D_f(i,a) := c(i,a) + \sum_{j \in S} J_\alpha(j, f) q(j|i,a) \tag{4.27}$$

and

$$E_f(i) := \{a \in A(i) : D_f(i,a) < \alpha J_\alpha(i, f)\}. \tag{4.28}$$

We then define an *improvement policy* $h \in F$ (depending on $f$) as follows:

$$h(i) \in E_f(i) \quad \text{if } E_f(i) \neq \emptyset \quad \text{and} \quad h(i) := f(i) \quad \text{if } E_f(i) = \emptyset. \tag{4.29}$$

The policy $h$ is indeed an "improvement" of $f$ because, as shown in the following lemma, the cost when using the policy $h$ does not exceed the cost when using $f$, that is, $J_\alpha(f) \succeq J_\alpha(h)$ if $h \neq f$.

**Lemma 4.17** *For any given $f \in F$, let $h \in F$ be defined as in* (4.29) *and suppose that $h \neq f$. Then $J_\alpha(f) \succeq J_\alpha(h)$.*

*Proof* From the definition of $h$ in (4.29) and $h \neq f$, by Lemma 4.16 we have

$$\alpha J_\alpha(f) \succeq c(h) + Q(h) J_\alpha(f), \tag{4.30}$$

which, together with Lemma 4.3, implies that $J_\alpha(f) \geq J_\alpha(h)$. By the uniqueness in Lemma 4.16 and (4.30), $J_\alpha(f) \succeq J_\alpha(h)$, and so the lemma follows. □

Using Lemma 4.17, we can state the following *policy iteration algorithm* for computing a discounted-cost optimal policy.

**The discounted-cost policy iteration algorithm:**

1. Pick an arbitrary $f \in F$. Let $k = 0$ and take $f_k := f$.
2. (Policy evaluation) Obtain $J_\alpha(f_k) = [\alpha I - Q(f_k)]^{-1} c(f_k)$ (by Lemma 4.16).
3. (Policy improvement) Obtain a policy $f_{k+1}$ from (4.29) (with $f_k$ and $f_{k+1}$ in lieu of $f$ and $h$, respectively).
4. If $f_{k+1} = f_k$, then stop because $f_{k+1}$ is discounted-cost optimal (by Theorem 4.18 below). Otherwise, increment $k$ by 1 and return to Step 2.

**Theorem 4.18** *For each fixed discount factor $\alpha > 0$, the discounted-cost policy iteration algorithm yields a discounted-cost optimal deterministic stationary policy in a finite number of iterations.*

*Proof* Let $\{f_k\}$ be the sequence of policies in the discounted-cost policy iteration algorithm above. Then, by Lemma 4.17, we have $J_\alpha(f_k) \succeq J_\alpha(f_{k+1})$. That is, as $k$

increases, $J_\alpha(f_k)$ decreases. Thus, each policy in the sequence of $\{f_k,\ k=0,1,\ldots\}$ is different. Since the number of policies is finite, the iterations must stop after a finite number. Suppose that the algorithm stops at a policy denoted by $f_\alpha^*$. Then, $f_\alpha^*$ satisfies (4.10). Thus, by Lemma 4.4 and Theorem 4.10, $f_\alpha^*$ is discounted-cost optimal. □

## 4.8 Examples

In this section, we apply the main results in this chapter to a class of controlled population processes.

*Example 4.1* (A controlled population process) Consider a controlled birth-and-death population process in which the state variable $i$ denotes the population size. Suppose that the birth and death rates of such a population process can be controlled by a decision-maker. We denote by $\lambda(i,a)\geq 0$ and $\mu(i,a)\geq 0$ the birth and death rates, respectively, which may depend on the population size $i\in S:=\{0,1,\ldots\}$ and also on decision variables $a$ taken by the decision-maker. For each population size $i\geq 0$, the decision-maker takes an action $a$ from a *finite* set $A(i)$ of available actions, which may increase or decrease $\lambda(i,a)$ and $\mu(i,a)$, and may incur a cost with rate $\tilde{c}(i,a)$. Moreover, suppose that the cost caused by each individual is represented by $p\geq 0$. Then the total cost in this system is $c(i,a)=pi+\tilde{c}(i,a)$ for each $i\in S$ and $a\in A(i)$. On the other hand, when there is no population in the system (i.e., $i=0$), it is natural to suppose that $\mu(0,a)\equiv 0$ for all $a\in A(0)$.

We now formulate this system as a continuous-time MDP. The corresponding transition rates $q(j|i,a)$ and cost $c(i,a)$ are given as follows.

For $i=0$ and each $a\in A(0)$,

$$\mu(0,a)=0,\qquad -q(0|0,a)=q(1|0,a):=\lambda(0,a)\geq 0,$$
$$q(j|0,a)=0\quad \text{for } j\geq 2.$$

For $i\geq 1$ and all $a\in A(i)$,

$$q(j|i,a):=\begin{cases}\mu(i,a) & \text{if } j=i-1,\\ -\mu(i,a)-\lambda(i,a) & \text{if } j=i,\\ \lambda(i,a) & \text{if } j=i+1,\\ 0 & \text{otherwise};\end{cases}\tag{4.31}$$

$$c(i,a):=pi+\tilde{c}(i,a)\quad \text{for } i\in S \text{ and } a\in A(i).$$

We aim to find conditions that ensure the existence of a discounted-cost optimal stationary policy. To do this, in the spirit of Assumption 2.2 (in Sect. 2.2) and Assumption 4.12, we consider the following conditions:

**B₁**. There exist a function $w \geq 1$ on $S$ and a constant $L > 0$ such that $\lambda(i,a)[w(i+1) - w(i)] - \mu(i,a)[w(i) - w(i-1)] \leq Lw(i)$ and $\mu(i,a) + \lambda(i,a) \leq Lw(i)$ for all $(i,a) \in K$, where $w(-1) := 0$.

**B₂**. $\tilde{c}(i,a)$ is bounded below in $(i,a) \in K$.

We can provide an alternative interpretation of Example 4.1 by means of the *special* case of *birth-and-death processes with controlled immigration*. Consider a pest population in a region which may be isolated to prevent immigration. There are "natural" birth and death rates denoted by (fixed) *positive* constants $\lambda$ and $\mu$, respectively, as well as an immigration parameter $b > 0$. Let a positive constant $C$ denote the cost when immigration is always prevented. The actions "1" and "0" denote that the immigration is admissible and prevented, respectively. Then, we have $A(i) := \{0, 1\}$ for each $i \geq 1$. However, when there is no pest in the region (i.e., $i = 0$), it is natural to let $A(0) := \{1\}$. (This can be explained as follows: for the ecological balance of the region, the pest is not permitted to become extinct, and so the immigration is admissible.) Suppose that the damage rate caused by a pest individual is represented by $p \geq 0$. Then, the cost is of the form $c(i,a) := pi + C(1-a)$. Using the notation in Example 4.1, for this model, we have $\lambda(i,a) = \lambda i + a$ and $\mu(i,a) = \mu i$ for all $i \geq 0$. Let $w(i) := i + 1$ for all $i \in S$; then we can easily verify that conditions $\mathbf{B}_1$ and $\mathbf{B}_2$ above are all satisfied.

Under these conditions, we obtain the following.

**Proposition 4.19** (On discount optimality) *Under conditions* $\mathbf{B}_1$ *and* $\mathbf{B}_2$, *the process in Example* 4.1 *satisfies Assumptions* 2.2 *and* 4.12. *Therefore* (*by Theorem* 4.14), *there exists a discounted-cost optimal stationary policy, which can be computed or at least can be approximated as a limit point of the sequence of stationary policies* $\{f_n\}$ *given in Theorem* 4.14(b).

*Proof* Under $\mathbf{B}_1$, by (2.4) , (2.5), and (4.31) we have

$$\sum_{j \in S} w(j) q(j|i, \pi_t) \leq Lw(i) \quad \text{and} \quad q^*(i) \leq Lw(i) \quad \forall (i,a) \in K.$$

These inequalities, together with Proposition C.9, give Assumption 2.2. Furthermore, from the description of Example 4.1 and (4.31) we see that Assumption 4.12 is also satisfied. □

## 4.9 Notes

In view of Lemma 4.3(a), the method for proving the existence of a discounted-cost $\varepsilon$-optimal stationary policy in Theorems 4.6 and 4.14 might be called a *minimum nonnegative solution approach*. This approach and the main results in this chapter are from Guo and Yin (2007) [61]. A similar approach has been used by several authors, for instance, Hernández-Lerma and Lasserre (1999) [74], Puterman (1994)

[129], and Sennnott (1999) [141] for nonnegative models of *discrete-time* MDPs. In fact, the main results in this chapter are analogous to those in [74, 129, 141] for the discrete-time case. It is important to note, however, that our results cannot be obtained from their discrete-time *counterparts* by the uniformization technique, because the transition rates here may *not* be bounded.

On the other hand, for a finite MDP model, we have provided in Sect. 4.7 a policy iteration algorithm for obtaining a discounted-cost optimal deterministic stationary policy. As in Chap. 6 of Puterman's (1994) [129] book (among many other references), such an algorithm can be seen as the "dual" of a linear programming problem associated to the MDP.

# Chapter 5
# Average Optimality for Nonnegative Costs

In this chapter, we turn our attention to the average-cost optimality problem, again for the nonnegative model in Chap. 4. First of all, it should be mentioned that an AR optimal policy may *not* exist when the state or the action space is *not* finite. Thus, the main purpose of this chapter is to find conditions ensuring the existence of AR optimal policies.

## 5.1 Introduction

In the previous chapter, we considered the discount optimality criterion for the nonnegative model. The main goal in this chapter is to prove the existence of an average-cost optimal stationary policy by using a *minimum nonnegative solution approach*, which has a similar idea to the approach we used for proving Theorems 4.6 and 4.14 in the discounted-cost case but requires a different technique.

In fact, we use this approach in combination with the *vanishing discount* (*or discount-factor*) *method*, which is so named because it studies some average-cost MDPs as the "limit" when the discount factor vanishes, that is, $\alpha \downarrow 0$, of some $\alpha$-discount problems. Here we use the vanishing discount approach to obtain, first, the average-cost *optimality inequality* (in Sect. 5.4), and then the *optimality equation* (in Sect. 5.5). The former requires, of course, weaker hypotheses than the latter, but still it suffices to obtain average-cost optimal stationary policies. The importance of the optimality equation, however, should not be underestimated because it is useful to obtain many classes of results, including approximation algorithms such as value iteration and policy iteration.

The remainder of this chapter is organized as follows. In Sect. 5.2, we recall the nonnegative MDP model and the definition of average-cost optimality. In Sect. 5.3, we prove Theorem 5.2, which presents two key results. The first, Theorem 5.2(a), gives a characterization of a finite-horizon expected cost as the minimal nonnegative solution of a certain inequality (see (5.3)). The second result, Theorem 5.2(b), shows how to obtain an upper bound for average costs of stationary policies. In

X. Guo, O. Hernández-Lerma, *Continuous-Time Markov Decision Processes*,
Stochastic Modelling and Applied Probability 62,
DOI 10.1007/978-3-642-02547-1_5, © Springer-Verlag Berlin Heidelberg 2009

Sects. 5.4 and 5.5, as already mentioned above, we introduce the average-cost optimality inequality and the optimality equation, respectively. Section 5.6 presents two examples illustrating our results, and it is worth mentioning that one of these examples, Example 5.2, concerns a continuous-time MDP that satisfies the optimality inequality, but the optimality equation does *not* hold. This shows, in particular, that there are (at least) *two* different ways to obtain average-cost optimal policies, namely, using the optimality inequality and using the optimality equation.

Finally, Sect. 5.7 presents a summary of some hypotheses and techniques to obtain average-cost optimal policies.

## 5.2 The Average-Cost Criterion

In this chapter we consider the *nonnegative* model of a continuous-time MDP

$$\{S, (A(i), i \in S), q(j|i,a), c(i,a)\},$$

which is the same as in (4.1).

To introduce the average-cost optimality criterion, we recall the notation in (4.2), namely: For all $i \in S$, $\pi = (\pi_t) \in \Pi$, $t \geq s \geq 0$, and $p_\pi(s,i,t,j)$, we have

$$c(i,\pi_t) := \int_{A(i)} c(i,a)\pi_t(da|i), \qquad E_i^\pi c(x(t),\pi_t) := \sum_{j\in S} c(j,\pi_t)p_\pi(0,i,t,j).$$

When $\pi = f \in F$, we write $c(i,\pi_t)$ as $c(i,f) := c(i,f(i))$.

For all $\pi \in \Pi$ and $i \in S$, we define the long-run expected *average cost*

$$J_c(i,\pi) := \limsup_{T\to\infty} \frac{1}{T}\int_0^T E_i^\pi c(x(t),\pi_t)\,dt. \tag{5.1}$$

(It should be noted that this expected average cost is defined by using the lim sup, instead of the lim inf in the average *reward* case.) The corresponding optimal average-cost function (or the optimal value function of the average-cost optimality problem) is

$$J_c^*(i) := \inf_{\pi\in\Pi} J_c(i,\pi) \quad \forall i \in S.$$

**Definition 5.1** For any fixed $\varepsilon \geq 0$, a policy $\pi^* \in \Pi$ is said to be average-cost $\varepsilon$-optimal if

$$J_c(i,\pi^*) \leq J_c^*(i) + \varepsilon \quad \forall i \in S.$$

An average-cost 0-optimal policy is simply called average-cost optimal.

## 5.3 The Minimum Nonnegative Solution Approach

For each $f \in F$, the associated transition function $p_f(s,i,t,j)$ is time homogeneous, that is, $p_f(s,i,s+t,j)$ is independent of $s \geq 0$. Thus, let $p_f(i,t,j) := p_f(s,i,s+t,j)$ for all $i, j \in S$ and $s, t \geq 0$. Recall from (2.22) the $t$-horizon expected cost defined as

$$J_t^c(i,f) := \int_0^t E_i^f c(x(s),f)\,ds = \int_0^t \sum_{j\in S} c(j,f)p_f(i,s,j)\,ds \tag{5.2}$$

for $i \in S$ and $t \geq 0$. We then have the following key result.

**Theorem 5.2** *Fix any $f \in F$ and $t > 0$.*

(a) *The function $J_s^c(i,f)$ $(i \in S, 0 \leq s \leq t)$ is the minimum nonnegative solution to the inequality*

$$u(i,s) \geq c(i,f)se^{-q_i(f)s} + \int_0^s e^{-q_i(f)y}\Big[q_i(f)c(i,f)y + \sum_{k\neq i} u(k,s-y)q(k|i,f)\Big]dy \quad \forall (i,s) \in S\times[0,t] \tag{5.3}$$

*and satisfies* (5.3) *with equality.*

(b) *Given a policy $f \in F$, if there exist a constant $g \geq 0$ and a real-valued function $u$ bounded below on $S$ such that*

$$g \geq c(i,f) + \sum_{j\in S} u(j)q(j|i,f) \quad \forall i \in S, \tag{5.4}$$

*then $g \geq J_c(i,f)$ for all $i \in S$.*

*Proof* (a) To prove (a), we first show that $J_s^c(i,f)$ for all $i \in S$ and $0 \leq s \leq t$ satisfies (5.3) with equality, that is,

$$J_s^c(i,f) = c(i,f)se^{-q_i(f)s} + \int_0^s e^{-q_i(f)y}\Big[q_i(f)c(i,f)y + \sum_{k\neq i} J_{s-y}^c(k,f)q(k|i,f)\Big]dy \quad \forall (i,s) \in S\times[0,t], \tag{5.5}$$

and then show that $u(i,s) \geq J_s^c(i,f)$ for all $(i,s) \in S\times[0,t]$ and any nonnegative function $u(i,s)$ satisfying (5.3).

To prove (5.5), we use the construction of the transition function $p_f(i,s,j)$: for $s \geq 0$, $i, j \in S$, and $n \geq 1$, let

$$q_{ij}^{(0)}(f,s) := \delta_{ij}e^{-q_i(f)s},$$

$$q_{ij}^{(n)}(f,s) := q_{ij}^{(0)}(f,s) + \sum_{k\neq i}\int_0^s e^{-q_i(f)y} q(k|i,f) q_{kj}^{(n-1)}(f,s-y)\,dy, \quad (5.6)$$

$$\begin{aligned} m_f^{(0)}(i,s) &:= \int_0^s \sum_{j\in S} c(j,f) q_{ij}^{(0)}(f,y)\,dy \\ &= c(i,f) s e^{-q_i(f)s} + \int_0^s q_i(f) e^{-q_i(f)y} c(i,f) y\,dy, \end{aligned} \quad (5.7)$$

$$m_f^{(n)}(i,s) := \int_0^s \sum_{j\in S} c(j,f) q_{ij}^{(n)}(f,y)\,dy. \quad (5.8)$$

Then, by Proposition C.5 (or see, for instance, Anderson (1991) [4, p. 71]) we have

$$q_{ij}^{(n)}(f,s) \uparrow p_f(i,s,j) \quad \text{as } n\to\infty.$$

Thus, by (5.2) and (5.8), as $n\to\infty$, we have

$$m_f^{(n)}(i,s) \uparrow J_s^c(i,f) \quad \forall i\in S \text{ and } s\geq 0. \quad (5.9)$$

On the other hand, for each $n\geq 1$, by (5.6) and (5.8) we have

$$\begin{aligned} m_f^{(n)}(i,s) &= m_f^{(0)}(i,s) \\ &\quad + \int_0^s\int_0^y e^{-q_i(f)z} \sum_{k\neq i}\left[\sum_{j\in S} c(j,f) q_{kj}^{(n-1)}(f,y-z)\right] q(k|i,f)\,dz\,dy \\ &= m_f^{(0)}(i,s) \\ &\quad + \int_0^s e^{-q_i(f)z} \sum_{k\neq i}\left[\int_z^s \sum_{j\in S} c(j,f) q_{kj}^{(n-1)}(f,y-z)\,dy\right] q(k|i,f)\,dz \\ &= m_f^{(0)}(i,s) + \int_0^s e^{-q_i(f)z} \sum_{k\neq i} m_f^{(n-1)}(k,s-z) q(k|i,f)\,dz, \end{aligned}$$

which, together with (5.7), gives

$$\begin{aligned} m_f^{(n)}(i,s) &= \int_0^s e^{-q_i(f)y}\left[q_i(f)c(i,f)y + \sum_{k\neq i} m_f^{(n-1)}(k,s-y) q(k|i,f)\right] dy \\ &\quad + c(i,f) s e^{-q_i(f)s}. \end{aligned} \quad (5.10)$$

Letting $n\to\infty$ in (5.10) and using (5.9), we obtain (5.5).

Now suppose that a nonnegative function $u(i,s)$ satisfies (5.3). Since $c(i,f)$ and $q(j|i,f)$ are nonnegative for all $j\neq i$, from (5.3) and (5.7) we have that $u(i,s)\geq m_f^{(0)}(i,s)$. Then, by induction and (5.10) we see that $u(i,s)\geq m_f^{(n)}(i,s)$ for all $n\geq 1$. This fact, together with (5.9), completes the proof of (a).

(b) Since the transition rates in model (4.1) are conservative, (5.4) still holds when the function $u$ is replaced with "$u+L$" for any constant $L$. Therefore, without loss of generality we may further assume that $u \geq 0$.

Let $\hat{u}(i,s) := u(i) + gs \geq 0$ for all $i \in S$ and $s \in [0,t]$, with $u \geq 0$ and $g$ as in (5.4). Then, we can rewrite (5.4) as

$$\sum_{k \neq i} u(k) q(k|i,f) \leq u(i) q_i(f) + g - c(i,f),$$

which, combined with $\sum_{k \neq i} q(k|i,f) = q_i(f)$, gives, after straightforward but laborious calculations,

$$\begin{aligned}
c(i,f)se^{-q_i(f)s} + \int_0^s e^{-q_i(f)y}\Big[q_i(f)c(i,f)y + \sum_{k\neq i}\big[u(k) + g(s-y)\big]q(k|i,f)\Big]dy \\
\leq c(i,f)se^{-q_i(f)s} + g\int_0^s e^{-q_i(f)y} q_i(f)(s-y)\,dy \\
+ \int_0^s e^{-q_i(f)y}\big[q_i(f)c(i,f)y + q_i(f)u(i) + g - c(i,f)\big]dy \\
= u(i) + gs - u(i)e^{-q_i(f)s} \\
\leq u(i) + gs.
\end{aligned}$$

This yields

$$\begin{aligned}
\hat{u}(i,s) \geq c(i,f)se^{-q_i(f)s} \\
+ \int_0^s e^{-q_i(f)y}\Big[q_i(f)c(i,f)y + \sum_{k\neq i}\hat{u}(k,s-y)q(k|i,f)\Big]ds.
\end{aligned}$$

Hence, $\hat{u}(i,s)$ is a nonnegative solution to (5.3). This implies, by (a),

$$J_s^c(i,f) \leq \hat{u}(i,s) = u(i) + gs \quad \forall (i,s) \in S \times [0,t] \text{ and } t > 0.$$

Multiplying this inequality by $1/s$ and letting $s \to \infty$, from (5.2) and (5.1) we obtain (b). □

*Remark 5.3* We call the method of proving Theorem 5.2 an *average-cost minimum nonnegative solution approach*. Obviously, this approach, which is possible by the assumption that $c(i,a)$ is *nonnegative*, is *different* from those in Guo and Cao (2005) [52] and in Guo and Hernández-Lerma (2003) [55], which require the *four* conditions (i)–(iv) in Sect. 5.7 below.

## 5.4 The Average-Cost Optimality Inequality

In this section, we show the existence of an average-cost optimal stationary policy under the following condition, which depends on the optimal $\alpha$-discounted cost function $J_\alpha^*(i)$ in Chap. 4.

**Assumption 5.4** For some decreasing sequence $\{\alpha_n, n \geq 1\}$ tending to zero (as $n \to \infty$) and some state $i_0 \in S$, there exist a nonnegative real-valued function $H$ on $S$ and a constant $L_1$ such that

(a) $\alpha_n J_{\alpha_n}^*(i_0)$ is bounded in $n$ (this implies that $|J_{\alpha_n}^*(i_0)| < \infty$, and so we may define the function $h_{\alpha_n} := J_{\alpha_n}^* - J_{\alpha_n}^*(i_0)$ on $S$ for each $n \geq 1$).
(b) $L_1 \leq h_{\alpha_n}(i) \leq H(i)$ for all $n \geq 1$ and $i \in S$.

*Remark 5.5* Assumption 5.4 implies that $J_\alpha^* < \infty$, and it is similar to the hypotheses used for *discrete-time* MDPs; see Cavazos-Cadena (1991) [25], Puterman (1994) [129, p. 415] and Sennott (1999) [141, p. 132], for instance.

To verify Assumption 5.4, we may use, among others, the following facts.

**Proposition 5.6**

(a) *Each of the following conditions* $(a_1)$ *and* $(a_2)$ *implies Assumption* 5.4(a):
$(a_1)$ *There exist* $\tilde{f} \in F$, *a function* $\tilde{w} > 0$ *on* $S$, *constants* $\tilde{\rho} > 0$, $\tilde{b} \geq 0$, *and* $\tilde{M} > 0$ *such that, for all* $i \in S$ *and* $a \in A(i)$, *we have*

$$c(i, \tilde{f}) \leq \tilde{M}\tilde{w}(i) \quad \textit{and} \quad \sum_{j \in S} \tilde{w}(j) q(j|i, \tilde{f}) \leq -\tilde{\rho}\tilde{w}(i) + \tilde{b}.$$

*(This condition* $(a_1)$ *is satisfied when* $c(i,a)$ *is bounded on* $K$.*)*
$(a_2)$ *There exist* $\tilde{f} \in F$ *and a state* $i^* \geq 0$ *such that*

$$q(j|i, \tilde{f}) = \begin{cases} 0 & \textit{if } j > i^* \textit{ and } i \leq i^*, \\ 0 & \textit{if } j \geq i+1 \textit{ when } i > i^*. \end{cases}$$

*(Here, without loss of generality we write the denumerable state space* $S$ *as* $\{0, 1, \ldots\}$.*)*
(b) *Suppose that Assumption* 4.12 *and the following conditions* $(b_1)$, $(b_2)$ *hold*:
$(b_1)$ *For every* $f \in F$, $c(i, f)$ *is nondecreasing in* $i \in S := \{0, 1, 2, \ldots\}$, *the set of nonnegative integers*; *and*
$(b_2)$ $\sum_{j \geq k} q(j|i, f) \leq \sum_{j \geq k} q(j|i+1, f)$ *for all* $f \in F$ *and all* $i, k \in S$ *with* $k \neq i+1$.
*Then* $J_\alpha^*(i) \geq J_\alpha^*(0)$ *for all* $i \in S$ *and* $\alpha > 0$.

*Proof* (a) Under condition $(a_1)$, by Theorem 6.5(a) below (in Sect. 6.2) we have

$$\alpha J_\alpha^*(i) \leq \alpha J_\alpha(i, \tilde{f}) \leq \frac{\alpha \tilde{M}}{\alpha + \tilde{\rho}} \tilde{w}(i) + \frac{\alpha \tilde{b} \tilde{M}}{\alpha + \tilde{\rho}} \leq \tilde{M}\tilde{w}(i) + \frac{\tilde{b}\tilde{M}}{\tilde{\rho}} \quad \forall i \in S,$$

which implies Assumption 5.4(a). Also, under condition ($a_2$), by (5.6)–(5.8) we have

$$p_{\tilde{f}}(i,t,j) = \lim_{n\to\infty} q_{ij}^{(n)}(f,t) = 0 \quad \forall j > \max\{i, i^*\} \text{ and } t \geq 0.$$

Thus, we have

$$J_\alpha^*(i) \leq J_\alpha(i,\tilde{f}) = \int_0^\infty \sum_{j=0}^{\max\{i^*,i\}} e^{-\alpha t} c(j,\tilde{f}) p_{\tilde{f}}(i,t,j)\,dt$$

$$\leq \frac{\sum_{j=0}^{\max\{i^*,i\}} c(j,\tilde{f})}{\alpha} < \infty \quad \forall i \in S,$$

which implies Assumption 5.4(a).

(b) By ($b_2$) and Proposition C.16, we see that $p_f(i,t,j)$ is stochastically monotone for all $f \in F$, and so $\sum_{j\geq k} p_f(i,t,j) \leq \sum_{j\geq k} p_f(m,t,j)$ for all $k \in S$, $i \leq m$, and $t \geq 0$. Let $c(-1,f) := 0$; then $c(k,f) - c(k-1,f) \geq 0$ for all $k \geq 0$. Thus, for each $m \geq i$, we have

$$\sum_{j=0}^\infty c(j,f) p_f(i,t,j) = \sum_{j=0}^\infty \sum_{k=0}^j \big[c(k,f) - c(k-1,f)\big] p_f(i,t,j)$$

$$= \sum_{k=0}^\infty \big[c(k,f) - c(k-1,f)\big] \sum_{j=k}^\infty p_f(i,t,j)$$

$$\leq \sum_{k=0}^\infty \big[c(k,f) - c(k-1,f)\big] \sum_{j=k}^\infty p_f(m,t,j)$$

$$= \sum_{j=0}^\infty c(j,f) p_f(m,t,j),$$

and so $\sum_{j\in S} c(j,f) p_f(i,t,j)$ is increasing in $i \in S$. Therefore, by (4.3), also $J_\alpha(i,f)$ is increasing in $i \in S$. Taking $f := f_\alpha$ in Theorem 4.10, we have $J_\alpha^*(i) = J_\alpha(i,f_\alpha) \geq J_\alpha(0,f_\alpha) = J_\alpha^*(0)$, and so (b) follows. □

The following theorem establishes an important stepping-stone to show the existence of average-cost optimal stationary policies, namely, the *average-cost optimality inequality* (ACOI) in (5.11).

**Theorem 5.7** *Suppose that Assumptions* 4.12(a), 4.12(b), *and* 5.4 *hold. Then,*

(a) *There exist a sequence* $\{\alpha_m\}$ *of discount factors, a policy* $f^* \in F$, *a constant* $g^*$, *and a real-valued function* $u^*$ *on* $S$ *such that, for all* $i \in S$,

$$f^*(i) = \lim_{k\to\infty} f_{\alpha_k}^*(i), \qquad g^* := \lim_{k\to\infty} \alpha_k J_{\alpha_k}^*(i_0), \quad \textit{and}$$

$$u^*(i) := \lim_{k\to\infty} h_{\alpha_k}(i) \geq L_1.$$

(b) $(g^*, f^*, u^*)$ *satisfy the following ACOI*:

$$\begin{aligned} g^* &\geq c\big(i, f^*\big) + \sum_{j\in S} u^*(j)q\big(j|i, f^*\big) \\ &\geq \inf_{a\in A(i)} \left\{ c(i,a) + \sum_{j\in S} u^*(j)q(j|i,a) \right\} \quad \forall i \in S. \end{aligned} \tag{5.11}$$

(c) *Any policy $f \in F$ realizing the minimum in (5.11) is average-cost optimal. Therefore, $f^*$ in (a) is an average-cost optimal stationary policy, and, moreover, the optimal average-cost function $J_c^*$ equals the constant $g^*$.*

*Proof* We first prove (a) and (b) together. By the boundedness conditions in Assumption 5.4, an elementary argument gives the existence of a subsequence $\{\alpha_m\}$ of $\{\alpha_n\}$ with $\alpha_m \downarrow 0$, a constant $g^*$, and a real-valued function $u^*$ on $S$ such that, for all $i \in S$,

$$g^* = \lim_{m\to\infty} \alpha_m J^*_{\alpha_m}(i_0) \geq 0, \quad \text{and} \quad u^*(i) = \lim_{m\to\infty} h_{\alpha_m}(i) \geq L_1. \tag{5.12}$$

Then, for all $\varepsilon > 0$ and $m \geq 1$, Theorem 4.6 and Assumption 5.4 give a policy $f_m \in F$ (depending on $\alpha_m$ and $\varepsilon$) such that

$$J^*_{\alpha_m}(i) \geq \frac{c(i, f_m)}{\alpha_m + q_i(f_m)} + \frac{1}{\alpha_m + q_i(f_m)} \sum_{j\neq i} J^*_{\alpha_m}(j)q(j|i, f_m) - \frac{\alpha_m \varepsilon}{\alpha_m + q_i(f_m)}.$$

Hence, (a) follows from (5.12). Moreover, since $|J^*_{\alpha_m}(i)q_i(f_m)| < \infty$, we see that

$$\alpha_m J^*_{\alpha_m}(i) \geq c(i, f_m) + \sum_{j\in S} J^*_{\alpha_m}(j)q(j|i, f_m) - \alpha_m\varepsilon \quad \forall i \in S \text{ and } m \geq 1. \tag{5.13}$$

Now, by Assumption 5.4, replace $J^*_{\alpha_m}(i)$ with $h_{\alpha_m}(i) + J^*_{\alpha_m}(i_0)$. Therefore, since $\sum_{j\in S} q(j|i, f_m) = 0$ for all $m \geq 1$ and $i \in S$, by (5.13) we have

$$\alpha_m J^*_{\alpha_m}(i_0) + \alpha_m h_{\alpha_m}(i) \geq c(i, f_m) + \sum_{j\in S} h_{\alpha_m}(j)q(j|i, f_m) - \alpha_m\varepsilon,$$

and so

$$\begin{aligned} \alpha_m J^*_{\alpha_m}(i_0) + \alpha_m h_{\alpha_m}(i) &\geq c(i, f_m) + \sum_{j\neq i} h_{\alpha_m}(j)q(j|i, f_m) \\ &\quad + h_{\alpha_m}(i)q(i|i, f_m) - \alpha_m\varepsilon. \end{aligned} \tag{5.14}$$

On the other hand, Assumptions 4.12(a) and 4.12(b) give the existence of a subsequence $\{f_k\}$ of $\{f_m\}$ and $f^* \in F$ such that, for all $i \in S$,

$$\lim_{k\to\infty} f_k(i) = f^*(i), \qquad \lim_{k\to\infty} c(i, f_k) = c\bigl(i, f^*\bigr),$$
$$\lim_{k\to\infty} q(j|i, f_k) = q\bigl(j|i, f^*\bigr).$$

These facts, together with Proposition A.3 (Generalized Fatou's Lemma) and (5.12)–(5.14), yield

$$\begin{aligned} g^* &\geq c\bigl(i, f^*\bigr) + \sum_{j\neq i} u^*(j)q\bigl(j|i, f^*\bigr) + u^*(i)q\bigl(i|i, f^*\bigr) \\ &= c\bigl(i, f^*\bigr) + \sum_{j\in S} u^*(j)q\bigl(j|i, f^*\bigr) \\ &\geq \inf_{a\in A(i)} \Bigl\{c(i,a) + \sum_{j\in S} u^*(i)q(j|i,a)\Bigr\} \quad \forall i \in S, \end{aligned}$$

and so (b) follows.

(c) Suppose that $f \in F$ realizes the minimum in (5.11), so that

$$g^* \geq c(i, f) + \sum_{j\in S} u^*(j)q(j|i, f) \quad \forall i \in S.$$

Thus, by Theorem 5.2(b),

$$g^* \geq J_c(i, f) \quad \forall i \in S. \tag{5.15}$$

On the other hand, by (5.12) and Assumption 5.4(b) we have

$$g^* = \lim_{m\to\infty} \alpha_m J^*_{\alpha_m}(i_0) = \lim_{m\to\infty} \alpha_m J^*_{\alpha_m}(i) \quad \forall i \in S.$$

This implies, by Proposition A.5 (the Tauberian Theorem), that for all $\pi \in \Pi$ and $i \in S$,

$$\begin{aligned} g^* &= \lim_{m\to\infty} \alpha_m J^*_{\alpha_m}(i) \\ &\leq \limsup_{m\to\infty} \alpha_m J_{\alpha_m}(i, \pi) \\ &= \limsup_{\alpha_m\downarrow 0} \alpha_m \int_0^\infty e^{-\alpha_m t} E_i^\pi c\bigl(x(t), \pi_t\bigr)\, dt \\ &\leq \limsup_{T\to\infty} \frac{1}{T} \int_0^T E_i^\pi c\bigl(x(t), \pi_t\bigr)\, dt = J_c(i, \pi). \end{aligned}$$

Therefore, by (5.15), $J_c(i, f) \leq g^* \leq J_c(i, \pi)$ for all $\pi \in \Pi$ and $i \in S$, which yields (c). □

*Remark 5.8*

(a) As was already noted, inequality (5.11) is called the *average-cost optimality inequality* (ACOI), and if it holds with *equality*, then it is called the *average-cost optimality equation*; see, for instance, Guo and Cao (2005) [52], Guo and Hernández-Lerma (2003), Guo and Liu (2001) [58], Guo and Zhu (2002) [63], Kakumanu (1971) [92], Kakumanu (1975) [94], Kitaev and Rykov (1995) [98], Miller (1968) [117], Puterman (1994) [129], Sennott (1999) [141].
(b) The conditions and results in Theorem 5.7 are *continuous-time versions* of results for discrete-time MDPs; see Puterman (1994) [129] and Sennott (1999) [141], for instance. Note that the proof of Theorem 5.7 is essentially an extension of the minimum nonnegative solution approach used to prove Theorem 5.2. (See Remark 5.3.)

It should be noted that Theorem 5.7 does *not* require part (c) of Assumption 4.12. If Assumption 4.12(c) is imposed, then we obtain the average-cost optimality *equation* (5.16) below, rather than the inequality (5.11).

## 5.5 The Average-Cost Optimality Equation

In this section we establish the average-cost optimality equation (ACOE) as follows.

**Theorem 5.9** *Under Assumptions* 4.12(a), 4.12(b), *and* 5.4, *suppose in addition that* $\sum_{j\in S} H(j)q(j|i,a)$ *is continuous in* $a \in A(i)$ *for any* $i \in S$. *Then*

(a) *There exist a constant* $g^* \geq 0$, *a stationary policy* $f^* \in F$, *and a real-valued function* $u^*$ *on* $S$ *satisfying the ACOE*

$$g^* = c\bigl(i, f^*\bigr) + \sum_{j\in S} u^*(j)q\bigl(j|i, f^*\bigr)$$

$$= \min_{a\in A(i)} \left\{c(i,a) + \sum_{j\in S} u^*(j)q(j|i,a)\right\} \quad \forall i \in S. \tag{5.16}$$

(b) *Any* $f \in F$ *realizing the minimum in* (5.16) *is average-cost optimal. Therefore,* $f^*$ *in (a) is an average-cost optimal stationary policy, and the optimal average-cost function equals the constant* $g^*$ *(i.e.,* $J_c^* \equiv g^*$*).*

*Proof* (a) Let $g^*$ and $u^*$ be as in (5.11). Then, since $J_\alpha^*(i_0) \geq 0$ with $i_0$ as in Assumption 5.4, by (5.12) we have $H \geq u^*$. Thus, it follows from Proposition A.4 (Generalized dominated convergence theorem) that the functions $\sum_{j\in S} u^*(j) \times q(j|i,a)$ and $\sum_{j\in S} J_\alpha^*(j)q(j|i,a)$ are both continuous in $a \in A(i)$ for all $i \in S$. Therefore, Theorem 4.6, together with Lemma 4.4, gives the existence of $f_\alpha \in F$

satisfying

$$\alpha J_\alpha^*(i) = c(i, f_\alpha) + \sum_{j\in S} J_\alpha^*(j) q(j|i, f_\alpha) \quad \forall i \in S \tag{5.17}$$

$$\le c(i, a) + \sum_{j\in S} J_\alpha^*(j) q(j|i, a) \quad \forall i \in S \text{ and } a \in A(i). \tag{5.18}$$

Then, as in the proof of Theorem 5.7(a), from (5.17) and (5.18) we can show the existence of $f^* \in F$ such that

$$g^* = c\big(i, f^*\big) + \sum_{j\in S} u^*(j) q\big(j|i, f^*\big) \quad \forall i \in S$$

$$\le \min_{a\in A(i)} \Big\{ c(i, a) + \sum_{j\in S} u^*(j) q(j|i, a) \Big\} \quad \forall i \in S.$$

The latter inequality, together with (5.11), yields (a).

Obviously, (b) follows from Theorem 5.7(b). □

## 5.6 Examples

In this section, we illustrate our results with a couple of examples.

*Example 5.1* Consider the controlled population processes in Example 4.1 (in Sect. 4.7). Let $S$, $q(j|i, a)$ and $c(i, a)$ be as in Example 4.1, except that each $A(i)$ here is generalized to be a *compact* set of available actions.

We aim to find conditions that ensure the existence of an average-cost optimal stationary policy. To do this, in the spirit of the conditions in Theorem 5.7, in addition to $\mathbf{B_1}$ and $\mathbf{B_2}$ in Example 4.1, now we also consider the following hypotheses:

$\mathbf{B_3}$. For each fixed $i \in S$, the functions $\lambda(i, a)$, $\mu(i, a)$, and $\tilde{c}(i, a)$ are all continuous on $A(i)$. (This condition is satisfied, of course, when the sets $A(i)$ are all finite.)

$\mathbf{B_4}$. There exists $a(i) \in A(i)$ such that $\lambda(i, a(i)) = 0$ for each $i \ge 0$, but $\mu(i, a(i)) > 0$ when $i \ge 1$. (This means that there exists $f \in F$ with $f(i) := a(i)$ such that the corresponding $[q(j|i, f)]$ is a pure-death matrix.)

$\mathbf{B_5}$. For each $f \in F$, the function $c(i, f)$ is increasing in $i \in S$. (This condition is satisfied, for example, when $\tilde{c}(i, a) := ai$ for all $i \in S$ and $a \in A(i)$, with $A(i) := [0, p]$ and $p$ as in Example 4.1.)

To further specialize Example 5.1, we consider again the birth-and-death process with *controlled immigration* in Example 4.1, in which, if $p \ge C$, conditions $\mathbf{B_3}$, $\mathbf{B_4}$, and $\mathbf{B_5}$ above are all satisfied.

Under conditions $\mathbf{B_1}$–$\mathbf{B_5}$, we obtain the following.

**Proposition 5.10** (On average cost optimality) *If conditions* $\mathbf{B_1}$ *and* $\mathbf{B_2}$ *in Example* 4.1 *and* $\mathbf{B_k}$ ($3 \le k \le 5$) *above are all satisfied, then Example* 5.1 *satisfies all the assumptions in Theorem* 5.9. *Therefore* (*by Theorem* 5.9), *there exists an average-cost optimal stationary policy.*

*Proof* Since Assumptions 2.2 and 4.12(a) and (b) have been verified in the proof of Proposition 4.19, we only need to verify Assumption 5.4 since the continuity of $\sum_{j\in S} H(j)q(j|i,a)$ in $a \in A(i)$ follows from the description of Example 4.1. Obviously, Proposition 5.6(a), together with $\mathbf{B_4}$, implies Assumption 5.4(a), and thus it only remains to verify Assumption 5.4(b). Under $\mathbf{B_5}$, by (4.31) and Proposition 5.6(b) we have

$$0 \le J_\alpha^*(i) - J_\alpha^*(0) = h_\alpha(i) < \infty \quad \forall i \in S. \tag{5.19}$$

Moreover, by Theorem 4.6(a) and (4.31) we have, for each $i \ge 1$,

$$\begin{aligned}
J_\alpha^*(i) &= \min_{a\in A(i)} \left\{ \frac{c(i,a)}{\alpha+\lambda(i,a)+\mu(i,a)} + \frac{1}{\alpha+\lambda(i,a)+\mu(i,a)} \sum_{j\neq i} J_\alpha^*(j)q(j|i,a) \right\} \\
&\le \frac{c(i,a(i))}{\alpha+\lambda(i,a(i))+\mu(i,a(i))} \\
&\quad + \frac{1}{\alpha+\lambda(i,a(i))+\mu(i,a(i))} \sum_{j\neq i} J_\alpha^*(j)q\big(j|i,a(i)\big) \\
&= \frac{c(i,a(i))}{\alpha+\mu(i,a(i))} + \frac{\mu(i,a(i))}{\alpha+\mu(i,a(i))} J_\alpha^*(i-1) \\
&\le \frac{c(i,a(i))}{\mu(i,a(i))} + J_\alpha^*(i-1),
\end{aligned}$$

i.e.,

$$J_\alpha^*(i) \le \frac{c(i,a(i))}{\mu(i,a(i))} + J_\alpha^*(i-1) \quad \forall i \ge 1.$$

Iteration of this inequality gives, for $i \ge 1$,

$$J_\alpha^*(i) \le \sum_{j=1}^{i} \frac{c(j,a(j))}{\mu(j,a(j))} + J_\alpha^*(0),$$

which, together with (5.19), gives Assumption 5.4(b) with $H(i) := \sum_{j=1}^{i} \frac{c(j,a(j))}{\mu(j,a(j))}$ for $i \ge 1$ and $H(0) := 0$. □

In the following example, we show that the conditions to guarantee the existence of an average-cost optimal stationary policy do not imply the existence of a solution to the average optimality equation.

*Example 5.2* Let $S := \{0, 1, \ldots\}$; $A(0) := \{1, 2\}$ and $A(i) \equiv \{0\}$ for all $i \geq 1$; $c(0, 1) := 0$, $c(0, 2) := 1$, and $c(i, 0) := 1$ for each $i \geq 1$; $q(i-1|i, 0) := 1$, $q(i|i, 0) := -1$ for each $i \geq 1$, $q(0|0, 1) := -1$, $q(0|0, 2) := -1$, $q(i|0, 1) := p_1(i)$, and $q(i|0, 2) = p_2(i)$ for all $i \geq 1$, where $\{p_k(i), i \geq 1\}$ is a probability distribution on $\{1, \ldots\}$ (for each $k = 1, 2$) such that $\sum_{i=1}^{\infty} i p_1(i) = \infty$.

**Proposition 5.11** (On average optimality) *For the continuous-time MDP in Example 5.2, there exists an average-cost optimal stationary policy, and the optimality average-cost inequality* (5.11) *is strict.*

*Proof* Since the transition rates and the cost function are bounded and each set of admissible actions is finite, Assumptions 2.2, 4.12, and 5.4(a) are satisfied. Hence, to prove the existence of a solution to the optimality inequality (5.11), by Theorem 5.7 it suffices to show that Assumption 5.4(b) is satisfied. To do so, let $f_1$ and $f_2$ be the stationary policies defined by $f_1(0) = 1$, $f_2(0) = 2$, and $f_1(i) = f_2(i) = 0$ for all $i \geq 1$. Then, from (4.3) we see that $J_\alpha(i, f_2) \equiv \frac{1}{\alpha}$ and $J_\alpha(i, f_1) \leq \frac{1}{\alpha}$ for all $i \in S$. Moreover, by (4.4) and the description of this example we have

$$\begin{aligned} \alpha J_\alpha(0, f_1) &= c(0, f_1) + \sum_{j \in S} q\big(j|0, f_1(0)\big) J_\alpha(j, f_1) \\ &= -J_\alpha(0, f_1) + \sum_{j \geq 1} p_1(j) J_\alpha(j, f_1), \end{aligned}$$

and thus,

$$J_\alpha(0, f_1) = \frac{1}{1+\alpha} \sum_{j \geq 1} p_1(j) J_\alpha(j, f_1) \leq \frac{1}{(1+\alpha)\alpha} < \frac{1}{\alpha} = J_\alpha(0, f_2),$$

which, together with Remark 4.15 and Theorem 4.14, implies that $f_1$ is discounted-cost optimal policy, that is, $J_\alpha^*(i) = J_\alpha(i, f_1)$ for all $i \in S$. Moreover, by (4.4) we have, for each $i \geq 1$,

$$J_\alpha^*(i) = \frac{1}{1+\alpha} + \frac{1}{1+\alpha} J_\alpha^*(i-1) = \cdots = \sum_{k=1}^{i} \frac{1}{(1+\alpha)^k} + \frac{1}{(1+\alpha)^i} J_\alpha^*(0),$$

and so $h_\alpha(i) := J_\alpha^*(i) - J_\alpha^*(0) \leq i$ for all $i \geq 0$, and

$$\begin{aligned} h_\alpha(i) &= \frac{1}{\alpha(1+\alpha)^i} \big[(1+\alpha)^{i+1} - 1 + \big[1 - (1+\alpha)^i\big]\alpha J_\alpha^*(0)\big] \\ &\geq \frac{1}{\alpha(1+\alpha)^i} \big[(1+\alpha)^{i+1} - 1 + \big[1 - (1+\alpha)^i\big]\big] \geq 0 \quad \forall i \geq 1. \end{aligned} \tag{5.20}$$

Therefore, $0 \leq h_\alpha(i) \leq i$ for all $i \in S$ and $\alpha > 0$, which implies that Assumption 5.4(b) holds, and thus Theorem 5.7 implies that $f_1$ is also average-cost optimal.

Let $g^*$ and $u^*$ be as in Theorem 5.7. Then, by (5.20) and $g^* := \lim_{m\to\infty} \alpha_m J^*_{\alpha_m}(0)$ for some sequence of $\{\alpha_m\}$ with $\lim_{m\to\infty} \alpha_n = 0$, we have

$$u^*(i) = \lim_{m\to\infty} h_{\alpha_m}(i) = 1 + i\big(1 - g^*\big), \quad \text{and} \quad J_c(i, f_1) = g^* \quad \forall i \in S. \qquad (5.21)$$

On the other hand, by Proposition C.12(b) and the description of this example we see that $p_{f_1}(i,t,j)$ is null recurrent when $\sum_{i\geq 1} i p_1(i) = \infty$, and thus $\lim_{t\to\infty} p_{f_1}(0,t,0) = 0$. Hence,

$$\lim_{t\to\infty} E_0^{f_1} c\big(x(t), f_1\big) = \lim_{t\to\infty} \big[1 - p_{f_1}(0,t,0)\big] = 1,$$

which, together with (5.1), gives $J_c(0, f_1) = 1$. Thus, by (5.21) we get $u^* = 1$, and

$$g^* = 1 > 0 = \min_{a\in A(0)} \left\{ c(0,a) + \sum_{j\in S} u^*(j) q(j|0,a) \right\}.$$

This shows that ACOI (5.11) is strict for $i = 0$. □

The reader may have noticed that the distribution $\{p_2(i), i \geq 1\}$ in Example 5.2 may be chosen arbitrarily, since it is not imposed by any additional condition in the discussions above, and we have $J_c(f_2) \equiv 1$. This implies that $f_2$ is also average-cost optimal. However, $f_2$ does not realize the minimum in (5.11) at $i = 0$. Thus, if we choose $\{p_2(i), i \geq 1\}$ such that $\sum_{i\geq 1} i p_2(i) < \infty$ which gives that $Q(f_2)$ is positive recurrent, then we can get an example for which there exists two average-cost optimal stationary policies: One is null recurrent, and the other is positive recurrent, and yet it fails to realize the minimum in ACOI (5.11).

## 5.7 Notes

Chapters 4 and 5 are, respectively, about the discounted- and the long-run average-cost criteria, which have been studied by many authors, including Guo and Hernández-Lerma (2003) [53, 54], Guo and Zhu (2002) [62], Howard (1960) [86], Kakumanu (1971) [92], Kitaev and Rykov (1995) [98], Miller (1968) [117], Puterman (1994) [129], and Sennott (1999) [141] for the discount criterion; and Feinberg (2004) [39], Guo and Hernández-Lerma (2003) [55], Guo and Liu [58], Guo and Rieder (2006) [59], Haviv and Puterman (1998) [68], Kakumanu (1972) [93], Lewis and Puterman (2001) [107] and (2002) [108], Lippman (1975) [110], Puterman (1994) [129], Sennott (1999) [141], Prieto-Rumeau and Hernández-Lerma (2005) [123, 124], and Yushkevich and Feinberg (1979) [165] for the average criterion. The main purpose of all of the literature concerns the existence of optimal stationary policies, and the characterization of the value function. When the state and the action spaces are both *finite*, the existence of an optimal stationary policy for either the discount or the average criterion is a well-known fact, and there is a "policy iteration" algorithm to obtain such an optimal policy; see the previous Sects. 3.5

and 4.7 and Guo, Song, and Zhang (2009) [66], Howard (1960) [86], Miller (1968) [117], Puterman (1994) [129], and Sennott (1999) [141], for instance. If, however, either the state or the action space is *not* finite, then the situation is a bit tricky. In fact, the uniformization technique by Lewis and Puterman (2001) [107] and (2002) [108], Lippman (1975) [110], Puterman (1994) [129], and Sennott (1999) [141] can be used to show that an AR optimal policy may *not* exist when the state or the action space is *not* finite. Thus, to ensure the existence of an optimal policy requires suitable conditions. These conditions include the following.

(i) *Regularity* conditions, which are used to guarantee the regularity of the transition probability function of the underlying continuous-time Markov processes
(ii) *Continuity* conditions for the class of randomized Markov policies, which are required to establish the existence of a transition function under a randomized Markov policy
(iii) An *expected-growth* condition, which is needed to ensure the finiteness of the corresponding optimality criterion when using possibly unbounded costs/rewards; and
(iv) An *absolute integrability* condition, which is used for the interchange of integrals and summations in some arguments.

To verify the regularity condition (i), the transition rates are assumed to be *bounded* by Feinberg (2004) [39], Haviv and Puterman (1998) [68], Kakumanu (1972) [93], Lewis and Puterman (2001) [107] and (2002) [108], Lippman (1975) [110], Puterman (1994) [129], Sennott (1999) [141], and Yushkevich and Feinberg (1979) [165]; or, alternatively, the transition rates are supposed to satisfy a "drift" condition in Guo and Hernández-Lerma (2003) [53, 54], and Prieto-Rumeau and Hernández-Lerma (2005) [123, 124].

There are several ways to show the existence of transition functions. In particular, the continuity conditions mentioned in (ii) are required when following an existence approach that can be traced back to Feller (1940) [42]. Those conditions, however, can be replaced with weaker measurability assumptions, as in Ye et al. (2008) [159].

To verify the expected growth condition (iii), the reward/cost rates are assumed to be bounded in Guo and Liu (2001) [58], Kakumanu (1971) [92], Howard (1960) [86], and Miller (1968) [117].

The *absolute integrability* condition (iv) is trivially satisfied when the reward/cost and the transition rates are both *bounded*; otherwise, it needs to be imposed, as in Guo and Hernández-Lerma (2003) [53–55], Guo and Liu (2001) [58], Guo and Rieder (2006) [59], and Prieto-Rumeau and Hernández-Lerma (2005) [123, 124]. Actually, to ensure the existence of an optimal policy, in addition to the above four conditions, we usually need a continuity–compactness assumption, as our Assumption 4.12 above (in Sect. 4.6), which is required by Guo and Hernández-Lerma (2003) [53–55] and Prieto-Rumeau and Hernández-Lerma (2005) [123, 124].

The question now is, assuming that we have sufficient conditions for the existence of average optimal policies, *how* do we prove such an existence result? The most common approaches to proving the existence of average optimal policies are the following. The first one is by means of *Kolmogorov's forward equation*, as in

Guo and Liu (2001) [58], Howard (1960) [86], Kakumanu (1971) [92], and Miller (1968) [117]. The second one is the *uniformization technique* used by Lewis and Puterman (2001) [107, 108], Lippman (1975) [110], Puterman (1994) [129], and Sennott (1999) [141], among other authors. The third one is using the *extended infinitesimal operator* of a Markov process, as in Guo and Hernández-Lerma (2003) [53–55], Guo and Rieder (2006) [59], and Prieto-Rumeau and Hernández-Lerma (2005) [123–125]. These three approaches require condition (iv) above.

In this chapter, we introduced a fourth approach: we *removed* condition (ii) and *avoided* conditions (iii) and (iv) by restricting ourselves to *nonnegative* costs. Under these circumstances, we provided a "minimum nonnegative solution approach" to proving the existence of $\varepsilon$-optimal stationary policies. The motivation for this chapter came from the arguments on nonnegative models for discrete-time MDPs by Puterman (1994) [129] and Sennott (1999) [141]. In fact, the main results in this chapter are analogous to those given by Puterman (1994) [129] and Sennott (1999) [141] in the discrete-time case, and they are from Guo and Yin (2007) [61].

Summarizing, we now have *four* approaches to deal with average-cost optimality: the average-cost minimum nonnegative solution approach; Kolmogorov's forward equation approach; the uniformization technique approach; and the extended infinitesimal operator approach. Each of them has advantages and disadvantages. For further discussion of these approaches, see Sect. 6.6 of Chap. 6 below.

# Chapter 6
# Discount Optimality for Unbounded Rewards

In Chap. 4, we have studied discounted-cost optimality for nonnegative MDP models, which means that the cost function $c(i,a)$ in (4.1) is nonnegative (or at least bounded below). Equivalently, the results in Chap. 4 can apply only to the case where the reward function $r(i,a) := -c(i,a)$ is bounded above. Therefore, the results in Chap. 4 are not applicable to the more general case where the reward function $r(i,a)$ has neither upper nor lower bounds. This general unbounded case is now our main subject.

*In this chapter, we consider the control model* (2.1).

## 6.1 Introduction

The discounted-reward criterion was introduced by Howard (1960) [86]. The main tool to study this optimality criterion is the *discounted-reward optimality equation* (DROE):

$$\alpha u(i) = \sup_{a\in A(i)} \left\{ r(i,a) + \sum_{j\in S} u(j) q(j|i,a) \right\} \quad \forall i \in S. \tag{6.1}$$

The DROE is also known as the discounted-reward *dynamic programming equation*. For some continuous-time Markov control processes (for example, controlled diffusions), the DROE is called the *Hamilton–Jacobi–Bellman equation*. (Compare (6.1) with the optimality equation (4.10) in the discounted-*cost* case.)

When $S$ and the action sets $A(i)$ are all finite, the existence of a solution to the DROE and of a discount optimal stationary policy was shown by Howard (1960) [86], Miller (1968) [117], Roykov (1966) [139], and Veinott (1969) [151] under various conditions. The main results in these references were extended by Kakumanu (1971) [92] to the case of a denumerable state space under the following conditions:

X. Guo, O. Hernández-Lerma, *Continuous-Time Markov Decision Processes*,
Stochastic Modelling and Applied Probability 62,
DOI 10.1007/978-3-642-02547-1_6, © Springer-Verlag Berlin Heidelberg 2009

(i) The transition rates and the reward function are both bounded, i.e.,

$$\|q\| := \sup_{i \in S} q^*(i) \quad \text{and} \quad \|r\| := \sup_{(i,a) \in K} \left| r(i,a) \right|$$

are finite (to avoid a trivial case, we suppose that $\|q\| > 0$ in the following arguments), with $K$ and $q^*(i)$ as in (2.2) and (2.4), respectively.
(ii) The action sets $A(i)$ are all denumerable.

It was shown by Kakumanu (1971) [92] that the optimal discounted-reward function $V_\alpha^*$ is the unique bounded solution to the DROE by approximating the denumerable MDP by finite MDPs. He also proved that any deterministic stationary policy realizing the maximum in (6.1) is discount optimal.

*Remark 6.1* Under the condition that $\|q\| < +\infty$ (see (i) above) the continuous-time MDP can be transformed into an equivalent *discrete-time* MDP. Indeed, let

$$\beta := \frac{\|q\|}{\alpha + \|q\|}, \qquad \bar{r}(i,a) := \frac{r(i,a)}{\alpha + \|q\|}, \quad \text{and} \quad \bar{p}(j|i,a) := \frac{q(j|i,a)}{\|q\|} + \delta_{ij} \tag{6.2}$$

for all $i, j \in S$ and $a \in A(i)$. Then it follows from our hypotheses in (2.3) that $[\bar{p}(j|i,a)]$ is a transition probability matrix and also that $0 < \beta < 1$. Moreover, a direct calculation shows that (6.1) can be rewritten as

$$u(i) = \sup_{a \in A(i)} \left\{ \bar{r}(i,a) + \beta \sum_{j \in S} u(j) \bar{p}(j|i,a) \right\} \quad \forall i \in S, \tag{6.3}$$

which is precisely the DROE for a *discrete-time* MDP with discount factor $\beta$, transition probabilities $\bar{p}(j|i,a)$, and reward function $\bar{r}(i,a)$. Therefore, when $\|q\| < +\infty$, the discounted-reward optimality equations for continuous- and discrete-time MDPs are *equivalent.* (Note that (6.2) was already introduced in (3.76) in connection with finite MDPs.)

The above approach (i.e., transforming a continuous-time MDP into an equivalent discrete-time MDP) is called the *uniformization technique*; see Bertsekas (2001) [12], Howard (1960) [86], Kakumanu (1977) [95], Lewis and Puterman (2001) [107], Puterman (1994) [120], Serfozo (1979) [142], Sennott (1999) [141], and Veinott (1969) [151], for instance. When this approach is applicable, many results about continuous-time MDPs (such as the existence of solutions to the DROE and the convergence of the value and the policy iteration algorithms) can be obtained from those for the discrete-time case; see Dynkin and Yushkevich (1979) [38], Feinberg and Shwartz (2002) [41], Hernández-Lerma and Lasserre (1999) [74], Hou and Guo (1998) [84], Howard (1960) [86], Puterman (1994) [129], and Sennott (1999) [141], for instance.

When the transition rates are *unbounded*, however, so that $\|q\| = +\infty$, we cannot guarantee that $\beta < 1$, and thus the uniformization technique is no longer applicable. Thus, to deal with the general case where $\|q\| \leq +\infty$, we need to follow a different approach.

This general case has been studied by several authors under different sets of hypotheses; see, for instance, Guo and Zhu (2002a) [62], Guo and Hernández-Lerma (2003a, 2003b) [53, 54], Hernández-Lerma and Govindan (2001) [72], Hou and Guo (1998) [84], Prieto-Rumeau and Hernández-Lerma (2006b) [125], Song (1987) [145], or Wu (1997) [158]. In this chapter, we will mainly discuss the results by Guo and Hernández-Lerma (2003a, 2003b) [53, 54], but it should be noted that the main arguments are only slightly different from those in the other mentioned references. In particular, to show the existence of optimal policies, here we use Kolmogorov's forward differential equation.

## 6.2 The Discounted-Reward Optimality Equation

First of all, we must find conditions ensuring that $V_\alpha(i,\pi)$ in (2.19) is finite for each $\pi \in \Pi$ and $i \in S$. To this end, we need the following key results in Lemmas 6.2 and 6.3.

**Lemma 6.2** *Let $\bar{w}$ be an arbitrary nonnegative function on $S$, and fix an arbitrary policy $\pi = (\pi_t) \in \Pi$ and $t > s \geq 0$. Then the function*

$$\sum_{j \in S} \bar{w}(j) p_\pi(z,i,t,j) \quad \text{of } (z,i) \in [s,t] \times S$$

*is the minimal nonnegative solution of the "backward" inequality*

$$h(z,i) \geq \sum_{k \neq i} \int_z^t e^{\int_z^y q(i|i,\pi_v)\,dv} h(y,k) q(k|i,\pi_y)\,dy + e^{\int_z^t q(i|i,\pi_v)\,dv} \bar{w}(i), \tag{6.4}$$

*and, moreover, it satisfies* (6.4) *with equality.*

*Proof* By Proposition C.4, the (minimum) transition probability function $p_\pi(s,i,t,j)$ can be constructed as follows: For all $i, j \in S$, $n \geq 0$, and $z \in [s,t]$, let

$$p_0^\pi(z,i,t,j) := \delta_{ij} e^{\int_z^t q(i|i,\pi_v)\,dv}, \tag{6.5}$$

$$p_{n+1}^\pi(z,i,t,j) := \sum_{k \neq i} \int_z^t e^{\int_z^y q(i|i,\pi_v)\,dv} q(k|i,\pi_y) p_n^\pi(y,k,t,j)\,dy. \tag{6.6}$$

Then

$$p_\pi(z,i,t,j) := \sum_{n=0}^\infty p_n^\pi(z,i,t,j). \tag{6.7}$$

From (6.5)–(6.7) we have that for all $z \in [s,t]$ and $i, j \in S$,

$$p_\pi(z,i,t,j) = \sum_{k \neq i} \int_z^t e^{\int_z^y q(i|i,\pi_v)\,dv} q(k|i,\pi_y) p_\pi(y,k,t,j)\,dy + \delta_{ij} e^{\int_z^t q(i|i,\pi_v)\,dv}. \tag{6.8}$$

Multiplying both sides of (6.8) by $\bar{w}(j)$ and then summing over $j \in S$, we see that the function $\sum_{j\in S} \bar{w}(j)p_\pi(z,i,t,j)$, with $(z,i) \in [s,t] \times S$, satisfies (6.4) with equality.

Suppose now that $h(z,i)$, $(z,i) \in [s,t] \times S$, is an arbitrary nonnegative solution of (6.4). Then from (6.4), using that $q(k|i,\pi_y) \geq 0$ $(k \neq i)$, we obtain

$$h(z,i) \geq e^{\int_z^t q(i|i,\pi_v)\,dv}\bar{w}(i) = \sum_{j\in S} \bar{w}(j)p_0^\pi(z,i,t,j) \quad \forall (z,i) \in [s,t] \times S.$$

Arguing by induction, let us now suppose that, for some $N \geq 0$,

$$h(z,i) \geq \sum_{j\in S} \bar{w}(j)\left[\sum_{n=0}^{N} p_n^\pi(z,i,t,j)\right] \quad \forall (z,i) \in [s,t] \times S.$$

Then from (6.4)–(6.6) and the induction hypothesis it follows that

$$\begin{aligned} h(z,i) &\geq \int_z^t e^{\int_z^y q(i|i,\pi_v)\,dv} \sum_{k\neq i} q(k|i,\pi_y)h(y,k)\,dy + e^{\int_z^t q(i|i,\pi_v)\,dv}\bar{w}(i) \\ &\geq \int_z^t e^{\int_z^y q(i|i,\pi_v)\,dv} \sum_{k\neq i} q(k|i,\pi_y)\left[\sum_{j\in S}\bar{w}(j)\sum_{n=0}^{N} p_n^\pi(y,k,t,j)\right]dy \\ &\quad + e^{\int_z^t q(i|i,\pi_v)\,dv}\bar{w}(i) \\ &= \sum_{j\in S}\bar{w}(j)\left[\sum_{n=0}^{N+1} p_n^\pi(z,i,t,j)\right], \end{aligned}$$

and so

$$h(z,i) \geq \sum_{j\in S} \bar{w}(j)\left[\sum_{n=0}^{N} p_n^\pi(z,i,t,j)\right] \quad \forall N \geq 0.$$

This inequality and (6.7) give that $h(z,i) \geq \sum_{j\in S} \bar{w}(j)p_\pi(z,i,t,j)$, and so Lemma 6.2 follows. □

The inequality in part (ii) of the following Lemma 6.3 is known as the *Lyapunov* (or *Foster–Lyapunov*) *stability* condition. In our present context, it is indeed a "stability" condition in the sense that it gives a bound on the expected growth of $\bar{w}(x(t))$ in part (i) of the lemma.

**Lemma 6.3** *Let $\bar{w}$ be an arbitrary nonnegative function on $S$, and $\bar{c}$ and $\bar{b}$ two constants such that $\bar{b} \geq 0$ and $\bar{c} \neq 0$. Then, for each fixed $\pi \in \Pi$, the following statements are equivalent*:

(i) $\sum_{j\in S} \bar{w}(j)p_\pi(s,i,t,j) \leq e^{-\bar{c}(t-s)}\bar{w}(i) + \frac{\bar{b}}{\bar{c}}[1 - e^{-\bar{c}(t-s)}]$ *for all $i \in S$ and a.e. $t \geq s \geq 0$.*

(ii) $\sum_{j\in S}\bar{w}(j)q(j|i,\pi_t)\le -\bar{c}\bar{w}(i)+\bar{b}$ *for all* $i\in S$ *and a.e.* $t\ge s\ge 0$.

*Proof* (i) $\Rightarrow$ (ii). Suppose that (i) holds. Then, for each $i\in S$, a.e. $t\ge 0$, and $\Delta t>0$, we have

$$\sum_{j\neq i}\bar{w}(j)\frac{p_\pi(t,i,t+\Delta t,j)}{\Delta t}\le\left[\frac{1-p_\pi(t,i,t+\Delta t,i)}{\Delta t}-\frac{1-e^{-\bar{c}\Delta t}}{\Delta t}\right]\bar{w}(i)$$
$$+\frac{\bar{b}(1-e^{-\bar{c}\Delta t})}{\bar{c}\Delta t}.$$

Letting $\Delta t\to 0$, the Fatou–Lebesgue Lemma yields

$$\sum_{j\neq i}\bar{w}(j)q(j|i,\pi_t)\le\bar{w}(i)\big[-q(i|i,\pi_t)-\bar{c}\big]+\bar{b},$$

which gives (ii).

(ii) $\Rightarrow$ (i). For any fixed $i\in S$ and a.e. $t\ge s\ge 0$, note that the right-hand side of (i) is nonnegative, i.e.,

$$h(z,i):=e^{-\bar{c}(t-z)}\bar{w}(i)+\frac{\bar{b}}{\bar{c}}\big[1-e^{-\bar{c}(t-z)}\big]\ge 0\quad\forall(z,i)\in[s,t]\times S.\tag{6.9}$$

Therefore, by Lemma 6.2, it suffices to show that $h$ satisfies inequality (6.4), that is, for all $z\in[s,t]$ and $i\in S$,

$$h(z,i)\ge\sum_{k\neq i}\int_z^t e^{\int_z^y q(i|i,\pi_v)\,dv}q(k|i,\pi_y)h(y,k)\,dy+e^{\int_z^t q(i|i,\pi_v)\,dv}\bar{w}(i).\tag{6.10}$$

Indeed, since $\sum_{j\neq i}q(j|i,\pi_y)=-q(i|i,\pi_y)$, by (6.9) and condition (ii), we have

$$\sum_{k\neq i}\int_z^t e^{\int_z^y q(i|i,\pi_v)\,dv}q(k|i,\pi_y)h(y,k)\,dy+e^{\int_z^t q(i|i,\pi_v)\,dv}\bar{w}(i)$$
$$\le\int_z^t e^{\int_z^y q(i|i,\pi_v)\,dv}e^{-\bar{c}(t-y)}\big[-q(i|i,\pi_y)\bar{w}(i)-\bar{c}\bar{w}(i)+\bar{b}\big]\,dy$$
$$+e^{\int_z^t q(i|i,\pi_v)\,dv}\bar{w}(i)+\frac{\bar{b}}{\bar{c}}\int_z^t e^{\int_z^y q(i|i,\pi_v)\,dv}q(i|i,\pi_y)e^{-\bar{c}(t-y)}\,dy$$
$$-\frac{\bar{b}}{\bar{c}}\int_z^t e^{\int_z^y q(i|i,\pi_v)\,dv}q(i|i,\pi_y)\,dy$$
$$=e^{-\bar{c}(t-z)}\bar{w}(i)+\bar{b}\int_z^t e^{\int_z^y q(i|i,\pi_v)\,dv}e^{-\bar{c}(t-y)}\,dy$$
$$+\frac{\bar{b}}{\bar{c}}\int_z^t e^{\int_z^y q(i|i,\pi_v)\,dv}q(i|i,\pi_y)e^{-\bar{c}(t-y)}\,dy$$

$$-\frac{\bar{b}}{\bar{c}}\int_z^t e^{\int_z^y q(i|i,\pi_v)\,dv} q(i|i,\pi_y)\,dy$$

$$= e^{-\bar{c}(t-z)}\bar{w}(i) - \frac{\bar{b}}{\bar{c}}e^{-\bar{c}t}\left[\int_z^t e^{-(-\bar{c}y-\int_z^y q(i|i,\pi_v)\,dv)}\big(-\bar{c} - q(i|i,\pi_y)\big)\,dy\right]$$

$$-\frac{\bar{b}}{\bar{c}}\big[e^{\int_z^t q(i|i,\pi_v)\,dv} - 1\big]$$

$$= e^{-\bar{c}(t-z)}\left[\bar{w}(i) - \frac{\bar{b}}{\bar{c}}\right] + \frac{\bar{b}}{\bar{c}} = h(z,i).$$

Hence, (6.10) holds. □

To guarantee the finiteness of the expected discounted reward $V_\alpha(i,\pi)$ in (2.19), by Assumption 2.2 and Lemma 6.3 we see that it is natural to require the following condition:

**Assumption 6.4**

(a) For every $(i,a) \in K$ and some constant $M > 0$, $|r(i,a)| \le Mw(i)$, where $w$ comes from Assumption 2.2.
(b) The positive discount factor $\alpha$ verifies that $\alpha > c_0$, where the constant $c_0$ is as in Assumption 2.2.

Under Assumptions 2.2 and 6.4, we see that the next result is a direct consequence of (2.19) and Lemma 6.3.

**Theorem 6.5** *Suppose that Assumptions* 2.2 *and* 6.4 *hold. Then*:

(a) *For all* $\pi \in \Pi$ *and* $i \in S$,

$$\big|V_\alpha(i,\pi)\big| \le \frac{b_0 M}{\alpha(\alpha - c_0)} + \frac{M}{\alpha - c_0}w(i)$$

*with* $c_0$ *and* $b_0$ *as in Assumption* 2.2. *In particular,* $V_\alpha(\cdot,\pi)$ *is in* $B_w(S)$.
(b) *The optimal discounted-reward function* $V_\alpha^*$ *is in* $B_w(S)$.
(c) $\lim_{t\to\infty} e^{-\alpha t}E_i^\pi u(x(t)) = 0$ *for all* $\pi \in \Pi$, $i \in S$, *and* $u \in B_w(S)$.

*Proof* Obviously, under Assumptions 2.2 and 6.4, by Lemma 6.3 we see that (a) and (b) are both true. To prove (c), by Assumptions 2.2 and 6.4(b), together with Lemma 6.3, we have

$$\big|e^{-\alpha t}E_i^\pi u\big(x(t)\big)\big| \le \|u\|_w e^{-\alpha t}\left[e^{c_0 t}w(i) - \frac{b_0}{c_0}\big(1 - e^{c_0 t}\big)\right]$$

$$= \|u\|_w\left[e^{-(\alpha-c_0)t}w(i) + \frac{b_0}{c_0}e^{-(\alpha-c_0)t} - \frac{b_0}{c_0}e^{-\alpha t}\right].$$

Letting $t \to \infty$, (c) follows. □

To establish the DROE (6.1) for the general case, we need to introduce some more notation. For each $i \in S$, let $m(i)$ be any positive number such that $m(i) \geq q^*(i) \geq 0$ (recall the notation in (2.4)). When $\|q\| < +\infty$, we can choose $m(i) := \|q\|$ for all $i \in S$. We define the operator $T : B_w(S) \to B_w(S)$ as

$$Tu(i) := \sup_{a\in A(i)} \left\{ \frac{r(i,a)}{\alpha + m(i)} + \frac{m(i)}{\alpha + m(i)} \sum_{j\in S} u(j) p(j|i,a) \right\} \tag{6.11}$$

for $u \in B_w(S)$ and $i \in S$, where

$$p(j|i,a) := \frac{q(j|i,a)}{m(i)} + \delta_{ij} \tag{6.12}$$

is a *probability measure* on $S$ for each $(i,a) \in K$. (Clearly, $p(j|i,a) = \bar{p}(j|i,a)$ when $\|q\| < +\infty$; recall (6.2).) Note that the operator $T$ is *monotone*, that is, $Tu \geq Tv$ if $u \geq v$.

Now, we recursively define the sequence $\{u_n\}$ $(n \geq 0)$ in $B_w(S)$ as

$$u_0(i) := -\frac{b_0 M}{\alpha(\alpha - c_0)} - \frac{M}{\alpha - c_0} w(i) \quad \text{for } i \in S \tag{6.13}$$

(where $w$, $c_0$, $b_0$, and $M$ are as in Assumptions 2.2 and 6.4), and for $n \geq 0$, define

$$u_{n+1} := Tu_n. \tag{6.14}$$

The choice of $u_0$ is inspired by Theorem 6.5(a), that is, $u_0$ is a lower bound for the $\alpha$-discounted rewards.

The next result proves the existence of solutions to the DROE.

**Theorem 6.6** *Suppose that Assumptions* 2.2 *and* 6.4 *hold, and let* $\{u_n\}$ *be defined as in* (6.11)–(6.14).

(a) *The sequence* $\{u_n\}_{n\geq 0}$ *is monotone nondecreasing, and the limit* $u^* := \lim_{n\to\infty} u_n$ *is in* $B_w(S)$.
(b) *The function* $u^*$ *in* (a) *satisfies the fixed-point equation* $u^* = Tu^*$, *or, equivalently,* $u^*$ *verifies the DROE* (6.15), *that is,*

$$\alpha u^*(i) = \sup_{a\in A(i)} \left\{ r(i,a) + \sum_{j\in S} u^*(j) q(j|i,a) \right\} \quad \forall i \in S. \tag{6.15}$$

*Proof* (a) To prove the monotonicity, we will first show that $u_0 \leq u_1$. Indeed, under Assumptions 2.2 and 6.4, a straightforward calculation yields

$$\begin{aligned} u_1(i) &\geq -\frac{Mw(i)}{\alpha + m(i)} - \frac{m(i)}{\alpha + m(i)} \left( \frac{b_0 M}{\alpha(\alpha - c_0)} + \frac{Mw(i)}{\alpha - c_0} + \frac{M(c_0 w(i) + b_0)}{(\alpha - c_0) m(i)} \right) \\ &= -\frac{b_0 M}{\alpha(\alpha - c_0)} - \frac{M}{\alpha - c_0} w(i) = u_0(i) \end{aligned} \tag{6.16}$$

for every $i \in S$. Hence, the monotonicity of $T$ gives

$$u_n = T^n u_0 \leq T^n u_1 = u_{n+1} \quad \text{for every } n \geq 1.$$

This establishes the monotonicity of $\{u_n\}$ and, therefore, the existence of the (pointwise) limit $u^*$.

Moreover, one can prove, by a calculation as in the proof of (6.16) and using an induction argument, that

$$|u_n(i)| \leq \frac{b_0 M}{\alpha(\alpha - c_0)} + \frac{M w(i)}{\alpha - c_0} \leq \frac{(\alpha + b_0)M}{\alpha(\alpha - c_0)} w(i) \quad \forall n \geq 0 \text{ and } i \in S,$$

which implies that $\sup_{n \geq 0} \|u_n\|_w$ is finite. Hence $u^*$ is in $B_w(S)$, as we wanted to prove.

(b) Note that, by the monotonicity of $T$, $Tu^* \geq Tu_n = u_{n+1}$ for all $n \geq 0$, and thus

$$Tu^* \geq u^*. \tag{6.17}$$

Next, we will show that (6.17) holds with equality. Indeed, recalling (6.11) and (6.14), we have

$$u_{n+1}(i) \geq \frac{r(i,a)}{\alpha + m(i)} + \frac{m(i)}{\alpha + m(i)} \sum_{j \in S} u_n(j) p(j|i,a)$$

for all $n \geq 0$ and $(i,a) \in K$. Letting $n \to \infty$ in the latter inequality and using Proposition A.4 (Generalized dominated convergence theorem), we obtain

$$u^*(i) \geq \frac{r(i,a)}{\alpha + m(i)} + \frac{m(i)}{\alpha + m(i)} \sum_{j \in S} u^*(j) p(j|i,a) \quad \forall (i,a) \in K,$$

which gives the reverse of the inequality (6.17), i.e., $u^* \geq Tu^*$. This shows that $Tu^* = u^*$, and thus DROE (6.15) follows. □

Observe that Theorem 6.6 establishes the existence of a solution to the DROE, but we have not yet proved that such a solution is precisely the optimal discounted-reward function $V_\alpha^*$. This is done in Theorem 6.9, below.

*Remark 6.7* It is important to note that for a continuous-time MDP, there are two different ways of defining the expected discounted-reward criteria. In the way we follow here, which is the most natural, the rewards are earned *continuously in time*, as in (2.19). In the second approach, the rewards are earned at the (*discrete*) *transition epochs only*, that is, at the times when the state process makes a transition from one state to a different one; see, for instance, Lippman (1975) [111], Puterman (1994) [129], and Sennott (1999) [141]. This explains why the DROE in the latter references is slightly different from our DROE (6.15) or (equivalently) $u^* = Tu^*$.

## 6.3 Discount Optimal Stationary Policies

To ensure the existence of a discount optimal stationary policy we will need the following additional conditions.

**Assumption 6.8**

(a) The action set $A(i)$ is compact for each $i \in S$.
(b) The functions $r(i,a)$, $q(j|i,a)$, and $\sum_{k\in S} w(k)q(k|i,a)$ are all continuous in $a \in A(i)$ for each fixed $i, j \in S$, with $w$ as in Assumption 2.2.
(c) There exists a nonnegative function $w'$ on $S$ and constants $c' > 0$, $b' \geq 0$, and $M' > 0$ such that

$$q^*(i)w(i) \leq M'w'(i) \quad \text{and} \quad \sum_{j\in S} w'(j)q(j|i,a) \leq c'w'(i) + b'$$

for all $(i,a) \in K$, with $K$ and $q^*(i)$ as in (2.2) and (2.4), respectively.

Assumptions 6.8(a) and 6.8(b) are standard continuity–compactness conditions (similar to those in Assumption 4.12). Assumption 6.8(c) is needed to use Kolmogorov's forward differential equation.

**Theorem 6.9** *Suppose that Assumptions* 2.2, 6.4, *and* 6.8(c) *hold, and let* $\pi = (\pi_t) \in \Pi$ *and* $u \in B_w(S)$. *The following hold*:

(a) *If*

$$\alpha u(i) \geq r(i,\pi_t) + \sum_{j\in S} u(j)q(j|i,\pi_t)$$

*for all* $i \in S$ *and a.e.* $t \geq 0$, *then* $u(i) \geq V_\alpha(i,\pi)$ *for all* $i \in S$.
(b) *If*

$$\alpha u(i) \leq r(i,\pi_t) + \sum_{j\in S} u(j)q(j|i,\pi_t)$$

*for all* $i \in S$ *and a.e.* $t \geq 0$, *then* $u(i) \leq V_\alpha(i,\pi)$ *for all* $i \in S$.
(c) *For each* $\pi \in \Pi_s$, $V_\alpha(\cdot,\pi)$ *is the unique solution in* $B_w(S)$ *to the equation*

$$\alpha u(i) = r(i,\pi) + \sum_{j\in S} u(j)q(j|i,\pi) \quad \forall i \in S.$$

*Proof* (a) Observe that, since $|u(j)| \leq \|u\|_w \, w(j)$ for all $j \in S$,

$$\begin{aligned}\sum_{j\in S} \big|u(j)q(j|k,\pi_s)\big| &\leq \|u\|_w \Big[\sum_{j\neq k} w(j)q(j|k,\pi_s) + w(k)q^*(k)\Big] \\ &\leq \|u\|_w \Big[\sum_{j\in S} w(j)q(j|k,\pi_s) + 2w(k)q^*(k)\Big]\end{aligned}$$

for all $s \geq 0$ and $k \in S$.

Therefore, by Lemma 6.3, Assumption 2.2(a), and Assumption 6.8(c), given $T \geq 0$ and $i \in S$, we have

$$\begin{aligned}
&\int_0^T \sum_{k\in S}\sum_{j\in S} \big|q(j|k,\pi_s)u(j)\big| p_\pi(0,i,s,k)\,ds \\
&\quad \leq \|u\|_w \int_0^T \sum_{k\in S}\big[c_0 w(k) + b_0 + 2q^*(k)w(k)\big] p_\pi(0,i,s,k)\,ds \\
&\quad \leq \|u\|_w \int_0^T \sum_{k\in S}\big[c_0 w(k) + b_0 + 2M'w'(k)\big] p_\pi(0,i,s,k)\,ds \\
&\quad \leq \|u\|_w T\bigg[|c_0|w(i)e^{|c_0|T} + 2M'w'(i)e^{c'T} + |b_0| + \frac{b'}{c'} + b_0\bigg]. \qquad (6.18)
\end{aligned}$$

Thus, by the current hypothesis and using Kolmogorov's forward differential equation (2.6) and Fubini's theorem, under the condition in (a), we have

$$\begin{aligned}
\int_0^T e^{-\alpha t} E_i^\pi r\big(x(t),\pi_t\big)\,dt &\leq \sum_{j\in S}\int_0^T \bigg[\alpha e^{-\alpha t}u(j)p_\pi(0,i,t,j) \\
&\qquad - e^{-\alpha t}u(j)\bigg(\sum_{k\in S} q(j|k,\pi_t)p_\pi(0,i,t,k)\bigg)\bigg]\,dt \\
&= u(i) - e^{-\alpha T}E_i^\pi u\big(x(T)\big). \qquad (6.19)
\end{aligned}$$

Finally, letting $T \to +\infty$ in (6.19) and using Theorem 6.5(c), we obtain (a).

(b) The proof of (b) is similar to that for (a).

(c) As in the proof of (a), by Fubini's theorem and Kolmogorov's backward equation (2.6) we obtain

$$\begin{aligned}
&r(i,\pi) + \sum_{j\in S} V_\alpha(j,\pi)q(j|i,\pi) \\
&\quad = r(i,\pi) + \sum_{j\in S}\bigg[\int_0^\infty e^{-\alpha t}\sum_{k\in S} r(k,\pi)p_\pi(0,j,t,k)\,dt\bigg] q(j|i,\pi) \\
&\quad = r(i,\pi) + \sum_{k\in S} r(k,\pi)\bigg[\int_0^\infty e^{-\alpha t}\sum_{j\in S} q(j|i,\pi)p_\pi(0,j,t,k)\,dt\bigg] \\
&\quad = r(i,\pi) + \sum_{k\in S} r(k,\pi)\bigg[\int_0^\infty e^{-\alpha t}\frac{\partial p_\pi(0,i,t,k)}{\partial t}\,dt\bigg] \\
&\quad = \alpha V_\alpha(i,\pi).
\end{aligned}$$

This implies that $V_\alpha(\cdot,\pi)$ satisfies the equation in (c). Finally, the uniqueness of solutions to the equation follows from (a) and (b). □

Let $\{u_n\}$ be the sequence recursively defined in (6.13) and (6.14) with $T$ as in (6.11). We say that a deterministic stationary policy $f \in F$ attains the maximum in the equation $u_{n+1} = Tu_n$ if (see (6.11) and (6.14))

$$u_{n+1}(i) = \frac{r(i,f)}{\alpha + m(i)} + \frac{m(i)}{\alpha + m(i)} \sum_{j \in S} u_n(j) p(j|i,f) \quad \forall i \in S.$$

Similarly, $f \in F$ attains the maximum in the DROE (6.15) if

$$\alpha u^*(i) = r(i,f) + \sum_{j \in S} u^*(j) q(j|i,f) \quad \forall i \in S.$$

We next state the main result of this section and first recall the concept: a policy $f \in F$ is said to be a *limit point* of a sequence $\{f_n\}$ in $F$ if there exists a subsequence $\{m\}$ of $\{n\}$ such that $f_m(i) \to f(i)$ as $m \to \infty$ for each $i \in S$.

**Theorem 6.10** *Suppose that Assumptions* 2.2, 6.4 *and* 6.8 *hold. Then*:

(a) *There exist deterministic stationary policies $f_n$ (for each $n \geq 0$) and $f_\alpha^*$ attaining the maximum in the equations $u_{n+1} = Tu_n$ and the DROE* (6.15), *respectively.*
(b) *The solution $u^*$ to the DROE* (6.15) *is $u^* = V_\alpha^*$, and the policy $f_\alpha^*$ in* (a) *is discounted-reward optimal.*
(c) *Every limit point in $F$ of the sequence $\{f_n\}$ in* (a) *is a discounted-reward optimal stationary policy.*

*Proof* (a) Since we have already shown (in Theorem 6.6) that $u_n$ and $u^*$ are in $B_w(S)$, from either Proposition A.4 (Generalized Dominated Convergence Theorem) or Lemma 8.3.7 in Hernández-Lerma and Lasserre (1999) [74] we see that the functions within brackets in (6.11) and (6.15) are continuous in $a \in A(i)$ for each $i \in S$. This fact and Assumption 6.8 give (a).

(b) For all $i \in S$ and $\pi = (\pi_t) \in \Pi$, it follows from Theorem 6.6(b), (2.4), and (2.20) that

$$\alpha u^*(i) \geq r(i,\pi_t) + \sum_{j \in S} u^*(j) q(j|i,\pi_t) \quad \text{for all } i \in S \text{ and } t \geq 0 \tag{6.20}$$

with equality if $\pi = f_\alpha^*$. Hence, (6.20), together with Theorem 6.9, yields that

$$V_\alpha\big(i, f_\alpha^*\big) = u^*(i) \geq V_\alpha(i,\pi)$$

for all $i \in S$ and $\pi \in \Pi$, thus proving (b).

(c) By part (a) and Proposition A.4 (Generalized Dominated Convergence Theorem), every limit point $f \in F$ of $\{f_n\}$ satisfies

$$u^*(i) = \frac{r(i,f)}{\alpha + m(i)} + \frac{m(i)}{\alpha + m(i)} \sum_{j \in S} u^*(j) p(j|i,f) \quad \forall i \in S,$$

which is equivalent to

$$\alpha u^*(i) = r(i, f) + \sum_{j\in S} u^*(j)q(j|i, f) \quad \forall i \in S.$$

Thus, by (b) and Theorem 6.9(c), $V_\alpha(i, f) = u^*(i) = V_\alpha^*(i)$ for every $i \in S$. □

## 6.4 A Value Iteration Algorithm

In this section, we summarize (6.11)–(6.14) in a value iteration algorithm to calculate (or at least to approximate) the optimal discounted-reward function $V_\alpha^*$. Throughout this section, we suppose that Assumptions 2.2, 6.4, and 6.8 hold.

From the arguments in Theorems 6.6 and 6.10 we see that the optimal discounted-reward function can be approximated as follows.

**A value iteration algorithm:**

1. Let

$$u_0(i) := -\frac{b_0 M}{\alpha(\alpha - c_0)} - \frac{M}{\alpha - c_0} w(i) \quad \text{for } i \in S,$$

   where $w$, $c_0$, $b_0$, and $M$ are as in Assumptions 2.2 and 6.4.
2. For $n \geq 0$, define

$$u_{n+1}(i) := \sup_{a\in A(i)} \left\{ \frac{r(i,a)}{\alpha + m(i)} + \frac{m(i)}{\alpha + m(i)} \sum_{j\in S} u_n(j)p(j|i,a) \right\} \tag{6.21}$$

   with $p(j|i, a)$ and $m(i)$ as in (6.12).
3. Choose $f_n \in F$ attaining the maximum in the right-hand side of (6.21).
4. $V_\alpha^*(i) = \lim_{n\to\infty} u_n(i)$ for all $i \in S$.
5. Every limit point in $F$ of the sequence $\{f_n\}$ is a discounted-reward optimal stationary policy.

## 6.5 Examples

In this section, we illustrate with some examples the main results in this chapter.

*Example 6.1* (Optimal control of birth-and-death systems) Consider a controlled birth-and-death system in which the state variable denotes the population size at any time $t \geq 0$. There are natural birth and death rates represented by nonnegative constants $\lambda$ and $\mu$, respectively. In addition, there is an emigration parameter $h_1$ and an immigration parameter $h_2$, which are assumed to be controlled by a decision-maker. When the state of the system is $i \in S := \{0, 1, \ldots\}$, the decision-maker takes an action $a$ from a given set $A(i)$, which may increase ($h_1(i, a) \geq 0$) or decrease ($h_1(i, a) \leq 0$) the emigration parameter, and also admit ($h_2(i, a) \geq 0$) or expel ($h_2(i,a) \le 0$) members of the population. This action incurs a cost $c(i,a)$. In addition, the decision-maker gets a reward $p_0 i$ for each unit of time during which the system remains in state $i$, where $p_0 > 0$ is a fixed reward parameter.

We now formulate this system as a continuous-time MDP. The corresponding transition rates $q(j|i,a)$ and reward function $r(i,a)$ are given as follows. For $i = 0$ and each $a \in A(0)$,

$$q(1|0,a) := -q(0|0,a) := h_2(0,a), \quad \text{and} \quad q(j|0,a) = 0 \quad \text{for } j \ge 2,$$

and for $i \ge 1$ and all $a \in A(i)$,

$$q(j|i,a) := \begin{cases} \mu i + h_1(i,a) & \text{if } j = i-1, \\ -(\mu+\lambda)i - h_1(i,a) - h_2(i,a) & \text{if } j = i, \\ \lambda i + h_2(i,a) & \text{if } j = i+1, \\ 0 & \text{otherwise;} \end{cases} \tag{6.22}$$

$$r(i,a) := p_0 i - c(i,a) \quad \text{for } (i,a) \in K := \{(i,a) : i \in S, a \in A(i)\}. \tag{6.23}$$

We aim to find conditions that ensure the existence of a discounted-reward optimal stationary policy.

To do this, we consider the following conditions:

**C$_1$**. $\alpha + \mu - \lambda > 0$ (where $\alpha$ is a given discount factor); $\mu i + h_1(i,a) \ge 0$ and $\lambda i + h_2(i,a) \ge 0$ for all $a \in A(i)$ and $i \ge 1$; and $h_1(0,a) = 0$ and $h_2(0,a) \ge 0$ for all $a \in A(0)$.

**C$_2$**. For each $i \in S$, $A(i)$ is a compact metric space.

**C$_3$**. The functions $c(i,a)$, $h_1(i,a)$, and $h_2(i,a)$ are continuous in $a \in A(i)$, and there exists a constant $\bar{M} > 0$ such that $\sup_{a\in A(i)} |c(i,a)| \le \bar{M}(i+1)$ for all $i \in S$, and $\|h_k\| := \sup_{(i,a)\in K} |h_k(i,a)| < \infty$ for $k = 1, 2$.

The model in Example 6.1 can have several interesting interpretations, as in the following *special* cases:

(i) *A queueing system with controlled service and arrival rates.* In Example 6.1, we take an action $a := (a_1, a_2)$ consisting of a *controlled* service rate $a_1$ and a *controlled* arrival rate $a_2$, and let $h_1(i,a) := a_1$, $h_2(i,a) := a_2$ for all $i \in S$ and $A(i) := [0, a_1^*] \times [0, a_2^*]$ with some positive constants $a_1^*$ and $a_2^*$. Then conditions **C$_1$**, **C$_2$**, and **C$_3$** above are all satisfied when $\alpha + \mu - \lambda > 0$, and $\sup_{a\in A(i)} |c(i,a)| \le \bar{M}(i+1)$ for all $i \in S$, with some $\bar{M} > 0$.

(ii) *A birth-and-death process with controlled immigration.* In Example 6.1, suppose that $h_1 \equiv 0$, $h_2(i,a) := a$. Then the model in Example 6.1 with such special data becomes the birth, death, and immigration process in Anderson (1991) [4] with birth, death, and immigration rates $\lambda$, $\mu$, and $a$, respectively. Thus, we can interpret the immigration rate $a$ as an action, and so $A(i)$ may take the form $A(i) := [0, a^*]$ for all $i \in S$, with some $a^* > 0$. Therefore, conditions **C$_1$**, **C$_2$**, and **C$_3$** above are all satisfied when $\alpha + \mu - \lambda > 0$ and $\sup_{a\in A(i)} |c(i,a)| \le \bar{M}(i+1)$ for all $i \in S$, with some $\bar{M} > 0$.

In any case, we obtain the following.

**Proposition 6.11** *Under conditions* $\mathbf{C_1}$, $\mathbf{C_2}$, *and* $\mathbf{C_3}$, *the above controlled birth-and-death system satisfies Assumptions* 2.2, 6.4, *and* 6.8. *Therefore (by Theorem* 6.10), *there exists a discounted-reward optimal stationary policy.*

*Proof* We shall first verify Assumption 2.2. Let $w(i) := i + 1$ for all $i \in S$, and $L_0 := \mu + \lambda + \|h_1\| + \|h_2\|$. Then Assumption 2.2(b) follows from (6.22) and $\mathbf{C_3}$. Moreover, from $\mathbf{C_1}$ and (6.22) we have

$$\sum_{j\in S} w(j)q(j|i,a) \le (\lambda - \mu)w(i) + L_0 \quad \forall i \ge 1. \tag{6.24}$$

For $i = 0$ and $a \in A(0)$, we also have

$$\sum_{j\in S} w(j)q(j|0,a) = h_2(0,a) \le (\lambda - \mu)w(0) + L_0. \tag{6.25}$$

From inequalities (6.24) and (6.25) we see that Assumption 2.2(a) holds with $c_0 := \lambda - \mu$ and $b_0 := L_0$, and so Assumption 2.2 follows.

Since $|r(i,a)| \le (p_0 + \bar{M})w(i)$ for all $i \in S$, from $\mathbf{C_1}$ and (6.24)–(6.25) we obtain Assumption 6.4.

Finally, to verify Assumption 6.8, let $w'(i) := (i+1)^2$ for all $i \in S$. Then $q^*(i)w(i) \le M'w'(i)$ for all $i \in S$, with $M' := L_0$. Thus, as in (6.24) and (6.25), we see that Assumption 6.8(c) is true. On the other hand, by (6.22) and by $\mathbf{C_2}$ and $\mathbf{C_3}$, we see that Assumptions 6.8(a)–(b) are both satisfied, and then Assumption 6.8 follows. □

*Example 6.2* (Optimal control of epidemic processes; see Anderson (1991) [4], Bailey (1975) [7], Ramsey (1981) [131], and Reuter (1961) [134]) The epidemic processes belong to the category of *population processes* and can be described as follows: We have a population of individuals who can contract a certain communicable disease. The individuals without the disease are called *susceptibles* (denoted by $m$), and those with the disease are called *infectives* (denoted by $n$). Thus, the state space for such a process is of the form $S := \{0, 1, 2, \ldots\}^2$, and its transition rates are given by

$$q\big(y|(m,n),a\big) := \begin{cases} \rho & \text{if } y = (m+1,n), \\ \beta & \text{if } y = (m,n+1), \\ \gamma m & \text{if } y = (m-1,n), \\ -(\rho+\beta+\gamma m+\delta n+\varepsilon mn) & \text{if } y = (m,n), \\ \delta n & \text{if } y = (m,n-1), \\ \varepsilon mn & \text{if } y = (m-1,n+1), \\ 0 & \text{otherwise,} \end{cases} \tag{6.26}$$

where $(m,n) \in S$, $a := (\rho, \beta, \gamma, \delta, \varepsilon)$. Here, the numbers $\rho, \beta, \gamma, \delta$, and $\varepsilon$ represent the birth rate of susceptibles, the birth rate of infectives, the *removal* rate of a susceptible from the population, the *removal* rate of an infective from the population, and the rate of new infectives, respectively. Many probability properties of such a process have been studied by several authors, including Anderson (1991) [4], Bailey (1975) [7], Ramsey (1981) [131], and Reuter (1961) [134].

Here, we are interested in the discounted-reward optimality control problem. Thus, we assume that these vectors $a = (\rho, \beta, \gamma, \delta, \varepsilon)$ can be controlled by a decision-maker, and so we may interpret $a$ as an action with values in some nonempty compact set $A(m,n) := [0, \rho^*] \times [0, \beta^*] \times [0, \gamma^*] \times [0, \delta^*] \times [0, \varepsilon^*]$ for all states $(m,n) \in S$. In addition, the reward function $r(m,n,a)$ is assumed to satisfy

$$\left|r(m,n,a)\right| \le M'(1+m+n)^2 \quad \forall (m,n) \in S \text{ and } a \in A(m,n)$$

for some constant $M' > 0$.

To verify Assumptions 2.2, 6.4, and 6.8, we let $w(m,n) := (1+m+n)^2$ and $w'(m,n) := (1+m+n)^4$. Thus, by (6.26) we have, for all $(m,n) \in S$ and $a \in A(m,n)$,

$$\begin{aligned}
\sum_{y\in S} w(y) q\big(y|(m,n),a\big) &= 3(\rho+\beta) + 2(m+n)(\rho+\beta) \\
&\quad - (\gamma m + \delta n)(1+2m+2n) \\
&\le \frac{1}{2}\alpha w(m,n) - \frac{1}{2}\alpha(m+n+1)^2 \\
&\quad + 3\big(\rho^* + \beta^*\big)(m+n+1) \\
&\le \frac{1}{2}\alpha w(m,n) + \frac{18(\beta^*+\rho^*)^2}{\alpha}; \\
\sum_{y\in S} w'(y) q\big(y|(m,n),a\big) &\le 15\big(\rho^*+\beta^*\big) w'(m,n),
\end{aligned}$$

which verify Assumptions 2.2(a) and 6.8(c). In addition, Assumptions 2.2(b), 6.4, and 6.8(a)–(b) follow from the model's description. □

Thus, we have the following fact:

**Proposition 6.12** *For each fixed discount factor $\alpha > 0$, there exists a discounted-reward optimal stationary policy in Example* 6.2.

*Example 6.3* (Optimal control of predator-prey processes; see Anderson (1991) [4], Hitchcock (1981) [77], and Ridler-Rowe (1988) [136], for instance) Suppose that the members of a certain population are either predators or their prey. Then the state space for such a predator–prey process is of the form $S := \{0, 1, 2, \ldots\}^2$, and the transition rates are given as follows:

$$q\big(y|(m,n),a\big) := \begin{cases} \lambda n & \text{if } y=(m,n+1), \\ \beta m & \text{if } y=(m-1,n), \\ -(\lambda n+\beta m+\varepsilon mn) & \text{if } y=(m,n), \\ \varepsilon mn & \text{if } y=(m-1,n+1), \\ 0 & \text{otherwise}, \end{cases} \tag{6.27}$$

where $(m,n)\in S$ and $a := (\lambda,\beta,\varepsilon)$. We assume that the numbers $\lambda$, $\beta$, and $\varepsilon$ can be controlled by a decision-maker, and so we interpret $a=(\lambda,\beta,\varepsilon)$ as a control action in $[0,\lambda^*]\times[0,\beta^*]\times[0,\varepsilon^*] =: A(m,n)$ at state $(m,n)\in S$, for some positive constants $\lambda^*$, $\beta^*$, and $\varepsilon^*$. In addition, the reward function $r(m,n,a)$ is assumed to satisfy

$$\big|r(m,n,a)\big| \le M'(1+m+n)^2 \quad \forall (m,n)\in S \text{ and } a\in A(m,n)$$

for some constant $M'>0$.

Then, as in the Example 6.2, we can see that Assumptions 2.2, 6.4, and 6.8 are satisfied when $\alpha-3\lambda^*>0$, where $\alpha$ is a given discount factor. In fact, let $w(m,n):=(1+m+n)^2$ and $w'(m,n):=(1+m+n)^4$. Then, by (6.27) we have, for every $(m,n)\in S$ and $a\in A(m,n)$,

$$\sum_{y\in S} w(y)q\big(y|(m,n),a\big) = \lambda n\big(1+2(1+m+n)\big)+\beta m\big(1-2(1+m+n)\big)$$

$$\le 3\lambda^* w(m,n);$$

$$\sum_{y\in S} w'(y)q\big(y|(m,n),a\big) = \lambda n(2+m+n)^4+\beta m(m+n)^4$$

$$-(\lambda n+\beta m)(1+m+n)^4$$

$$\le 15\lambda^* w'(m,n),$$

which verify Assumptions 2.2(a) and 6.8(c). Finally, Assumptions 2.2(b), 6.4, and 6.8(a)–(b) follow from the model's description. □

Thus, we have the following fact:

**Proposition 6.13** *For each fixed discount factor $\alpha>3\lambda^*$ (with $\lambda^*$ as in Example* 6.3), *there exists a discounted-reward optimal stationary policy in Example* 6.3.

## 6.6 Notes

*An Open Problem* Note that each policy in this book is history-independent. Thus, an interesting (and so far unsolved) problem would be to study discount optimality in the class of *history-dependent* policies. In fact, such continuous-time MDPs have been studied by Yushkevich (1977) [162] and Yushkevich and Feinberg (1979) [165]. In those papers, however, the analysis is restricted to very particular classes of MDPs (namely, MDPs with transition rates that are uniformly bounded in both states and actions). Yushkevich (1977) [162] himself mentions that the issue of dropping the boundedness condition in his Lemma 5.1 and Theorems 5.1 and 5.2 in Yushkevich (1977) [162] is an *unsolved problem*. This suggests that for our model, in which the transition rates and the reward function can be *unbounded*, the above-mentioned problem is perhaps quite challenging.

Also, we should mention that some of the results in this section can be extended to several classes of continuous-time MDPs, for instance, MDPs with Polish state spaces, controlled stochastic differential equations, and so forth; see Guo (2007) [50], Hernández-Lerma (1994) [70], Hernández-Lerma and Govindan (2001) [72], and Prieto-Rumeau and Hernández-Lerma (2006b) [126], for instance.

Finally, up to now we have mentioned *four* approaches to showing the existence of discount optimal stationary policies, namely: the minimum nonnegative solution approach, Kolmogorov's forward equation approach, the uniformization technique approach, and the extended infinitesimal operator approach. Each of them has an advantage over the others. Roughly speaking, the conditions for the minimum nonnegative solution approach are very weak, except that the cost function is required to be bounded below; the Kolmogorov forward equation approach can be regarded as a special case of the extended infinitesimal operator approach, which can deal with reward functions that may have neither upper nor lower bounds; the uniformization technique approach can simplify many arguments by using results on discrete-time MDPs, but it only deals with the case in which the transition rates are bounded.

# Chapter 7
# Average Optimality for Unbounded Rewards

Average optimality has been studied in Chaps. 3 and 5 for finite and nonnegative models, respectively. However, these results are not applicable to the more general case where the reward function $r(i, a)$ has neither upper nor lower bounds. The average optimality for this general case will be studied in this chapter.

*As in Chap.* 6, *this chapter deals with the control model* (2.1).

## 7.1 Introduction

In Chap. 3, we have established the existence of an expected average reward (AR) optimal stationary policy for *finite* MDPs. The finiteness of the state and the action spaces is crucial, because, as in Puterman (1994) [129] or Sennott (1999) [141], we can provide an example for which no AR optimal policy exists when either the state space or the action space is infinite. Thus, to guarantee the existence of AR optimal policies for models with infinitely many states or actions, some conditions need to be imposed on the models. In short, the question dealt with in this chapter is: for non-finite MDPs, what conditions ensure the existence of an AR optimal stationary policy?

The first thing to do is to guarantee that the expected AR is indeed well defined. To this end, in Sect. 7.2, we introduce the concept of uniform exponential ergodicity under which the expected AR of a stationary policy is a constant independent of the initial state. In Sect. 7.3, we first give conditions for the existence of solutions to the *average-reward optimality equation* (AROE) (7.4) below. It is then shown that a stationary policy is AR optimal if and only if it attains the maximum in the AROE. Section 7.4 presents a policy iteration algorithm for the computation of the optimal AR function (i.e., the AR optimal value) and an AR optimal policy; in fact, this algorithm gives a solution to the AROE. Finally, in Sect. 7.5, we introduce some examples, and we conclude in Sect. 7.6 with some comments on the different approaches to obtain AR optimal policies.

X. Guo, O. Hernández-Lerma, *Continuous-Time Markov Decision Processes*,
Stochastic Modelling and Applied Probability 62,
DOI 10.1007/978-3-642-02547-1_7, © Springer-Verlag Berlin Heidelberg 2009

## 7.2 Exponential Ergodicity Conditions

Since the reward function $r(i,a)$ may be unbounded, the expected average reward $\bar{V}(i,\pi)$ of a policy $\pi \in \Pi$, defined in (2.21), may be infinite. To guarantee the finiteness of $\bar{V}(i,\pi)$, in the spirit of Lemma 6.3, we will impose the following "drift condition." For the simplicity of statements in this chapter, we will suppose that the state space $S$ is the set of all nonnegative integers, that is, $S := \{0, 1, \ldots\}$.

**Assumption 7.1**

(a) (Drift condition) There exist a nondecreasing function $w \geq 1$ on $S$ and constants $c_1 > 0$ and $b_1 \geq 0$ such that $\sum_{j\in S} w(j)q(j|i,a) \leq -c_1 w(i) + b_1\delta_{i0}$ for all $(i,a) \in K$.
(b) $q^*(i) \leq L_0 w(i)$ for all $i \in S$, with $L_0 > 0$ and $q^*(i)$ as Assumption 2.2(b).
(c) $|r(i,a)| \leq M w(i)$ for all $(i,a) \in K$, with some $M > 0$.
(d) Assumption 6.8 holds with $w$ there replaced by the nondecreasing function in (a) above.

The drift condition in Assumption 7.1(a) is also known as a Lyapunov or Foster–Lyapunov condition, and it is a key part of standard ergodicity hypothesis; see Chen (2000) [27], Lund et al. (1996) [113] and Meyn and Tweedie (1993) [116], among others.

Obviously, Assumption 7.1 implies Assumptions 2.2, 6.4, and 6.8. Hence, under Assumption 7.1, the transition function $p_\pi(s,i,t,j)$ is regular (see (2.8)) for every $\pi \in \Pi$.

As a consequence of (2.21)–(2.22) and Lemma 6.3, we have the following.

**Lemma 7.2** *Suppose that Assumption* 7.1 *holds. Then the expected average reward is uniformly bounded, i.e.,*

$$\left|\bar{V}(i,\pi)\right| \leq Mb_1/c_1$$

*for all $i \in S$ and $\pi \in \Pi$.*

The next result will be used to show the existence of an expected AR optimal policy.

**Proposition 7.3** *Suppose that Assumption* 7.1 *holds, and consider $\pi \in \Pi$, $u \in B_w(S)$, and a real number $g$. Then the following hold.*

(a) *If for all $i \in S$ and a.e. $t \geq 0$,*

$$g \geq r(i,\pi_t) + \sum_{j\in S} u(j)q(j|i,\pi_t),$$

*then $g \geq \bar{V}(i,\pi)$ for all $i \in S$.*

(b) *If for all $i \in S$ and a.e. $t \geq 0$,*

$$g \leq r(i, \pi_t) + \sum_{j \in S} u(j) q(j|i, \pi_t),$$

*then $g \leq \bar{V}(i, \pi)$ for all $i \in S$.*

*Proof* (a) As in the proof of (6.19), under the condition in (a), we have

$$gT \geq \int_0^T E_i^\pi r\big(x(t), \pi_t\big)\, dt + E_i^\pi u\big(x(T)\big) - u(i). \tag{7.1}$$

On the other hand, from Lemma 6.3 and Assumption 7.1(a) it is easily deduced that

$$\lim_{T \to \infty} \frac{1}{T} E_i^\pi u\big(x(T)\big) = 0 \quad \text{for all } \pi \in \Pi \text{ and } i \in S.$$

This fact, together with (2.22) and (7.1), gives (a).

(b) The proof of (b) is similar. □

Now we focus on the issue of the existence of AR optimal stationary policies (recall Definition 2.8). To do so, in addition to Assumption 7.1, we impose the following irreducibility condition.

**Assumption 7.4** For each $f \in F$, the corresponding Markov process $\{x(t)\}$ with transition function $p_f(i, t, j)$ is irreducible, which means that, for any two states $i \neq j$, there exists a set of distinct states $i = i_1, \ldots, i_m$ such that

$$q(i_2|i_1, f) \cdots q(j|i_m, f) > 0.$$

Under Assumptions 7.1(a) and 7.4, for each $f \in F$, Propositions C.11 and C.12 yield that the Markov chain $\{x(t)\}$ has a unique invariant probability measure, denoted by $\mu_f$, which satisfies that $\mu_f(j) = \lim_{t \to \infty} p_f(i, t, j)$ (independent of $i \in S$) for all $j \in S$. Thus, by Assumption 7.1(a) and Lemma 6.3(i), we have

$$\mu_f(w) := \sum_{j \in S} w(j) \mu_f(j) \leq \frac{b_1}{c_1},$$

which shows that the $\mu_f$-expectation of $w$ (i.e., $\mu_f(w)$) is finite. Therefore, for all $f \in F$ and $u \in B_w(S)$, the inequality $|u(i)| \leq \|u\|_w w(i)$ for all $i \in S$ gives that the expectation

$$\mu_f(u) := \sum_{i \in S} u(i) \mu_f(i) \tag{7.2}$$

exists and is finite.

**Assumption 7.5** The control model (2.1) is uniformly $w$-exponentially ergodic, which means the following: there exist constants $\delta > 0$ and $L_2 > 0$ such that (using the notation in (7.2))

$$\sup_{f \in F} \left| E_i^f u\big(x(t)\big) - \mu_f(u) \right| \le L_2 e^{-\delta t} \|u\|_w w(i)$$

for all $i \in S$, $u \in B_w(S)$, and $t \ge 0$.

It is worth noting that, under our current assumptions, the gain of a deterministic stationary policy $f \in F$ is constant, i.e., $\bar{V}(i, f)$ does not depend on the initial state $i$ and equals the $\mu_f$-expectation of $r(f)$. More precisely, let

$$g(f) := \sum_{j \in S} r(j, f) \mu_f(j).$$

Then, by (2.22),

$$V_T(i, f) = T g(f) + E_i^f \int_0^T \left[ r\big(x(t), f\big) - g(f) \right] dt.$$

Hence, multiplying by $1/T$ and letting $T \to \infty$, from Assumption 7.5 we obtain

$$\bar{V}(i, f) = g(f) = \sum_{j \in S} r(j, f) \mu_f(j) \quad \forall i \in S. \tag{7.3}$$

Among Assumptions 7.1, 7.4, and 7.5 made so far on the control model, Assumption 7.5 seems to be the most difficult to verify in practice. Hence, we next propose sufficient conditions for uniform $w$-exponential ergodicity.

**Proposition 7.6** *In addition to Assumptions* 7.1(a) *and* 7.4, *suppose that, for each fixed* $f \in F$,

(i) (*Stochastic monotonicity condition*) $\sum_{j \ge k} q(j|i, f) \le \sum_{j \ge k} q(j|i+1, f)$ *for all* $i, k \in S$ *such that* $k \ne i+1$.
(ii) *For each* $j > i > 0$, *there exist nonzero distinct states* $i_1, i_2, \ldots, i_m \ge j$ *such that*

$$q(i_1|i, f) \cdots q(i_m|i_{m-1}, f) > 0.$$

*Then Assumption* 7.5 *holds with* $\delta := c_1$ *and* $L_2 := 2(1 + \frac{b_1}{c_1})$, *where* $c_1$ *and* $b_1$ *are the constants in Assumption* 7.1.

This proposition obviously follows from Proposition C.17.

Condition (i) in Proposition 7.6 is a variant of the "monotonicity conditions" in Anderson (1991) [4, p. 249]. Condition (ii) requires that, for any two states $j > i > 0$, the process $\{x(t)\}$ can travel with positive probability from the state $i$ to the set $\{j, j+1, \ldots\}$ without passing through the state $0 \in S$.

Other sufficient conditions for uniform $w$-exponential ergodicity are given by Guo, Hernández-Lerma, and Prieto-Rumeau (2006) [65] and Prieto-Rumeau and Hernández-Lerma (2006) [126], for instance.

## 7.3 The Existence of AR Optimal Policies

We begin with introducing the *average-reward optimality equation* (AROE) (7.4) below.

A pair $(g^*, u) \in \mathbb{R} \times B_w(S)$ is said to be a solution to the AROE if

$$g^* = \sup_{a \in A(i)} \left\{ r(i,a) + \sum_{j \in S} u(j) q(j|i,a) \right\} \quad \forall i \in S. \tag{7.4}$$

Under some assumptions, the supremum in (7.4) can be attained for every $i \in S$. In such a case, we say that $f \in F$ attains the maximum in the AROE (7.4), that is,

$$g^* = r(i,f) + \sum_{j \in S} u(j) q(j|i,f) \quad \forall i \in S. \tag{7.5}$$

**Lemma 7.7** *Suppose that Assumptions* 7.1, 7.4, *and* 7.5 *are satisfied. Consider an arbitrary fixed state* $i_0 \in S$. *Then, for all* $f \in F$ *and discount factors* $\alpha > 0$, *the relative differences of the discounted-reward function* $V_\alpha(f)$, *namely,*

$$u_\alpha^f(i) := V_\alpha(i,f) - V_\alpha(i_0,f) \quad \textit{for } i \in S, \tag{7.6}$$

*are uniformly* $w$*-bounded in* $\alpha > 0$ *and* $f \in F$. *More precisely, we have*

$$\left\| u_\alpha^f \right\|_w \le \frac{L_2 M}{\delta} \left[ 1 + w(i_0) \right] \quad \forall \alpha > 0 \textit{ and } f \in F.$$

*Proof* Choose any $\alpha > 0$ and $f \in F$. By Assumption 7.1(c), $|r(i,f)| \le M w(i)$ for all $i \in S$. Recalling the notation in (7.3), from Assumptions 7.4 and 7.5 we have

$$\left| E_i^f r\big(x(t), f\big) - g(f) \right| \le L_2 M e^{-\delta t} w(i) \quad \forall i \in S. \tag{7.7}$$

Thus, for each $i \in S$, by (7.7) and (2.19) we have

$$\begin{aligned}
\left| u_\alpha^f(i) \right| &= \left| E_i^f \left[ \int_0^\infty e^{-\alpha t} r\big(x(t), f\big)\, dt \right] - E_{i_0}^f \left[ \int_0^\infty e^{-\alpha t} r\big(x(t), f\big)\, dt \right] \right| \\
&\le L_2 M \int_0^\infty e^{-(\alpha+\delta)t} \big( w(i) + w(i_0) \big)\, dt \\
&= \frac{L_2 M}{\alpha + \delta} \big( w(i) + w(i_0) \big) \\
&\le \frac{L_2 M}{\delta} \big( 1 + w(i_0) \big) w(i),
\end{aligned}$$

which completes the proof. □

Now we present the main result of this section.

**Theorem 7.8** *Suppose that Assumptions* 7.1, 7.4, *and* 7.5 *hold. Then*:

(a) *There exists a solution* $(g^*, \bar{u}) \in \mathbb{R} \times B_w(S)$ *to AROE* (7.4). *Moreover, the constant* $g^*$ *coincides with the optimal average reward function* $\bar{V}^*$, *i.e.*,

$$g^* = \bar{V}^*(i) \quad \forall i \in S,$$

*and* $\bar{u}$ *is unique up to additive constants.*

(b) *A deterministic stationary policy is AR optimal if and only if it attains the maximum in AROE* (7.4).

*Proof* The proof proceeds in several steps. First, in (7.6) we take $f$ as $f^*_\alpha$, the $\alpha$-discounted reward optimal stationary policy in Theorem 6.10; hence $V_\alpha(i, f^*_\alpha) = V^*_\alpha(i)$, the optimal discounted reward function, and instead of the function in (7.6), we now take $u_\alpha(i) := u_\alpha^{f^*_\alpha}(i)$. Then following the vanishing discount approach already used in Sects. 5.4 and 5.5, we will show the existence of a solution to the AROE.

Lemma 7.7 and Proposition A.7 give the existence of a sequence $\{\alpha_n\}$ of discount factors such that $\alpha_n \downarrow 0$, a constant $g^*$, and a function $\bar{u} \in B_w(S)$ such that

$$\lim_{n\to\infty} \alpha_n V^*_{\alpha_n}(i_0) = g^* \quad \text{and} \quad \lim_{n\to\infty} u_{\alpha_n}(i) = \bar{u}(i) \tag{7.8}$$

for all $i \in S$. (Observe the analogy between (7.8) and (5.12).) On the other hand, for all $n \geq 1$ and $i \in S$, by Theorem 6.10(b) we have (using the notation in (6.11) and (7.6))

$$\frac{\alpha_n V^*_{\alpha_n}(i_0)}{m(i)} + \frac{\alpha_n u_{\alpha_n}(i)}{m(i)} + u_{\alpha_n}(i) \geq \frac{r(i,a)}{m(i)} + \sum_{j\in S} u_{\alpha_n}(j)\left[\frac{q(j|i,a)}{m(i)} + \delta_{ij}\right]$$

for all $(i, a) \in K$, which, together with (7.8), gives

$$\frac{g^*}{m(i)} + \bar{u}(i) \geq \frac{r(i,a)}{m(i)} + \sum_{j\in S} \bar{u}(j)\left[\frac{q(j|i,a)}{m(i)} + \delta_{ij}\right] \quad \forall (i,a) \in K.$$

Thus,

$$g^* \geq \sup_{a\in A(i)} \left\{ r(i,a) + \sum_{j\in S} \bar{u}(j) q(j|i,a) \right\} \quad \forall i \in S. \tag{7.9}$$

To prove that $(g^*, \bar{u})$ is a solution to the AROE (7.4), it remains to show the reverse inequality in (7.9). As a consequence of Theorem 6.10, for each $n \geq 1$, there exists $f_n \in F$ such that

$$\frac{\alpha_n V^*_{\alpha_n}(i_0)}{m(i)} + \frac{\alpha_n u_{\alpha_n}(i)}{m(i)} + u_{\alpha_n}(i) = \frac{r(i,f_n)}{m(i)} + \sum_{j\in S} u_{\alpha_n}(j)\left[\frac{q(j|i,f_n)}{m(i)} + \delta_{ij}\right] \tag{7.10}$$

for all $i \in S$. By Assumption 7.1(d), $F$ is compact, and thus we may suppose that there exists a policy $f' \in F$ such that

$$\lim_{n\to\infty} f_n(i) = f'(i) \quad \forall i \in S.$$

Letting $n \to \infty$ in (7.10) and applying Proposition A.4, we obtain

$$\frac{g^*}{m(i)} + \bar{u}(i) = \frac{r(i,f')}{m(i)} + \sum_{j\in S} \bar{u}(j)\left[\frac{q(j|i,f')}{m(i)} + \delta_{ij}\right] \quad \forall i \in S,$$

which can be rewritten as

$$\begin{aligned} g^* &= r(i,f') + \sum_{j\in S} \bar{u}(j) q(j|i,f') \\ &\le \sup_{a\in A(i)} \left\{ r(i,a) + \sum_{j\in S} \bar{u}(j) q(j|i,a) \right\} \quad \forall i \in S. \end{aligned} \tag{7.11}$$

Hence, from (7.9) and (7.11) it follows that $(g^*, \bar{u})$ is a solution of the AROE.

Next, we are going to prove that $g^* = \bar{V}^*(i)$ for every $i \in S$.

Pick an arbitrary policy $\pi \in \Pi$. It follows from the AROE (7.4), together with (2.5) and (2.20), that

$$g^* \ge r(i,\pi_t) + \sum_{j\in S} \bar{u}(j) q(j|i,\pi_t) \quad \forall i \in S \text{ and } t \ge 0. \tag{7.12}$$

Then, by Proposition 7.3 we have $g^* \ge \bar{V}(i,\pi)$, and so, since $\pi \in \Pi$ is arbitrary, $g^* \ge \bar{V}^*(i)$ for all $i \in S$.

Observe now that our assumptions ensure the existence of a policy $f^* \in F$ attaining the maximum in the AROE, that is,

$$g^* = r(i,f^*) + \sum_{j\in S} \bar{u}(j) q(j|i,f^*) \quad \forall i \in S.$$

Therefore, Proposition 7.3 gives $g^* = \bar{V}(i,f^*)$ for all $i \in S$. As a consequence, $g^* = \bar{V}^*(i)$ for every $i \in S$, and, moreover, $f^*$ is AR optimal.

Finally, note that, by (7.3),

$$\bar{V}(i,f) = \sum_{j\in S} r(j,f)\mu_f(j) = g(f) \quad \forall f \in F \text{ and } i \in S. \tag{7.13}$$

Our next step in the proof of Theorem 7.8 is to show that a necessary and sufficient condition for a deterministic stationary policy to be AR optimal is that it attains the maximum in the AROE.

In fact, we have already proved the sufficiency part. We will prove the necessity by contradiction. Thus, suppose that $f^* \in F$ is an AR optimal policy that does not attain the maximum in the AROE (7.4). Then there exist $i' \in S$ and a constant $\beta > 0$ (depending on $i'$ and $f^*$) such that

$$g^* \ge r(i, f^*) + \beta \delta_{i'i} + \sum_{j \in S} \bar{u}(j) q(j|i, f^*) \quad \forall i \in S. \tag{7.14}$$

By the irreducibility in Assumption 7.4, the invariant probability measure $\mu_{f^*}$ of $p_{f^*}(i, t, j)$ is supported on all of $S$, meaning that $\mu_{f^*}(j) > 0$ for every $j \in S$. Then, as in the proof of (7.13), from (7.14) and Proposition 7.3 we have

$$g^* \ge g(f^*) + \beta \mu_{f^*}(i') > g(f^*), \tag{7.15}$$

which contradicts the fact that $f^*$ is AR optimal.

Therefore, to complete the proof of this theorem, it only remains to show that the function $\bar{u}$ in the AROE is unique up to additive constants since the constant in the AROE equals the optimal average-reward function. To this end, for each $f \in F$, let us define an operator $Q^f$ on $B_w(S)$ as

$$(Q^f h)(i) := \sum_{j \in S} h(j) q(j|i, f) \quad \text{for } h \in B_w(S) \text{ and } i \in S. \tag{7.16}$$

Suppose now that $(g^*, \bar{u})$ and $(g^*, u')$ are two solutions to the AROE and that $f^* \in F$ is AR optimal. Then (by part (b)) $f^*$ attains the maximum in the *two* AROEs. Hence, subtracting the two AROEs gives

$$\sum_{j \in S} (\bar{u}(j) - u'(j)) q(j|i, f^*) = (Q^{f^*}(\bar{u} - u'))(i) = 0 \quad \forall i \in S. \tag{7.17}$$

Then, as in the proof of (7.1), by (7.17) and (6.19) (with $f^*$ in lieu of $\pi$) we can derive that

$$E_i^{f^*}[\bar{u}(x(t)) - u'(x(t))] = \bar{u}(i) - u'(i) \quad \forall i \in S.$$

Letting $t \to +\infty$ in this equality, it follows from Assumption 7.5 that

$$\mu_{f^*}(\bar{u} - u') = \bar{u}(i) - u'(i) \quad \forall i \in S,$$

showing that the functions $\bar{u}$ and $u'$ differ by the constant $\mu_{f^*}(\bar{u} - u')$. □

From Proposition 7.3 and the proof of Theorem 7.8 we obtain the following.

**Corollary 7.9** *Suppose that Assumptions* 7.1, 7.4, *and* 7.5 *hold, and let* $f \in F$, $g \in \mathbb{R}$, *and* $u \in B_w(S)$. *Then the following facts hold*:

(a) *If*

$$g \ge r(i, f) + \sum_{j \in S} u(j) q(j|i, f) \quad \forall i \in S,$$

*then* $g \ge \bar{V}(i, f) = \sum_{j \in S} r(j, f) \mu_f(j)$ *for all* $i \in S$.

(b) *If*

$$g = r(i, f) + \sum_{j \in S} u(j) q(j|i, f) \quad \forall i \in S,$$

*then* $g = \bar{V}(i, f) = \sum_{j \in S} r(j, f) \mu_f(j)$ *for all* $i \in S$.

*Remark 7.10* The AROE (7.4) is obviously equivalent to

$$\frac{g^*}{m(i)} + \bar{u}(i) = \sup_{a \in A(i)} \left\{ \frac{r(i, a)}{m(i)} + \sum_{j \in S} \bar{u}(j) p(j|i, a) \right\} \quad \forall i \in S$$

(recall the notation in (6.12)), which is different from the discrete-time AROE (see, e.g., Hernández-Lerma and Lasserre (1996) [73], Puterman (1994) [129], and Sennott (1999) [141]) because of the denominator $m(i)$. This difference can also be explained as in Remark 6.7.

We denote by $\mathbf{F}_{\mathrm{ao}}$ the family of AR optimal deterministic stationary policies, and by $\mathbf{F}_{\mathrm{ca}}$ the set of *canonical policies* defined as the policies in $F$ attaining the maximum in the AROE (7.4). Theorem 7.8(b) above shows that, in fact,

$$\mathbf{F}_{\mathrm{ao}} = \mathbf{F}_{\mathrm{ca}}.$$

If we drop the irreducibility hypothesis in Assumption 7.4, then $\mathbf{F}_{\mathrm{ca}} \subseteq \mathbf{F}_{\mathrm{ao}}$ still holds, but the reverse relationship, $\mathbf{F}_{\mathrm{ao}} \subseteq \mathbf{F}_{\mathrm{ca}}$, may fail. That is, we may have an AR optimal policy that is not canonical, which is in fact a situation that we already encountered in Example 5.2 and Proposition 5.11.

## 7.4 The Policy Iteration Algorithm

In this section, we give a policy iteration algorithm to obtain an AR optimal policy. *Throughout this section, we suppose that Assumptions* 7.1, 7.4, *and* 7.5 *are all satisfied.*

*The Bias of a Stationary Policy* Let $f \in F$. We say that a pair $(g, h) \in \mathbb{R} \times B_w(S)$ is a solution to the *Poisson equation* for $f \in F$ if

$$g = r(i, f) + \sum_{j \in S} h(j) q(j|i, f) \quad \forall i \in S.$$

Now recalling (7.13), the expected average reward (or gain) of $f$ is

$$\bar{V}(i, f) = \mu_f\bigl(r(\cdot, f)\bigr) = g(f) \quad \forall i \in S.$$

We define the *bias* (or "potential"—see Remark 3.2) of $f$ as

$$h_f(i) := \int_0^\infty \left[E_i^f r\big(x(t), f\big) - g(f)\right] dt \quad \text{for } i \in S. \tag{7.18}$$

By (7.7), $h_f$ is finite and in $B_w(S)$. Moreover, (7.7) yields that the bias is uniformly bounded in the $w$-norm because

$$\sup_{f\in F} \|h_f\|_w \le L_2 M/\delta. \tag{7.19}$$

**Proposition 7.11** *For every $f \in F$, the solutions to the Poisson equation for $f$ are of the form*

$$\big(g(f), h_f + z\big) \quad \textit{with } z \textit{ any real number.}$$

*Moreover, $(g(f), h_f)$ is the unique solution to the Poisson equation*

$$g(f) = r(i, f) + \sum_{j\in S} h_f(j) q(j|i, f) \quad \forall i \in S \tag{7.20}$$

*for which $\mu_f(h_f) = 0$.*

*Proof* First of all, we will prove that $(g(f), h_f)$ is indeed a solution to the Poisson equation for $f$. Our assumptions (in particular, Assumptions 7.1 and 7.5) allow us to interchange the sums and integrals in the following equations:

$$\begin{aligned}
\sum_{j\in S} h_f(j) q(j|i, f) &= \sum_{j\in S} \left[\int_0^\infty \big(E_j^f r\big(x(t), f\big) - g(f)\big)\, dt\right] q(j|i, f) \quad \big[\text{by (7.18)}\big] \\
&= \int_0^\infty \sum_{j\in S} E_j^f r\big(x(t), f\big) q(j|i, f)\, dt \\
&\qquad \left[\text{since } \sum_{j\in S} q(j|i, f) = 0 \text{ by (2.3)}\right] \\
&= \sum_{j\in S} \int_0^\infty r(j, f) \frac{d}{dt} p_f(i, t, j)\, dt \quad \big[\text{by (2.6)}\big] \\
&= \sum_{j\in S} \Big[r(j, f) \lim_{t\to+\infty} p_f(i, t, j)\Big] - r(i, f) \\
&= \sum_{j\in S} r(j, f) \mu_f(j) - r(i, f) = g(f) - r(i, f),
\end{aligned}$$

as we wanted to prove.

Suppose now that $(g,h)$ is a solution to the Poisson equation for $f$, that is, $(g,h) \in \mathbb{R} \times B_w(S)$, and

$$g = r(i,f) + \sum_{j \in S} h(j) q(j|i,f) \quad \forall i \in S.$$

Therefore, by Corollary 7.9(b), we have $g = g(f)$ because $\mu_f(r(\cdot,f)) = g(f)$.

Suppose now that $(g(f),h)$ and $(g(f),h')$ are two solutions to the Poisson equation for $f \in F$. Subtracting the corresponding Poisson equations, we get (cf. (7.17))

$$\sum_{j \in S} \big(h(j) - h'(j)\big) q(j|i,f) = 0 \quad \forall i \in S.$$

Then, as in the proof of Theorem 7.8(a), we can prove that $h$ and $h'$ are equal up to an additive constant.

To complete the proof, it remains to show that $\mu_f(h_f) = 0$, but this is obtained by taking the $\mu_f$-expectation in (7.18). □

*Remark 7.12* Given $f \in F$, we can determine the gain and the bias of $f$ by solving the following system of linear equations.

First, determine the i.p.m. (invariant probability measure) $\mu_f$ as the unique nonnegative solution (by Proposition C.12) to

$$\begin{cases} \sum_{i \in S} q(j|i,f) \mu_f(i) = 0 \quad \forall j \in S, \\ \sum_{j \in S} \mu_f(j) = 1. \end{cases}$$

Then, as a consequence of Proposition 7.11, the gain $g(f) = \sum_{j \in S} r(j,f) \times \mu_f(j) \in \mathbb{R}$ and the bias $h_f \in B_w(S)$ of $f$ form the unique solution to the system of linear equations

$$\begin{cases} g = r(i,f) + \sum_{j \in S} h(j) q(j|i,f) \quad \text{for } i \in S, \\ \sum_{i \in S} h(i) \mu_f(i) = 0. \end{cases}$$

**Policy Iteration Algorithm 7.1** This algorithm is a standard tool to analyze MDPs. It works as follows:

*Step I.* Choose an arbitrary policy $f \in F$.
*Step II.* Determine the gain $g(f)$ and the bias $h_f$ of $f$ as in Remark 7.12.
*Step III.* Define a policy $f' \in F$ in the following way: for each $i \in S$, if

$$r(i,f) + \sum_{j \in S} h_f(j) q(j|i,f) = \max_{a \in A(i)} \left\{ r(i,a) + \sum_{j \in S} h_f(j) q(j|i,a) \right\}, \tag{7.21}$$

let $f'(i) := f(i)$; otherwise (i.e., if in (7.21) we have a strict inequality), choose $f'(i) \in A(i)$ such that

$$r(i, f') + \sum_{j\in S} h_f(j)q(j|i, f') = \max_{a\in A(i)} \left\{ r(i,a) + \sum_{j\in S} h_f(j)q(j|i,a) \right\}. \tag{7.22}$$

*Step IV.* If $f' = f$ (or, equivalently, if (7.21) holds for every $i \in S$), then $f$ is an AR optimal policy, and the algorithm stops. Otherwise, replace $f$ with $f'$ and return to Step II.

Let $f_0 \in F$ be the initial policy in the policy iteration algorithm (see Step I), and let $\{f_n\}$ be the sequence of stationary policies obtained by the repeated application of the algorithm.

If $f_n = f_{n+1}$ for some $n$, then it follows from Proposition 7.11 that the pair $(g(f_n), h_{f_n})$ is a solution to the AROE, and thus, by Theorem 7.8, $f_n$ is AR optimal.

Hence, to analyze the convergence of the policy iteration algorithm, we will consider the case

$$f_n \neq f_{n+1} \quad \text{for every } n \geq 0. \tag{7.23}$$

Define, for $n \geq 1$ and $i \in S$,

$$\begin{aligned} \varepsilon(f_n, i) &:= r(i, f_n) + \sum_{j\in S} h_{f_{n-1}}(j)q(j|i, f_n) \\ &\quad - \left[ r(i, f_{n-1}) + \sum_{j\in S} h_{f_{n-1}}(j)q(j|i, f_{n-1}) \right], \end{aligned}$$

which by Proposition 7.11 can be expressed as

$$\varepsilon(f_n, i) = r(i, f_n) + \sum_{j\in S} h_{f_{n-1}}(j)q(j|i, f_n) - g(f_{n-1}). \tag{7.24}$$

Observe (by Step III above) that $\varepsilon(f_n, i) = 0$ if $f_n(i) = f_{n-1}(i)$, whereas $\varepsilon(f_n, i) > 0$ if $f_n(i) \neq f_{n-1}(i)$. Hence, $\varepsilon(f_n, i)$ can be interpreted as the "improvement" of the $n$th iteration of the algorithm.

**Lemma 7.13** *Suppose that* (7.23) *is satisfied. Then the following statements hold.*

(a) *The sequence* $\{g(f_n)\}$ *is strictly increasing and it has a finite limit.*
(b) *For every* $i \in S$, $\varepsilon(f_n, i) \to 0$ *as* $n \to \infty$.

*Proof* (a) As in the proof of (7.3), from (7.24) we have

$$\sum_{i\in S} \varepsilon(f_n, i)\mu_{f_n}(i) = g(f_n) - g(f_{n-1}) \quad \forall n \geq 1. \tag{7.25}$$

On the other hand, the hypothesis (7.23) implies that, for every $n \geq 1$, there exists some $i \in S$ with $\varepsilon(f_n, i) > 0$, and, besides, by the irreducibility condition in Assumption 7.4, $\mu_{f_n}(i) > 0$ for every $i \in S$. Hence, $g(f_n) > g(f_{n-1})$, as we wanted to prove. Moreover, as a consequence of Lemma 7.2, the sequence $\{g(f_n)\}$ is bounded above and, therefore, converges to some finite limit.

(b) Let $\mu(i) := \inf_{f \in F} \mu_f(i)$ for all $i \in S$. We will show that $\mu(i) > 0$ for all $i \in S$. To this end, first *fix* an arbitrary state $i \in S$. Since $F$ is compact, there exist a sequence $\{f_m\}$ and $f$ (depending on $i$) in $F$ such that

$$\mu(i) = \lim_{m \to \infty} \mu_{f_m}(i) \quad \text{and} \quad \lim_{m \to \infty} f_m(j) = f(j) \quad \forall j \in S. \tag{7.26}$$

On the other hand, we can prove that $\lim_{m \to \infty} p_{f_m}(i, t, j) = p_f(i, t, j)$ for all $i, j \in S$ and $t \geq 0$. To do so, let

$$\gamma_{ij}^{\alpha}(h) := \int_0^{\infty} e^{-\alpha t} p_h(i, t, j)\, dt \quad (\text{for each } h \in F, i, j \in S, \alpha > 0)$$

be the Laplace transform of $p_h(i, t, j)$. Then, we have

$$\gamma_{ij}^{\alpha}(h) \geq 0, \quad \alpha \sum_{j \in S} \gamma_{ij}^{\alpha}(h) = 1 \quad \forall h \in F, i, j \in S \text{ and } \alpha > 0,$$

and it follows from Theorem 6.9(c) that, for all $m \geq 1$ and $j, k \in S$,

$$\lim_{m \to \infty} q(k|j, f_m) = q(k|j, f), \qquad \alpha \gamma_{ij}^{\alpha}(f_m) = \delta_{ij} + \sum_{k \in S} \gamma_{ik}^{\alpha}(f_m) q(k|j, f_m). \tag{7.27}$$

We now fix $i, j \in S$ and $\alpha > 0$, and then choose an arbitrary subsequence $\{\gamma_{ij}^{\alpha}(f_{m_k})\}$ of $\{\gamma_{ij}^{\alpha}(f_m)\}$ converging to a nonnegative number $v_{ij}^{\alpha}$. Since $S$ is denumerable, there exists a subsequence $\{l\}$ of $\{m_k\}$ such that $\{\gamma_{i'j'}^{\alpha}(f_l)\}$ converges to a nonnegative number $v_{i'j'}^{\alpha}$ for each $i', j' \in S$ (as $l \to \infty$). Hence, by (7.26)–(7.27) and Proposition A.4 we have

$$\alpha v_{ij}^{\alpha} = \delta_{ij} + \sum_{k \in S} v_{ik}^{\alpha} q(k|j, f). \tag{7.28}$$

By the uniqueness of solution to (7.28) in Theorem 6.9(c) we have $v_{ij}^{\alpha} = \gamma_{ij}^{\alpha}(f)$. Then, as the above subsequence $\{\gamma_{ij}^{\alpha}(f_{m_k})\}$ was arbitrarily chosen and all such subsequences have the same limit $\gamma_{ij}^{\alpha}(f)$, we obtain $\lim_{m \to \infty} \gamma_{ij}^{\alpha}(f_m) = \gamma_{ij}^{\alpha}(f)$. This means that $\gamma_{ij}^{\alpha}(f)$ is continuous on $F$, and so is $p_f(i, t, j)$ in $f \in F$. Thus, replacing $f$ and the function $r(j, f)$ in (7.7) with $f_n$ and the function $\delta_{ij}$, respectively, and then letting $m \to \infty$, by (7.26) we obtain

$$\left| p_f(i, t, i) - \mu(i) \right| \leq L_2 e^{-\delta t} w(i).$$

Consequently, $\mu(i) = \lim_{t \to \infty} p_f(i, t, i) = \mu_f(i) > 0$. Therefore, since $i \in S$ was arbitrary, $\mu(i) := \inf_{f \in F} \mu_f(i)$ is positive for all $i \in S$. Hence, recalling that

$\varepsilon(f_n, i) \geq 0$ for all $n \geq 1$ and $i \in S$, by (7.25) we see that, for any $i \in S$,

$$0 \leq \varepsilon(f_n, i)\mu(i) \leq \varepsilon(f_n, i)\mu_{f_n}(i) \leq g(f_n) - g(f_{n-1}).$$

This implies (since $\mu(i) > 0$ and recalling that $\{g(f_n)\}$ is converging) that $\lim_{n\to\infty} \varepsilon(f_n, i) = 0$. □

**Proposition 7.14** *Let $\{f_n\}$ be the sequence obtained from the policy iteration algorithm* 7.1. *Then $g(f_n)$ converges to $g^*$, the optimal average reward function of the continuous-time MDP, and any limit point $f \in F$ of the sequence $\{f_n\}$ is an AR optimal stationary policy.*

*Proof* First of all, without loss of generality we suppose that the sequence $\{f_n\}$ satisfying (7.23) has limit points since $F$ is compact.

By (7.19), there exists a subsequence $\{f_m\}$ of $f_n$ such that $h_{f_m}$ converges pointwise to some $h \in B_w(S)$. Therefore, we have

$$\begin{aligned} g(f_m) &\to g \quad \bigl[\text{Lemma 7.13(a)}\bigr], \qquad f_m \to f, \quad \text{and} \\ h_{f_m} &\to h \quad [\text{pointwise}]. \end{aligned} \tag{7.29}$$

Now, by Proposition 7.11 and the definition of the improvement term $\varepsilon(f_{m+1}, i)$ in (7.24), we have

$$\begin{aligned} g(f_m) &= r(i, f_m) + \sum_{j\in S} h_{f_m}(j) q(j|i, f_m) \\ &= \max_{a\in A(i)} \Bigl\{ r(i,a) + \sum_{j\in S} h_{f_m}(j) q(j|i,a) \Bigr\} - \varepsilon(f_{m+1}, i) \end{aligned} \tag{7.30}$$

for all $i \in S$. As in the proof of Theorem 7.8(a), letting $m \to \infty$ in (7.30), by (7.29) and Lemma 7.13(b) we have

$$\begin{aligned} g &= r(i, f) + \sum_{j\in S} h(j) q(j|i, f) \\ &= \max_{a\in A(i)} \Bigl\{ r(i,a) + \sum_{j\in S} h(j) q(j|i,a) \Bigr\} \end{aligned} \tag{7.31}$$

for all $i \in S$. This shows (by Theorem 7.8(b)) that $f$ is AR optimal and also that $g$ equals the optimal AR function. □

We summarize our results in the following theorem.

**Theorem 7.15** *Suppose that Assumptions* 7.1, 7.4, *and* 7.5 *hold, and let $f_1 \in F$ be an arbitrary initial policy for the policy iteration algorithm* 7.1. *Let $\{f_n\} \subseteq F$ be the sequence of policies obtained by the policy iteration algorithm* 7.1. *Then one of the following results hold.*

(a) *Either*
   (i) *the algorithm converges in a finite number of iterations to an AR optimal policy, or*
   (ii) *as $n \to \infty$, the sequence $\{g(f_n)\}$ converges to the optimal AR function value $g^*$, and any limit point of $\{f_n\}$ is an AR optimal stationary policy.*

(b) *There exists a subsequence $\{f_m\} \subset \{f_n\}$ for which* (7.29) *holds, and, in addition, the limiting triplet $(g, f, h) \in \mathbb{R} \times F \times B_w(S)$ satisfies* (7.31), *so that $(g, h)$ satisfies the AROE, and $f$ is a canonical policy.*

As a consequence of Theorems 7.15 and 7.8(b), if there exists a unique AR optimal stationary policy $f^* \in F$, then the whole sequence $\{f_n\}$ converges pointwise to $f^*$.

## 7.5 Examples

This section presents some applications of the main results in this chapter.

*Example 7.1* (Average-optimal control of a birth-and-death system) Consider a controlled birth-and-death system in which the state variable denotes the population size at any time $t \geq 0$. The birth rate is assumed to be a *fixed* constant $\lambda > 0$, but the death rates $\mu$ are assumed to be controlled by a decision-maker; hence, we interpret a death rate $\mu$ as an *action a* (i.e., $\mu =: a$). When the system's state is $i \in S := \{0, 1, \ldots\}$, the decision-maker takes an action $a$ from a given set $A(i) \equiv [\mu_1, \mu_2]$ with $\mu_2 > \mu_1 > 0$, which increases or decreases the death rates given by (7.33)–(7.34) below. This action incurs a cost $c(i, a)$. In addition, suppose that there is a benefit represented by $p > 0$ for each unit of time, and then the decision-maker gets a reward $pi$ for each unit of time during which the system remains in state $i$.

We now formulate this system as a continuous-time MDP. The corresponding transition rates $q(j|i, a)$ are given as follows: for each $a \in [\mu_1, \mu_2]$,

$$q(1|0, a) = -q(0|0, a) := \lambda, \quad \text{and} \quad q(j|0, a) = 0 \quad \text{for } j \geq 2, \tag{7.32}$$

$$q(0|1, a) := a, \qquad q(1|1, a) = -a - \lambda, \qquad q(2|1, a) := \lambda, \qquad q(j|1, a) = 0 \tag{7.33}$$

for all $j \geq 3$. For all $i \geq 2$ and $a \in A(i) = [\mu_1, \mu_2]$,

$$q(j|i, a) := \begin{cases} p_1 ai & \text{if } j = i - 2, \\ p_2 ai & \text{if } j = i - 1, \\ -(a + \lambda)i & \text{if } j = i, \\ \lambda i & \text{if } j = i + 1, \\ 0 & \text{otherwise}, \end{cases} \tag{7.34}$$

where $p_1 \geq 0$ and $p_2 \geq 0$ are fixed constants such that $p_1 + p_2 = 1$.

By the model's description we see that the reward rates $r(i,a)$ are of the form

$$r(i,a) := pi - c(i,a) \quad \text{for } (i,a) \in K := \{(i,a) : i \in S, a \in A(i)\}. \tag{7.35}$$

We wish to find conditions that ensure the existence of an AR optimal stationary policy. To do this, we consider the following assumptions:

**D$_1$**. $\mu_1 - \lambda > 0$.
**D$_2$**. $p_1 \le \frac{\mu_1}{2\mu_2}$ with $p_1$ as in (7.34). (This condition trivially holds when $p_1 = 0$.)
**D$_3$**. The function $c(i,a)$ is continuous in $a \in A(i) = [\mu_1, \mu_2]$ for each fixed $i \in S$, and $\sup_{a\in A(i)} |c(i,a)| < \tilde{M}(i+1)$ for all $i \in S$, for some constant $\tilde{M} \ge 0$.

Under these conditions, we obtain the following.

**Proposition 7.16** *Under conditions* **D$_1$**, **D$_2$**, *and* **D$_3$**, *the above controlled birth-and-death system satisfies Assumptions* 7.1, 7.4, *and* 7.5. *Therefore (by Theorem* 7.8), *there exists an AR optimal stationary policy.*

*Proof* We shall first verify Assumption 7.1. Let $c_1 := \frac{1}{2}(\mu_1 - \lambda) > 0$ (by **D$_1$**), and $w(i) := i + 1$ for all $i \in S$. Then, from (7.32) and (7.33) we have

$$\sum_{j\in S} w(j)q(j|0,a) = \lambda \le -c_1 w(0) + \mu_1 + \lambda \quad \forall a \in A(i); \tag{7.36}$$

$$\sum_{j\in S} w(j)q(j|1,a) = -(a-\lambda) \le -c_1 w(1) \quad \forall a \in A(i). \tag{7.37}$$

Moreover, for all $i \ge 2$ and $a \in [\mu_1, \mu_2]$, from (7.34) we have

$$\begin{aligned}\sum_{j\in S} w(j)q(j|i,a) &= -(a + ap_1 - \lambda)i \\ &\le -\frac{2}{3}(a + ap_1 - \lambda)w(i) \le -c_1 w(i).\end{aligned} \tag{7.38}$$

By inequalities (7.36)–(7.38) we have, for all $i \in S$ and $a \in A(i)$,

$$\begin{aligned}\sum_{j\in S} w(j)q(j|i,a) &\le -c_1 w(i) + (\mu_1 + \lambda)\delta_{i0} \\ &\le -c_1 w(i) + \mu_1 + \lambda,\end{aligned} \tag{7.39}$$

which verifies Assumption 7.1(a). On the other hand, by (7.32)–(7.34), we have $q^*(i) \le (\mu_2 + \lambda)(i+1) = (\mu_2 + \lambda)w(i)$, and so Assumption 7.1(b) follows. By (7.35) and **D$_3$** we have $|r(i,a)| \le (p + \tilde{M})w(i)$ for all $i \in S$, which implies Assumption 7.1(c). We now verify Assumption 7.1(d). By (7.32)–(7.34) and **D$_3$** we see that Assumptions 6.8(a) and 6.8(b) hold. To verify Assumption 6.8(c), let

$$w'(i) := (i+1)(i+2) \quad \text{for each } i \in S.$$

Then by (7.32)–(7.34) we have

$$q^*(i)w(i) \le (\mu_2 + \lambda)w'(i) \quad \forall i \in S, \quad \text{and}$$

$$\sum_{j\in S} w'(j)q(j|i,a) \le 6\lambda w'(i) \quad \forall a \in [\mu_1, \mu_2] \text{ and } i \in S,$$

which imply Assumption 6.8(c) with $M' := (\mu_2 + \lambda)$, $c' := 6\lambda$, $b' := 0$. Thus, Assumption 7.1(d) is verified. Hence, Assumption 7.1 holds.

Obviously, Assumption 7.4 follows from the description of the model.

Finally, we verify Assumption 7.5. Since $0 \le p_1 \le \frac{\mu_1}{2\mu_2}$, by (7.32)–(7.34) we have, for each fixed $f \in F$,

$$\sum_{j\ge k} q\big(j|i, f(i)\big) \le \sum_{j\ge k} q\big(j|i+1, f(i+1)\big) \quad \forall i, k \in S \text{ such that } k \ne i+1,$$

which, together with Proposition C.16, implies that the corresponding Markov process $x(t)$ is stochastically ordered. Thus, Assumption 7.5 follows from (7.34), (7.39), and Proposition 7.6. □

*Example 7.2* (Average-optimal control of upwardly skip-free processes) We may recall from Example 1.3 that the upwardly skip-free processes, also known as birth-and-death processes with *catastrophes*, belong to the category of *population processes* (see Anderson (1991) [4], Chap. 9, p. 292) with the state space $S := \{0, 1, 2, \ldots\}$. Here we are interested in the AR optimal control problem for such processes with catastrophes of *two* sizes, and so the transition rates are of the form

$$q(j|i,a) := \begin{cases} \lambda i + a_1 & \text{if } j = i+1, \\ -(\lambda i + \mu i + d(i,a_2) + a_1) & \text{if } j = i, \\ \mu i + d(i,a_2)\gamma_i^1 & \text{if } j = i-1, \\ d(i,a_2)\gamma_i^2 & \text{if } j = i-2, \\ 0 & \text{otherwise,} \end{cases} \tag{7.40}$$

where $i \ge 2$, $a := (a_1, a_2)$, and the birth rate $\lambda > 0$ and death rate $\mu > 0$ are fixed. Moreover, the *immigration* rates $a_1 \ge 0$, and the $d(i, a_2)$ are nonnegative numbers representing the rates at which the "catastrophes" occur and which are assumed to be controlled by decisions $a_2$ in some compact set $B(i)$ when the process is in state $i \ge 2$. The numbers $\gamma_i^1$ and $\gamma_i^2$ are nonnegative and such that $\gamma_i^1 + \gamma_i^2 = 1$ for all $i \ge 2$, with $\gamma_i^k$ denoting the probability that the process makes a transition from $i$ to $i - k$ ($k = 1, 2$), given that a catastrophe occurs when the process is in state $i \ge 2$. When the state $i$ is 0 or 1, it is natural to let $q(1|0,a) = a_1$, $q(0|0,a) =$

$-a_1, q(j|0,a) = 0$ for $j \geq 2$, and

$$q(j|1,a) = \begin{cases} \lambda + a_1 & \text{if } j = 2, \\ -\lambda - \mu - a_1 - d(1,a_2) & \text{if } j = 1, \\ \mu + d(1,a_2) & \text{if } j = 0, \\ 0 & \text{otherwise,} \end{cases}$$

with $d(1,a_2)$ having the similar meaning as $d(i,a_2)$ above. On the other hand, we suppose that the immigration rates $a_1$ can also be controlled, and so we interpret $a := (a_1, a_2)$ as an action. Thus, we consider the admissible action sets $A(0) := [0,b]$ and $A(i)$ that are finite subsets of $[0,b] \times B(i)$ for $i \geq 1$, for some constant $b > 0$. In addition, we suppose that the damage caused by a catastrophe is represented by $p > 0$ for each unit of time and that it incurs a cost at rate $c(i,a_2)$ to take decision $a_2 \in B(i)$ at state $i \geq 1$. Let $c(0,\cdot) \equiv 0$. Also, we assume that the benefits obtained by the transitions to $i-1$ and $i-2$ from $i$ ($\geq 2$) are given by positive constants $q_1$ and $q_2$, respectively, and the benefit earned by each $a_1 \in [0,b]$ is denoted by $\tilde{r}(a_1)$. Hence, the reward function is of the form

$$r(i,a) := \tilde{r}(a_1) - c(i,a_2) - pd(i,a_2) + q_1\gamma_i^1 d(i,a_2) + q_2\gamma_i^2 d(i,a_2)$$

for all $a = (a_1,a_2) \in A(i)$, where $\gamma_0^1 = \gamma_0^2 := 0, \gamma_1^1 := 1, \gamma_1^2 := 0, d(0,a_2) := 0$.

As in Example 7.1, it can be verified that, under the following conditions $\mathbf{E_k}$ ($k = 1,2,3$), the above controlled upwardly skip-free process satisfies Assumptions 7.1, 7.4, and 7.5.

$\mathbf{E_1}$. $\mu - \lambda > 0$; $\gamma_{i+1}^2 \leq \inf_{\{a_2 \in B(i)\}} \frac{d(i,a_2)+\mu i}{d(i+1,a_2)}$ for all $i \geq 1$.

$\mathbf{E_2}$. $b \leq \lambda - \mu + \inf_{\{i \geq 1, a_2 \in B(i)\}} \{d(i,a_2) + \gamma_i^2 d(i,a_2)\}$.

$\mathbf{E_3}$. For each $i \in S$, the functions $\tilde{r}(a_1)$, $d(i,a_2)$, and $c(i,a_2)$ are continuous in $(a_1,a_2) \in A(i)$, and

$$\sup_{a_2 \in B(i)} |d(i,a_2)| \leq L_1'(i+1), \qquad \sup_{a_2 \in B(i)} |c(i,a_2)| < L_2'(i+1)$$

for some constants $L_1' > 0$ and $L_2' > 0$.

In particular, these conditions $\mathbf{E_k}$ ($k = 1,2,3$) hold when $\lambda < \mu \leq b + \lambda$, $\tilde{r}(a_1) := \tau a_1$, $d(i,a_2) := 2a_2 i$, $\gamma_i^2 \leq \frac{1}{2} + \frac{\mu}{4\beta}$, and $B(i) := [b,\beta]$ for all $i \geq 1$, for some constants $\tau > 0$ and $\beta > b$.

Therefore (by Theorem 7.8), we have the following fact.

**Proposition 7.17** *Under conditions* $\mathbf{E_1}$, $\mathbf{E_2}$, *and* $\mathbf{E_3}$, *the above controlled upwardly skip-free process satisfies Assumptions* 7.1, 7.4, *and* 7.5. *Therefore* (*by Theorem* 7.8), *there exists an AR optimal stationary policy.*

We next provide two examples about queueing systems which also satisfy Assumptions 7.1, 7.4, and 7.5. Therefore (by Theorem 7.8), there exists an AR optimal stationary policy for each of the two examples.

*Example 7.3* (Average-optimal control of a pair of $M/M/1$ queues in tandem) Consider a tandem queueing system consisting of two servers in series. Customers arrive as a Poisson stream with *unit* rate to the first queue where they are served with mean service time $a_1^{-1}$. After service is completed at the first queue, each customer immediately departs and joins the second queue, where the mean service time is $a_2^{-1}$. After service is completed at the second queue, the customers leave the system. The state space is $S := \{0, 1, 2, \ldots\}^2$. We interpret a given pair $(a_1, a_2) =: a$ of mean service times as an action taken from the action set $A(i_1, i_2) \equiv [\mu_1, \mu_1^*] \times [\mu_2, \mu_2^*]$, with positive constants $\mu_1^* > \mu_1$, $\mu_2^* > \mu_2$. Let

$$w(i_1, i_2) := \sigma_1^{i_1-1} + \sigma_2^{i_1+i_2-1} + \gamma \sigma_1^{-\beta_1(i_1-1)} \sigma_2^{-\beta_2(i_1+i_2-1)},$$

where $\sigma_1 = 1.06$, $\sigma_2 = 1.03$, $\gamma = 0.4$, $\beta_1 = 1.5$, $\beta_2 = 0.3$, $\mu_1 \geq 3$, and $\mu_2 \geq 2$. Suppose that $r(i_1, i_2, a)$ is *bounded* in all $(i_1, i_2, a)$ and *continuous* in $a \in A(i_1, i_2)$ for each fixed $(i_1, i_2) \in S$. Then by a direct calculation and Proposition 7.6, we see that Assumption 7.1 is satisfied with $c_1 := 0.002$ and a constant $b_1 > 0$. Moreover, under these parameter values which are found by the computer program Mathematica, a straightforward calculation yields Assumptions 7.1, 7.4, and 7.5. (If necessary, see Lund, Meyn, and Tweedie (1996) [113] for details.) Therefore, there exists an AR optimal stationary policy for our tandem queueing system.

*Example 7.4* (Average-optimal control of $M/M/N/0$ queue systems) Here the state space is $S := \{0, 1, \ldots, N\}$ for some integer $N \geq 1$. Suppose that the arrival rate $\lambda$ is fixed but the service rates can be controlled. Therefore, we interpret the service rates $a$ as actions which may depend on the current states $i \in S$. We denote by $A(i)$ the action set at state $i \in S$. When the system is empty, we may suppose that $A(0) := \{0\}$. For each $i \geq 1$, let $A(i) := [\mu_1, \mu_2]$ with constants $\mu_2 > \mu_1 > 0$. The transition rates are given as follows:

$$q(0|0,0) = -\lambda = -q(1|0,0), \qquad q(j|0,0) = 0 \quad \text{for } 2 \leq j \leq N;$$

$$q(N|N,a) = -Na = -q(N-1|N,a), \qquad q(j|N,a) = 0$$

for all $0 \leq j \leq N-2$ and $a \in A(N)$; moreover, for all $1 \leq i \leq N-1$ and $a \in A(i)$,

$$q(j|i,a) := \begin{cases} \lambda & \text{if } j = i+1, \\ -(\lambda + ai) & \text{if } j = i, \\ ai & \text{if } j = i-1, \\ 0 & \text{otherwise.} \end{cases}$$

Finally, suppose that $\mu_1 > \lambda$, and that the given reward function $r(i, a)$ is continuous in $a \in A(i)$ for all $i \in S$. Then, as in the Example 7.1, we can see that this controlled $M/M/N/0$ system satisfies Assumptions 7.1, 7.4, and 7.5.

*Remark 7.18* In the verification of Assumptions 7.1, 7.4, and 7.5 for our four examples, a key step is the verification of Assumption 7.5 by means of Proposition 7.6.

This is due to the advantageous feature of Proposition 7.6 of being expressed in terms of the primitive data of the model; this allows a direct verification of conditions (a) and (b) in the proposition. We should also note that these conditions have to be *uniform* with respect to the actions. In fact, this uniformity is crucial to show that the exponential convergence rate $\delta$ and the constant $L_2$ in Assumption 7.5 are *independent* of all the stationary policies. To conclude, we mention that other examples and approaches that yield Assumption 7.5 can be seen in Down, Meyn, and Tweedie (1995) [37], Lund, Meyn, and Tweedie (1996) [113], and Tweedie (1981) [149], for instance.

## 7.6 Notes

As we already mentioned in Sect. 7.1, an AR optimal policy may not exist for a *non*-finite MDP. Thus, the main aim of the study on the average-reward optimality criterion is to find conditions ensuring the existence of AR optimal policies, and many conditions have indeed been proposed; see, for instance, Hou and Guo (1998) [84] and Kakumanu (1972) [93] for bounded transition rates and rewards; Haviv and Puterman (1998) [68], Lewis and Puterman (2001) [107], Puterman (1994) [129], Sennott (1999) [141], Serfozo (1981) [143], and Yushkevich and Feinberg (1979) [165] for bounded transition rates but unbounded rewards; Bather (1976) [9], Guo and Liu (2001), Hou and Guo (1998) [84], and Song (1987) [145] for unbounded transition rates but bounded rewards; and Guo and Zhu (2002b) [63], and Guo and Hernández-Lerma (2003c) [55] for unbounded transition rates and unbounded rewards. For the case of a Polish state space (that is, where $S$ is a complete separable metric space), the reader is referred to Doshi (1976) [36], Guo and Rieder (2006) [59], and Hernández-Lerma (1994) [69]. We also note that when the transition rates are bounded, some results for continuous-time MDPs can be obtained from those for discrete-time MDPs by using the uniformization technique; see, e.g., Haviv and Puterman (1998) [68], Lembersky (1974) [106], Lewis and Puterman (2001) [107], Puterman (1994) [129], or Veinott (1969) [151].

This chapter concerns MDPs with *unbounded transition rates*, *unbounded reward functions*, and a *denumerable state space*, as in Guo and Liu (2001) [58], Guo and Zhu (2002b) [63], and Guo and Hernández-Lerma (2003c) [55]. The main results in this chapter are from Guo and Hernández-Lerma (2003c) [55].

On the other hand, the existing approaches used to showing the existence of AR optimal policies include *Kolmogorov's forward equation approach* by Guo and Liu (2001) [58], Guo and Zhu (2002a, 2002b) [62, 63], Kakumanu (1971) [92], and Miller (1968) [117], for instance; the *uniformization technique* by Lewis and Puterman (2001) [107], Puterman (1994) [129], and Sennott (1999) [141]; the *extended generator approach* by Guo and Cao (2005) [52] and Guo and Hernández-Lerma (2003) [53, 54, 56]; the *average-cost minimum nonnegative solution approach* (that is, the optimality inequality approach associated to the vanishing discount method) provided in Sect. 5.4; and the *convex analytic approach* in Piunovskiy (2004) [120].

Now, in view of Theorem 7.15 (in Sect. 7.4), we can add another approach, namely the policy iteration algorithm.

Each of these approaches has its own advantages. Roughly speaking, Kolmogorov's forward equation and the extended generator approaches, as well as the policy iteration algorithm, can deal with the case that the reward function may have neither upper nor lower bounds, but of course other conditions are also required. The uniformization technique (see Remark 6.1) is applicable only when the transition rates are bounded. The minimum nonnegative solution approach can show that an AR optimal policy may exist even when the AROE approach fails; however, it cannot deal with the case of unbounded from below rewards. Finally, the convex analytic approach can also be used to study multi-criteria and multi-constrained problems, but it is not very common because it mainly deals with the problem of *existence* of optimal policies; it is not obvious at all that it can be used for computational or approximation purposes, such as, for instance, the policy iteration algorithm in Sect. 7.4.

# Chapter 8
# Average Optimality for Pathwise Rewards

In Chaps. 3, 5, and 7, we have studied the optimality problem under the *expected average reward* $\bar{V}(i,\pi)$. However, the sample-path reward $r(x(t),\pi_t)$ corresponding to an average-reward optimal policy that maximizes an expected average reward may have fluctuations from its expected value. To take these fluctuations into account, we next consider the *pathwise average-reward* (PAR) criterion.

*The model considered in this chapter is the same as that in Chap. 7.*

## 8.1 Introduction

In Chaps. 5 and 7, under suitable conditions, we have studied the long-run *expected* average reward (EAR) (recall (2.21))

$$\bar{V}(i,\pi) := \liminf_{T\to\infty} \frac{1}{T}\int_0^T E_i^\pi r\bigl(x(t),\pi_t\bigr)\,dt. \tag{8.1}$$

Hence, $\bar{V}(i,\pi)$ concerns the long-run average of the expected reward $E_i^\pi r(x(t),\pi_t)$. In contrast, the *pathwise* average reward (PAR) $V_p(\cdot,\cdot)$, defined by

$$V_p(i,\pi) := \liminf_{T\to\infty} \frac{1}{T}\int_0^T r\bigl(x(t),\pi_t\bigr)\,dt, \tag{8.2}$$

concerns the long-run *average of the pathwise reward* $r(x(t),\pi_t)$ and, of course, it may have fluctuations from its expected value, as shown in the following example.

*Example 8.1* Let $S := \{1,2\}$. For some $\hat{\pi} = (\hat{\pi}_t)$ and $f \in \Pi$, suppose that

$$Q(\hat{\pi}_t) := \begin{pmatrix} -1+t & 1-t \\ 2-2t & -2+2t \end{pmatrix} \quad \text{and} \quad Q(f) := \begin{pmatrix} 0 & 0 \\ 0 & 0 \end{pmatrix} \tag{8.3}$$

X. Guo, O. Hernández-Lerma, *Continuous-Time Markov Decision Processes*, Stochastic Modelling and Applied Probability 62,
DOI 10.1007/978-3-642-02547-1_8, © Springer-Verlag Berlin Heidelberg 2009

for $0 \le t \le t_0 := 1$, where $Q(\hat{\pi}_t)$ is arbitrarily given for all $t > 1$. Moreover, define

$$Q(\tilde{\pi}_t) := \begin{cases} Q(\hat{\pi}_t) & \text{for } 0 \le t \le t_0, \\ Q(f) & \text{for } t > t_0. \end{cases} \tag{8.4}$$

By (8.3), (8.4), and Definition 2.1, we see that the associated policy $\tilde{\pi}$ belongs to $\Pi$. (Recall that for any $\pi \in \Pi$, the associated transition probability function $p_\pi(s,i,t,j)$ can be constructed as in (6.5)–(6.7).) Hence, for all $i, j \in S$, by (8.3), (8.4), and (6.5)–(6.7), we have $p_{\hat{\pi}}(0,i,t_0,j) > 0$. Moreover, $0 < p_{\hat{\pi}}(0,i,t_0,2) < 1$ and

$$p_{\tilde{\pi}}(s,i,t,j) = \begin{cases} p_{\hat{\pi}}(s,i,t,j) & \text{for } 0 \le s \le t \le t_0, \\ p_f(s,i,t,j) & \text{for } t \ge s > t_0. \end{cases} \tag{8.5}$$

Let $r(1,a) = 0$ and $r(2,a) = 1$ for all $a \in A(i)$ with $i = 1,2$. Then, by (8.3) and (8.4) we see that states 1 and 2 are absorbing after time $t_0$. In fact, by (8.3), (8.5), and (6.7), we have

$$p_{\tilde{\pi}}(t_0,i,t,i) = 1 \quad \forall i \in S \text{ and } t \ge t_0. \tag{8.6}$$

Noting that $r(1,\tilde{\pi}_t) = 0$ and $r(2,\tilde{\pi}_t) = 1$ for all $t \ge 0$, by (8.6) and (8.2) we have that for each $i \in S$,

$$V_p(i,\tilde{\pi}) = 1 \quad \text{for any sample path in } \{x(t) = 2, t \ge t_0\}. \tag{8.7}$$

On the other hand, by (8.6) and the Chapman–Kolmogorov equation (2.7), we have

$$\begin{aligned} p_{\tilde{\pi}}(0,i,t,2) &= p_{\tilde{\pi}}(0,i,t_0,1)p_{\tilde{\pi}}(t_0,1,t,2) + p_{\tilde{\pi}}(0,i,t_0,2)p_{\tilde{\pi}}(t_0,2,t,2) \\ &= p_{\tilde{\pi}}(0,i,t_0,2) < 1 \quad \forall t_0 < t. \end{aligned} \tag{8.8}$$

Using again $r(1,\tilde{\pi}_t) = 0$ and $r(2,\tilde{\pi}_t) = 1$ for all $t \ge 0$, by (8.8) and (8.1) we see that

$$\begin{aligned} \bar{V}(i,\tilde{\pi}) &= \liminf_{T\to\infty} \frac{\int_0^T p_{\tilde{\pi}}(0,i,t,2)\,dt}{T} \\ &= \liminf_{T\to\infty} \frac{\int_{t_0}^T p_{\tilde{\pi}}(0,i,t,2)\,dt}{T} \\ &= p_{\tilde{\pi}}(0,i,t_0,2) < 1 \quad \forall i \in S. \end{aligned} \tag{8.9}$$

This fact, together with (8.7) and $P_i^{\tilde{\pi}}(\{x(t) = 2, t \ge t_0\}) = p_{\tilde{\pi}}(0,i,t_0,2) \in (0,1)$, shows the difference between the pathwise AR (PAR) and the expected AR (EAR).

To take these fluctuations into account, we next study the *pathwise average reward* criterion.

## 8.2 The Optimal Control Problem

Recall from (8.2) that the PAR criterion $V_p(\cdot,\cdot)$ is defined as follows: for all $\pi = (\pi_t) \in \Pi$ and $i \in S$,

$$V_p(i,\pi) := \liminf_{T\to\infty} \frac{1}{T} \int_0^T r\bigl(x(t),\pi_t\bigr)\,dt. \tag{8.10}$$

Note that $V_p(i,\pi)$ has been defined by the *sample-path rewards* $r(x(t),\pi_t)$, $t \geq 0$; therefore, it is a *random variable* rather than a number as in the EAR criterion defined by (2.21). Thus, the following definition of an optimal policy for the PAR criterion is different from that for the EAR criterion.

**Definition 8.1** For a given $\varepsilon \geq 0$, a policy $\pi^* \in \Pi$ is said to be $\varepsilon$-PAR-optimal if there exists a constant $g^*$ such that

$$P_i^{\pi^*}\bigl(V_p\bigl(i,\pi^*\bigr) \geq g^* - \varepsilon\bigr) = 1 \quad \text{and} \quad P_i^{\pi}\bigl(V_p(i,\pi) \leq g^*\bigr) = 1$$

for all $i \in S$ and $\pi \in \Pi$. For $\varepsilon = 0$, a 0-PAR-optimal policy is simply called a PAR-optimal policy.

The main goal of this chapter is to give conditions that ensure the existence and calculation of $\varepsilon$-PAR-optimal stationary policies.

## 8.3 Optimality Conditions and Preliminaries

In this section, we present some basic optimality conditions and preliminary facts that are needed to prove our main results. In fact, perhaps not surprisingly, we will impose the same assumptions used for Theorems 7.8 and 7.15. To be more specific, to ensure the existence of a PAR-optimal stationary policy, we will suppose the following.

**Assumption 8.2** Assumptions 6.8(a), 6.8(b), 7.1(a)–7.1(c), 7.4, and 7.5 are all satisfied.

In Chap. 7, we have established the AROE (7.4) by using the *policy iteration algorithm*; see Theorem 7.15. For ease of reference, we recall here the policy iteration algorithm 7.1:

*Step I.* Take $n = 0$ and $f_n \in F$.
*Step II.* Solve

$$\begin{cases} 0 = \sum_{i\in S} q(j|i,f_n)\mu_{f_n}(i) & \text{for } j \in S,\\ 1 = \sum_{j\in S} \mu_{f_n}(j), \end{cases}$$

for $\mu_{f_n}$, then calculate $g(f_n) = \sum_{j\in S} r(j, f_n)\mu_{f_n}(j)$ and, finally, $h_{f_n}$ from the system of linear equations

$$\begin{cases} g(f_n) = r(i, f_n) + \sum_{j\in S} h(j)q(j|i, f_n) & \text{for } i \in S, \\ 0 = \sum_{i\in S} h(i)\mu_{f_n}(i). \end{cases}$$

*Step III.* Define the new stationary policy $f_{n+1}$ in the following way:
Set $f_{n+1}(i) := f_n(i)$ for all $i \in S$ for which

$$r\big(i, f_n(i)\big) + \sum_{j\in S} h_{f_n}(j)q\big(j|i, f_n(i)\big) = \max_{a\in A(i)} \Big\{ r(i, a) + \sum_{j\in S} h_{f_n}(j)q(j|i, a) \Big\}; \tag{8.11}$$

otherwise (i.e., when (8.11) does not hold), choose $f_{n+1}(i) \in A(i)$ such that

$$r\big(i, f_{n+1}(i)\big) + \sum_{j\in S} h_{f_n}(j)q\big(j|i, f_{n+1}(i)\big) = \max_{a\in A(i)} \Big\{ r(i, a) + \sum_{j\in S} h_{f_n}(j)q(j|i, a) \Big\}. \tag{8.12}$$

*Step IV.* If $f_{n+1}(i)$ satisfies (8.11) for all $i \in S$, then stop because (by Theorem 8.5 below) $f_{n+1}$ is PAR-optimal; otherwise, replace $f_n$ with $f_{n+1}$ and go back to Step II.

To prove the existence of a PAR-optimal stationary policy, in addition to Assumption 8.2, we propose the following conditions.

**Assumption 8.3** Let $w \geq 1$ be as in Assumption 7.1(a). For $k = 1, 2$, there exist nonnegative functions $w_k^* \geq 1$ on $S$ and constants $c_k^* > 0$, $b_k^* \geq 0$, and $M_k^* > 0$ such that, for all $i \in S$ and $a \in A(i)$,

(a) $w^2(i) \leq M_1^* w_1^*(i)$ and $\sum_{j\in S} w_1^*(j)q(j|i, a) \leq -c_1^* w_1^*(i) + b_1^*$.
(b) $[q^*(i)w(i)]^2 \leq M_2^* w_2^*(i)$ and $\sum_{j\in S} w_2^*(j)q(j|i, a) \leq -c_2^* w_k^*(i) + b_2^*$.

*Remark 8.4*

(a) Assumption 8.3(a) allows us to use the martingale stability theorem (see Proposition A.15); however, it is not required if a solution $u^*$ to (8.14) below and the transition rates are both uniformly bounded.
(b) Assumption 8.3(b) is slightly different from Assumption 6.8(c). Nevertheless, all the conclusions in Chap. 7 still hold if Assumption 6.8(c) is replaced by Assumption 8.3(b), because the former is used only for the finiteness of

$E_i^\pi[q^*(x(t))w(x(t))]$, which also follows from Assumption 8.3(b). In fact, by Lemma 6.3 and Assumption 8.3(b) we have

$$\left[E_i^\pi q^*\big(x(t)\big)w\big(x(t)\big)\right]^2 \le E_i^\pi\left[q^*\big(x(t)\big)w\big(x(t)\big)\right]^2 \le M_2^* E_i^\pi w_2^*\big(x(t)\big) < \infty.$$

For each $n \ge 1$, take $f_n$ as the policy obtained in the Policy Iteration Algorithm above, and let (as in (7.24))

$$\varepsilon(f_n, i) := r\big(i, f_n(i)\big) + \sum_{j \in S} h_{f_{n-1}}(j) q\big(j|i, f_n(i)\big) - g(f_{n-1}) \tag{8.13}$$

for all $i \in S$.

Then, by Lemma 7.13, we have

$$g(f_{n+1}) > g(f_n) \quad \text{if } f_{n+1} \neq f_n, \quad \text{and} \quad \varepsilon(f_n, i) \to 0 \quad (\text{as } n \to \infty),$$

which were used to establish the AROE (7.31). This equation is restated in (8.14), below, because now it will be used to obtain PAR-optimal policies.

## 8.4 The Existence of PAR Optimal Policies

In this section, we state and prove our main result, Theorem 8.5.

**Theorem 8.5** *Under Assumptions* 8.2 *and* 8.3, *the following statements hold.*

(a) *There exist a unique constant $g^*$, a function $u^* \in B_w(S)$, and a stationary policy $f^* \in F$ satisfying the average-reward optimality equation (AROE)*

$$\begin{aligned} g^* &= r\big(i, f^*(i)\big) + \sum_{j \in S} u^*(j) q\big(j|i, f^*(i)\big) \\ &= \max_{a \in A(i)} \left\{ r(i,a) + \sum_{j \in S} u^*(j) q(j|i,a) \right\} \quad \forall i \in S. \end{aligned} \tag{8.14}$$

(b) *The policy $f^*$ in* (a) *is PAR-optimal, and $P_i^{f^*}(V_p(i, f^*) = g^*) = 1$ for all $i \in S$, with $g^*$ as in* (a).

(c) *A policy $f$ in $F$ is PAR-optimal if and only if it realizes the maximum in* (8.14).

(d) *For given $\varepsilon \ge 0$, $f \in F$, and $g^*$ as in* (a) *above, if there is a function $u' \in B_w(S)$ such that*

$$g^* \le r\big(i, f(i)\big) + \sum_{j \in S} u'(j) q\big(j|i, f(i)\big) + \varepsilon \quad \forall i \in S, \tag{8.15}$$

*then $f$ is $\varepsilon$-PAR-optimal.*

*Proof* (a) Indeed, part (a) has been obtained in Theorem 7.8. (Recall our Remark 8.4(b) above.) The proof is based on the fact that if $f_n$, $g(f_n)$, and $h_{f_n}$ are as in (8.11)–(8.13), then there exist a subsequence $\{f_{n_k}\}$ of $\{f_n\}$, $f^* \in F$, $u^* \in B_w(S)$, and a constant $g^*$ such that for each $i \in S$,

$$\lim_{k\to\infty} h_{f_{n_k}}(i) =: u^*(i), \qquad \lim_{k\to\infty} f_{n_k}(i) = f^*(i), \quad \text{and} \quad \lim_{k\to\infty} g(f_{n_k}) = g^*. \tag{8.16}$$

The triplet $(g^*, u^*, f^*)$ satisfies (8.14).

(b) To prove (b), for all $i \in S$, $\pi = (\pi_t) \in \Pi$, and $t \geq 0$, let

$$\Delta(i, \pi_t) := r(i, \pi_t) + \sum_{j\in S} u^*(j) q(j|i, \pi_t) - g^*, \tag{8.17}$$

$$\bar{h}(i, \pi_t) := \sum_{j\in S} u^*(j) q(j|i, \pi_t), \tag{8.18}$$

and let

$$\mathcal{F}_t := \sigma\{x(s), 0 \leq s \leq t\}$$

be the $\sigma$-algebra generated by $\{x(s), 0 \leq s \leq t\}$.

Observe that, in particular,

$$\Delta(i, f(i)) = r(i, f(i)) + \sum_{j\in S} u^*(j) q(j|i, f(i)) - g^*$$

for all $f \in F$.

We now define the (continuous-time) stochastic process

$$M(t, \pi) := \int_0^t \bar{h}(x(y), \pi_y)\, dy - u^*(x(t)) \quad \text{for } t \geq 0. \tag{8.19}$$

Then $\{M(t, \pi), \mathcal{F}_t, t \geq 0\}$ is a continuous-time $P_i^\pi$-martingale (for each fixed $i \in S$), that is,

$$E_i^\pi[M(t, \pi)|\mathcal{F}_s] = M(s, \pi) \quad \forall t \geq s \geq 0. \tag{8.20}$$

Indeed, for all $t \geq s \geq 0$, by (8.19) and the Markov property (see (B.1) in Appendix B.1) we have

$$\begin{aligned} E_i^\pi[M(t, \pi)|\mathcal{F}_s] = M(s, \pi) + E_i^\pi\left[\int_s^t \bar{h}(x(y), \pi_y)\, dy|\mathcal{F}_s\right] \\ + u^*(x(s)) - E^\pi_{x(s),s} u^*(x(t)). \end{aligned} \tag{8.21}$$

Since $u^*$ is in $B_w(S)$ (by (8.16)), it follows from Assumption 8.2 (see Assumption 7.1(a)) that

$$\begin{aligned}\sum_{j\in S}\left|u^*(j)q(j|i,a)\right| &\leq \|u^*\|_w\left[\sum_{j\in S}w(j)q(j|i,a)-2w(i)q(i|i,a)\right]\\ &\leq \|u^*\|_w\left[-c_1w(i)+b_1+2q^*(i)w(i)\right]\\ &\leq \|u^*\|_w\left[b_1+2q^*(i)w(i)\right] \qquad (8.22)\end{aligned}$$

for all $i \in S$ and $a \in A(i)$. Therefore, by (8.18) and (8.22) we obtain

$$\left|\bar{h}(i,\pi_y)\right| \leq \|u^*\|_w\left[b_1+2q^*(i)w(i)\right] \quad \forall i \in S \text{ and } y \geq 0. \qquad (8.23)$$

On the other hand, by the Markov property we have

$$E_i^\pi\left[\int_s^t \bar{h}\big(x(y),\pi_y\big)\,dy|\mathcal{F}_s\right] = E_{x(s),s}^\pi\left[\int_s^t \bar{h}\big(x(y),\pi_y\big)\,dy\right],$$

which, together with (8.23), Assumption 8.3, Lemma 6.3(i) and Fubini's theorem, gives

$$E_i^\pi\left[\int_s^t \bar{h}\big(x(y),\pi_y\big)\,dy|\mathcal{F}_s\right] = \int_s^t \left[E_{x(s),s}^\pi \bar{h}\big(x(y),\pi_y\big)\right]dy. \qquad (8.24)$$

By (2.6), (8.24), and (8.18) we obtain

$$E_i^\pi\left[\int_s^t \bar{h}\big(x(y),\pi_y\big)\,dy|\mathcal{F}_s\right] = E_{x(s),s}^\pi u^*\big(x(t)\big) - u^*\big(x(s)\big).$$

This equality and (8.21) give (8.20).

As a consequence of (8.20), the process $\{M(n,\pi),\mathcal{F}_n, n=1,2,\ldots\}$ is a discrete-time $P_i^\pi$-martingale. Moreover, by Assumption 8.3 and Lemma 6.3 we have

$$E_i^\pi w_k^*\big(x(t)\big) \leq w_k^*(i) + \frac{b_k^*}{c_k^*} \quad \forall t \geq 0 \text{ and } k = 1,2. \qquad (8.25)$$

Thus, since "$(x+y)^2 \leq 2(x^2+y^2)$", by (8.19)–(8.23) and the Hölder inequality, we have

$$\begin{aligned}&E_i^\pi\left[M(n+1,\pi)-M(n,\pi)\right]^2\\ &\quad= E_i^\pi\left[\int_n^{n+1}\bar{h}\big(x(y),\pi_y\big)\,dy + u^*\big(x(n)\big) - u^*\big(x(n+1)\big)\right]^2\\ &\quad\leq 2E_i^\pi\left[\int_n^{n+1}\bar{h}\big(x(y),\pi_y\big)\,dy\right]^2 + 2E_i^\pi\left[u^*\big(x(n+1)\big)-u^*\big(x(n)\big)\right]^2\end{aligned}$$

$$\leq 2E_i^\pi\left[\int_n^{n+1} \bar{h}^2(x(y),\pi_y)\,dy\right] \quad \text{(by the Hölder inequality)}$$

$$+4\|u^*\|_w^2 E_i^\pi\left[w^2(x(n+1))+w^2(x(n))\right]$$

$$\leq 2E_i^\pi\left[\int_n^{n+1} \|u^*\|_w^2\left[b_1+2q^*(x(y))w(x(y))\right]^2 dy\right] \quad \text{(by (8.23))}$$

$$+4M_1^*\|u^*\|_w^2 E_i^\pi\left[w_1^*(x(n+1))+w_1^*(x(n))\right] \quad \text{(by Assumption 8.3(a))}$$

$$\leq 4\|u^*\|_w^2 E_i^\pi\left[\int_n^{n+1} \left(b_1^2+4[q^*(x(y))w(x(y))]^2\right) dy\right]$$

$$+4M_1^*\|u^*\|_w^2 E_i^\pi\left[w_1^*(x(n+1))+w_1^*(x(n))\right]$$

$$\leq 4\|u^*\|_w^2 E_i^\pi\left[\int_n^{n+1} \left(b_1^2+4M_2^* w_2^*(x(y))\right) dy\right] \quad \text{(by Assumption 8.3)}$$

$$+4M_1^*\|u^*\|_w^2 E_i^\pi\left[w_1^*(x(n+1))+w_1^*(x(n))\right]$$

$$\leq 16\|u^*\|_w^2\left[b_1^2+M_2^*\left(w_2^*(i)+\frac{b_2^*}{c_2^*}\right)+M_1^*\left(w_1^*(i)+\frac{b_1^*}{c_1^*}\right)\right] \quad \text{(by (8.25))}. \tag{8.26}$$

This means that $E_i^\pi[M(n+1,\pi)-M(n,\pi)]^2$ is bounded in $n \geq 1$. Thus, by the martingale stability theorem (see Proposition A.15) we have

$$\lim_{n\to\infty} \frac{M(n,\pi)}{n} = 0 \quad P_i^\pi\text{-a.s.} \tag{8.27}$$

Now, for any $T \geq 1$, let $[T]$ be the unique integer such that $[T] \leq T < [T]+1$. By (8.19) we obtain

$$\frac{M(T,\pi)}{T} = \frac{M([T],\pi)}{T} + \frac{\int_{[T]}^T \bar{h}(x(y),\pi_y)\,dy}{T} - \frac{u^*(x(T))}{T} + \frac{u^*(x([T]))}{T}. \tag{8.28}$$

Pick an arbitrary $\varepsilon > 0$. Then, as in the proof of (8.26), Chebyshev's inequality gives

$$P_i^\pi\left(\frac{|\int_{[T]}^T \bar{h}(x(y),\pi_y)\,dy|}{[T]} > \varepsilon\right) \leq \frac{E_i^\pi[\int_{[T]}^T |\bar{h}(x(y),\pi_y)|\,dy]^2}{\varepsilon^2[T]^2}$$

$$\leq \frac{16\|u^*\|_w^2[b_1^2+M_2^*(w_2^*(i)+\frac{b_2^*}{c_2^*})]}{\varepsilon^2[T]^2}. \tag{8.29}$$

Since $\sum_{[T]=1}^{\infty} \frac{1}{[T]^2} < \infty$, by (8.29) and the Borel–Cantelli lemma (see Proposition A.17), we have

$$P_i^\pi\left(\limsup_{[T]}\left\{\frac{|\int_{[T]}^T \bar{h}(x(y),\pi_y)\,dy|}{[T]} > \varepsilon\right\}\right) = 0.$$

Now let

$$D_{[T]} := \left\{\frac{|\int_{[T]}^T \bar{h}(x(y),\pi_y)\,dy|}{[T]} > \varepsilon\right\} \in \mathcal{B}(\Omega)$$

(recalling Theorem 2.3), $D := \limsup_{[T]} D_{[T]} \in \mathcal{B}(\Omega)$, and let $D^c$ denote the complement of any set $D$. Then $P_i^\pi(D^c) = 1$. Let $z \in D^c$, which means that $z$ is in finitely many sets $D_{[T]}$. Hence, there exists an integer $N_0(z)$ (depending on $z$) such that $z \notin D_{[T]}$ for all $[T] \geq N_0(z)$, i.e.,

$$\frac{|\int_{[T]}^T \bar{h}(x(y),\pi_y)\,dy|}{[T]} \leq \varepsilon \quad \text{for all } [T] \geq N_0(z) \text{ and } z \in D^c,$$

which, together with $P_i^\pi(D^c) = 1$, yields

$$\lim_{[T]\to\infty} \frac{\int_{[T]}^T \bar{h}(x(y),\pi_y)\,dy}{[T]} = 0 \quad P_i^\pi\text{-a.s.} \tag{8.30}$$

Similarly, we have

$$\lim_{[T]\to\infty} \frac{u^*(x(T))}{[T]} = \lim_{[T]\to\infty} \frac{u^*(x([T]))}{[T]} = 0 \quad P_i^\pi\text{-a.s.} \tag{8.31}$$

Since $\lim_{T\to\infty} \frac{[T]}{T} = 1$, by (8.27), (8.28), (8.30), and (8.31), we have

$$\lim_{T\to\infty} \frac{M(T,\pi)}{T} = 0 \quad P_i^\pi\text{-a.s.} \tag{8.32}$$

Moreover, it follows from (8.17)–(8.19) that

$$M(t,\pi) = -\int_0^t r\big(x(y),\pi_y\big)\,dy + \int_0^t \Delta\big(x(y),\pi_y\big)\,dy - u^*\big(x(t)\big) + tg^*. \tag{8.33}$$

By (8.14), (8.17), and (2.5) we have $\Delta(i,\pi_t) \leq 0$ and $\Delta(i, f^*(i)) = 0$ for all $t \geq 0$ and $i \in S$. Thus, by (8.31), (8.32), and (8.33) we obtain

$$P_i^\pi\big(V_p(i,\pi) \leq g^*\big) = 1 \quad \text{and} \tag{8.34}$$

$$P_i^{f^*}\big(V_p(i,f^*) = g^*\big) = 1, \tag{8.35}$$

which, together with the arbitrariness of $\pi$ and $i$, give (b).

(c) By (b), it suffices to prove that $f$ ($\in F$) realizes the maximum in (8.14) if $f$ is PAR-optimal. To prove this by contradiction, suppose that $f$ is PAR-optimal but does not realize the maximum in (8.14). Then there exist some $i' \in S$ and a constant $\alpha(i', f) > 0$ (depending on $i'$ and $f$) such that

$$g^* \geq \left[r(i, f) + \alpha(i', f)\delta_{i'i}\right] + \sum_{j\in S} u^*(j)q(j|i, f) \quad \forall i \in S. \tag{8.36}$$

On the other hand, since $f$ is PAR-optimal, by (b) and (8.35) we have $V_p(i, f) = g^*$ $P_i^{\pi}$-a.s. for all $i \in S$. Moreover, as in the proof of (8.35), from the Poisson equation for $f$ (that is, $g(f) = r(i, f) + \sum_{j\in S} h_f(j)q(j|i, f)$), we also have $V_p(i, f) = g(f)$ $P_i^f$-a.s., and so

$$g^* = g(f) = \sum_{j\in S} r(j, f)\mu_f(j) \quad P_i^f\text{-a.s.} \tag{8.37}$$

Hence (as in the proof of (7.15)), integration of both sides of (8.36) with respect to $\mu_f$ yields

$$g^* \geq \sum_{j\in S}\left[r(j, f) + \alpha(i', f)\delta_{i'j}\right]\mu_f(j) = g^* + \alpha(i', f)\mu_f(i'),$$

which gives a contradiction because $\mu_f(i')$ and $\alpha(i', f)$ are both positive.

(d) Let $\Delta_{u'}(i, f(i)) := r(i, f) + \sum_{j\in S} u'(j)q(j|i, f) - g^*$, so that (8.15) becomes $\Delta_{u'}(i, f(i)) \geq -\varepsilon$ for all $i \in S$. Thus, as in the proof of (8.34), we have

$$P_i^f\left(V_p(i, f) \geq g^* - \varepsilon\right) = 1,$$

which, together with (b), gives (d). □

Recalling the terminology introduced at the end of Sect. 7.3, Theorem 8.5(a) establishes the AROE (8.14) and the existence of a *canonical policy* $f^*$, that is, $f^*$ is in $F_{ca}$. Furthermore, Theorem 8.5(c) shows that a policy $f \in F$ is canonical if and only if $f$ is PAR-optimal.

*Remark 8.6*

(a) Under Assumptions 8.2 and 8.3, it follows from Theorems 8.5 and 7.8 that the canonical policy $f^*$ in Theorem 8.5(a) is also optimal for the *expected* AR criterion. However, it has been shown in Example 5.2 that an optimal stationary policy for the expected AR criterion may *not* be canonical. Therefore, it is natural to *guess* that a PAR-optimal stationary policy may *not* be canonical either. This issue, however, remains unsolved to this date.
(b) In the proof of the "only if part" of Theorem 8.5(c), we cannot obtain (8.37) by using the dominated convergence theorem because $V(i, f)$ is defined as a "lim inf" (see (8.2)) instead of "lim".

When the transition rates and the reward function are *uniformly bounded*, we need to impose conditions only on the discrete-time *embedded* Markov chains to guarantee the existence of PAR-optimal stationary policies. This is stated in the following corollary.

**Corollary 8.7** *Suppose that the following conditions* (i)–(iii) *are satisfied.*

(i) $\|q\| := \sup_{i\in S} q^*(i) < \infty$ *and* $\|r\| := \sup_{i\in S, a\in A(i)} |r(i,a)| < \infty$.
(ii) *For each* $i \in S$, $A(i)$ *is compact, and* $r(i,a)$ *and* $q(j|i,a)$ *are continuous in* $a \in A(i)$ *for any fixed* $i, j \in S$.
(iii) *Either* $\inf_{i\neq j_0, a\in A(i)} q(j_0|i,a) > 0$ *for some* $j_0 \in S$ *or* $\sum_{j\in S} \sup_{i\in S, a\in A(i)} \times (\frac{q(j|i,a)}{\|q\|} + \delta_{ij}) < 2$.

*Then,*

(a) *There exists a PAR-optimal stationary policy.*
(b) *For each* $\varepsilon > 0$, *an* $\varepsilon$*-PAR-optimal stationary policy can be obtained in a finite number of steps of the Policy Iteration Algorithm* 7.1.

*Proof* For $k = 1, 2$, define the maps $T_k$ on the set $B(S)$ of *bounded functions* on $S$ as

$$T_k u(i) := \sup_{a\in A(i)} \left\{ \frac{r(i,a)}{\|q\|+1} + \sum_{j\in S} u(j) \left[ \left( \frac{q(j|i,a)}{\|q\|+1} + \delta_{ij} \right) - \mu_k(j) \right] \right\} \tag{8.38}$$

for all $i \in S$ and $u \in B(S)$, where the measures $\mu_k$ on $S$ are given by

$$\mu_1(j) := \inf_{i\in S, a\in A(i)} \left[ \frac{q(j|i,a)}{\|q\|+1} + \delta_{ij} \right], \quad \text{and} \quad \mu_2(j) := \sup_{i\in S, a\in A(i)} \left[ \frac{q(j|i,a)}{\|q\|+1} + \delta_{ij} \right]$$

for $j \in S$, which are related, respectively, to the first and the second hypotheses in condition (iii). Thus, the maps $T_1$ and $T_2$ are both contractive with the contraction factors $\beta_1$ and $\beta_2$ given by

$$0 < \beta_1 := 1 - \sum_{j\in S} \mu_1(j) < 1 \quad \text{and} \quad 0 < \beta_2 := \sum_{j\in S} \mu_2(j) - 1 < 1. \tag{8.39}$$

Hence, Banach's fixed point theorem and condition (ii) give the existence of $u^* \in B(S)$, $f^* \in F$, and a unique constant $g^*$ satisfying (8.14). Then, as in the proof of Theorem 8.5(b) and (d), we obtain Corollary 8.7. □

*Remark 8.8* The two hypotheses in condition (iii) of Corollary 8.7 are variants of the ergodicity condition in Hernández-Lerma (1989) [69] for discrete-time MDPs. Each of these conditions, together with Assumption 7.4, implies that the embedded Markov chain with transition probability $(\frac{q(j|i,f(i))}{1+\|q\|} + \delta_{ij})$ has a unique invariant probability measure; see Hernández-Lerma (1989) [69, p. 56], for instance. Observe that the "stochastic monotonicity condition" in Proposition 7.6(i) and condition (iii) in Corollary 8.7 are quite different, although both yield some form of ergodicity.

## 8.5 Policy and Value Iteration Algorithms

From Assumption 7.4 and Theorem 8.5(c) we immediately obtain the following fact.

**Proposition 8.9** *Suppose that Assumptions* 8.2 *and* 8.3 *hold. Then any limit point* $f^*$ *of the sequence* $\{f_n\}$ *obtained by Policy Iteration Algorithm* 7.1 *is PAR-optimal.*

Under the conditions in Corollary 8.7, we now provide a *value iteration algorithm* to compute $\varepsilon$ ($>0$)-PAR-optimal stationary policies. It should be mentioned that, as in the proof of Corollary 8.7, the choice of $k=1$ (or 2) corresponds to the first (or the second) hypothesis in condition (iii) of Corollary 8.7. Thus, we will understand that $k$ in this algorithm is *fixed.*

**Value Iteration Algorithm 8.1** Let $T_k$ be as in (8.38).

*Step I.* Fix an arbitrary $\varepsilon>0$ and take arbitrarily $u_0 \in B(S)$.
*Step II.* If $T_k u_0 = u_0$, then obtain a policy $f$ in $F$ satisfying that

$$r\bigl(i, f(i)\bigr) + \sum_{j\in S} u_0(j) q\bigl(j|i, f(i)\bigr) = \sup_{a\in A(i)} \biggl\{ r(i,a) + \sum_{j\in S} u_0(j) q(j|i,a) \biggr\} \quad \forall i \in S$$

and stop; $f$ is PAR-optimal (by Theorem 8.5(b)); otherwise, calculate a positive integer $N \ge \frac{1}{\beta_k} \ln \frac{\varepsilon(1-\beta_k)}{4(1+\|q\|)\|u_1-u_0\|} + 1$, with $\beta_k$ as in (8.39), and let $u_N := T_k^N u_0 = T_k(T_k^{N-1} u_0)$.
*Step III.* Choose $f_\varepsilon(i) \in A(i)$ such that for each $i \in S$,

$$r\bigl(i, f_\varepsilon(i)\bigr) + \sum_{j\in S} u_N(j) q\bigl(j|i, f_\varepsilon(i)\bigr) \ge \sup_{a\in A(i)} \biggl\{ r(i,a) + \sum_{j\in S} u_N(j) q(j|i,a) \biggr\} - \frac{\varepsilon}{2}.$$

From Corollary 8.7 (which shows the existence of a solution to the AROE (8.14)) and Theorem 8.5(d) we have the following fact.

**Proposition 8.10** *Under conditions* (i)–(iii) *in Corollary* 8.7, *the policy* $f_\varepsilon$ *obtained by Value Iteration Algorithm* 8.1 *is* $\varepsilon$*-PAR-optimal.*

With respect to Policy Iteration Algorithm 7.1, if we luckily choose an initial policy such that this algorithm stops after a *finite* number of iterations, then

Proposition 8.9 shows that a PAR-optimal stationary policy can be computed. Otherwise, since the policy space $F$ may be infinite, the algorithm may not stop in any finite number of iterations. In this case, Proposition 8.9 shows that a PAR-optimal stationary policy can be approximated. On the other hand, Proposition 8.10 implies that, under the conditions in Corollary 8.7, an $\varepsilon$-PAR-optimal stationary policy can indeed be computed in a finite number of iterations for any given $\varepsilon > 0$.

## 8.6 An Example

In this section, we apply our results to a birth-and-death system.

*Example 8.2* (PAR optimality for a controlled birth-and-death system) Consider a controlled birth-and-death system in which the state variable denotes the population size at any time $t \geq 0$. There are natural birth and death rates denoted by *positive* constants $\lambda$ and $\mu$, respectively, as well as *nonnegative* emigration and immigration parameters. These two parameters are assumed to be controlled by a decision-maker and denoted by $h_1(i, a_1)$ and $h_2(i, a_2)$, respectively, which may depend on the system's states $i$ and decision variables $a_1$ and $a_2$ taken by the decision-maker. When the system is at state $i \in S := \{0, 1, \ldots\}$, the decision-maker takes an action $a := (a_1, a_2)$ from a *compact* set $A(i) := A_1(i) \times A_2(i)$ of available actions. These actions may increase/decrease the emigration parameter $h_1(i, a_1)$ and incur a cost $c(i, a_1)$, and also increase/decrease the immigration parameter $h_2(i, a_2)$ and give a reward $\bar{r}(i, a_2)$. We also suppose that the benefit by an individual is represented by $p > 0$. Then the *net* income in this system is $r(i, a) := pi + \bar{r}(i, a_2) - c(i, a_1)$ for all $i \in S$ and $a = (a_1, a_2) \in A(i)$. On the other hand, if there are no individuals in the system (i.e., $i = 0$), it is impossible to decrease/increase the emigration rate, and so we have $h_1(0, a_1) \equiv 0$ for all $a_1 \in A_1(0)$. Also, in this case, $i = 0$, we may assume that the decision-maker hopes to increase the immigration rate, and then $h_2(0, a_2) > 0$ for all $a_2 \in A_2(0)$. (The latter assumption guarantees the irreducibility condition in Assumption 7.4.)

We now formulate this system as a continuous-time MDP. The corresponding transition rates $q(j|i, a)$ and reward function $r(i, a)$ are given as follows.

For $i = 0$ and each $a = (a_1, a_2) \in A(0)$,

$$q(1|0, a) = -q(0|0, a) := h_2(0, a_2) > 0, \qquad q(j|0, a) = 0 \quad \text{for } j \geq 2,$$

and, for $i \geq 1$ and each $a = (a_1, a_2) \in A(i)$,

$$q(j|i, a) := \begin{cases} \mu i + h_1(i, a_1) & \text{if } j = i - 1, \\ -(\mu + \lambda)i - h_1(i, a_1) - h_2(i, a_2) & \text{if } j = i, \\ \lambda i + h_2(i, a_2) & \text{if } j = i + 1, \\ 0 & \text{otherwise;} \end{cases} \tag{8.40}$$

$$r(i,a) := pi + \bar{r}(i,a_2) - c(i,a_1) \quad \text{for } i \in S \text{ and } a = (a_1,a_2) \in A(i). \tag{8.41}$$

We aim to find conditions that ensure the existence of a PAR-optimal stationary policy. Since we wish to obtain Assumptions 8.2 and 8.3, we consider the following conditions:

**F₁**. (i) $\mu - \lambda > 0$.
(ii) Either $\kappa := \mu - \lambda + h_2^* - h_{1*} \le 0$, or $\mu - \lambda > |h_2^* - h_{1*}|$ when $\kappa > 0$, where $h_2^* := \sup_{i \ge 1, a_2 \in A_2(i)} h_2(i,a_2)$, $h_{1*} := \inf_{i \ge 1, a_1 \in A_1(i)} h_1(i,a_1)$.

**F₂**. For each fixed $i \in S$, the functions $h_1(i,\cdot), h_2(i,\cdot), c(i,\cdot)$, and $\bar{r}(i,\cdot)$ are all continuous.

**F₃**. (i) There exist positive constants $\bar{L}_k$ $(k = 1,2)$ such that $|c(i,a_1)| \le \bar{L}_1(i+1)$ and $|\bar{r}(i,a_2)| \le \bar{L}_2(i+1)$ for all $i \in S$ and $(a_1,a_2) \in A_1(i) \times A_2(i)$.
(ii) $\|h_k\| := \sup_{i \in S, a_k \in A_k(i)} |h_k(i,a_k)| < \infty$ for $k = 1,2$.

To further explain Example 8.2, we consider the special case of *birth-and-death processes with controlled immigration*. Consider a pest population in a region that may be isolated to prevent immigration. Let $\hat{L}$ denote the cost when immigration is always prevented, $\hat{a}$ denote the immigration rate without any control, and an action $a \in [0,1]$ denote the level of immigration prevented, where $\hat{L}$ and $\hat{a}$ are fixed positive constants. When the population size is $i \in S := \{0,1,\ldots\}$, an action $a$ from a set $A(i)$ consisting of available actions is taken. Then the cost $\hat{L}a$ is incurred, the immigration rate $(1-a)\hat{a}$ is permitted, and the evolution of the population depends on birth, death, and immigration with parameters $\lambda$, $\mu$, and $(1-a)\hat{a}$, respectively, where $\lambda$ and $\mu$ are given positive constants. Suppose that the damage caused by an individual is represented by $p > 0$. Then the reward function is of the form $r(i,a) := -pi - \hat{L}a$ for all $i \in S$ and $a \in A(i)$. Obviously, we have $A(i) := [0,1]$ for each $i \ge 1$. However, when there is no pest in the region (i.e., $i = 0$), to guarantee the irreducibility condition in Assumption 7.4, we need that $A(0) := [0,\beta]$ for some $\beta \in (0,1)$. (This, however, can be explained as follows: for the ecological balance of the region, the pest is not permitted to become extinct, and so the immigration rate $(1-\beta)\hat{a} > 0$ is left.) Using the notation in Example 8.2, for this model, we have $h_1 \equiv 0$ and $h_2(i,a_2) = (1-a)\hat{a}$ with $a_2 := a$ here. Hence, when $\mu - \lambda > \hat{a}$, conditions **F₁**, **F₂**, and **F₃** above are all satisfied, and we obtain the following.

**Proposition 8.11** *Under conditions* **F₁**, **F₂**, *and* **F₃**, *the above controlled birth-and-death system satisfies Assumptions* 8.2 *and* 8.3. *Therefore* (*by Theorem* 8.5), *there exists a PAR-optimal stationary policy, which can be computed* (*or at least approximated*) *by the Policy Iteration Algorithm* 7.1.

*Proof* We shall first verify Assumption 8.2. Let $w(i) := i + 1$ for all $i \in S$, and let

$$\rho := \frac{\mu - \lambda - h_2^* + h_{1*}}{2} = \mu - \lambda - \frac{\kappa}{2} > 0 \quad \text{when } \mu - \lambda > |h_2^* - h_{1*}|.$$

Then, for each $a = (a_1, a_2) \in A(i)$ with $i \geq 1$, by $\mathbf{F_1}$ and (8.40) we have

$$\begin{aligned}\sum_{j\in S} w(j)q(j|i,a) &= (\lambda-\mu)(i+1)+\mu-\lambda-h_1(i,a_1)+h_2(i,a_2)\\ &\leq -(\mu-\lambda)w(i)+\kappa\\ &\leq \begin{cases}-(\mu-\lambda)w(i) & \text{when } \kappa\leq 0,\\ -\rho w(i) & \text{when } \kappa>0 \text{ (and so } \rho>0).\end{cases}\end{aligned} \tag{8.42}$$

In particular, for $i = 0$ and each $a = (a_1, a_2) \in A(0)$, we obtain

$$\sum_{j\in S} w(j)q(j|0,a) = h_2(0,a_2) \leq -(\mu-\lambda)w(0)+b' = -\rho w(0)+b'-\frac{\kappa}{2}, \tag{8.43}$$

where $b' := \mu - \lambda + \|h_2\| > 0$.

By inequalities (8.42) and (8.43) we see that Assumption 7.1(a) holds with $c_1 := \mu - \lambda$ and $b_1 := b'$ when $\kappa \leq 0$, or $c_1 := \rho$ and $b_1 := b'$ when $\kappa > 0$. Furthermore, since $h_2(0, a_2) > 0$ for all $a_2 \in A_2(0)$, by (8.40) and $\mathbf{F_3}$(ii) we see that Assumption 7.1(b) is true. Also, by $\mathbf{F_3}$(i) and (8.41), we have

$$\left|r(i,a)\right| \leq pi+\bar{L}_1(i+1)+\bar{L}_2(i+1) \leq (p+\bar{L}_1+\bar{L}_2)w(i) \quad \forall i\in S \text{ and } a\in A(i),$$

which verifies Assumption 7.1(c). On the other hand, by Proposition 7.6 and the model's description, we see that Assumptions 6.8(a), 6.8(b), 7.4, and 7.5 are also true, and so Assumption 8.2 follows.

To verify Assumption 8.3, let

$$w_1^*(i) := i^2+1, \quad w_2^*(i) := i^4+1 \quad \forall i\in S. \tag{8.44}$$

Then

$$w^2(i) \leq M_1^* w_1^*(i), \quad \left[q^*(i)w(i)\right]^2 \leq M_2^* w_2^*(i) \quad \forall i\in S, \tag{8.45}$$

with $M_1^* := 3$ and $M_2^* := 8(\lambda + \mu + \|h_1\| + \|h_2\|)$. Moreover, for all $i \geq 1$ and $a = (a_1, a_2) \in A(i)$, by (8.40), (8.44), and $\mathbf{F_3}$(ii), we have

$$\begin{aligned}\sum_{j\in S} w_1^*(j)q(j|i,a) &= -2i\left[\mu i+h_1(i,a_1)\right]+\mu i+h_1(i,a_1)\\ &\quad +2i\left[\lambda i+h_2(i,a_2)\right]+\lambda i+h_2(i,a_2)\\ &\leq -2(\mu-\lambda)\left(i^2+1\right)+3\left(\mu+\lambda+\|h_1\|+\|h_2\|\right)i\\ &\leq -(\mu-\lambda)w_1^*(i)\end{aligned} \tag{8.46}$$

when $i \geq \frac{3(\mu+\lambda+\|h_1\|+\|h_2\|)}{\mu-\lambda} + 1 =: i_*$.

On the other hand, since $A(i)$ is assumed to be compact for each $i \in S$, by (8.40) and (8.44) we see that $\sum_{j\in S} w_1^*(j)q(j|i,a)$ and $(\mu - \lambda)w_1^*(i)$ are both bounded in

$a \in A(i)$ and $i \leq i_*$. Thus, by (8.46) there exists a positive constant $b_1^*$ such that

$$\sum_{j\in S} w_1^*(j)q(j|i,a) \leq -(\mu-\lambda)w_1^*(i) + b_1^* \quad \forall i \in S \text{ and } a \in A(i). \tag{8.47}$$

Also, for all $i \geq 1$ and $a \in A(i)$, by (8.40) and (8.44) we have

$$\begin{aligned}\sum_{j\in S} w_2^*(j)q(j|i,a) &\leq -2(\mu-\lambda)\bigl(i^4+1\bigr) - (\mu-\lambda)i^4 \\ &\quad + \varepsilon_3 i^3 + \varepsilon_2 i^2 + \varepsilon_1 i + \varepsilon_0,\end{aligned} \tag{8.48}$$

where the constants $\varepsilon_k$ $(k = 0, 1, 2, 3)$ are completely determined by $\lambda$, $\mu$, $\|h_1\|$, and $\|h_2\|$. Similarly, by (8.48) and (8.40), there exists a positive constant $b_2^*$ such that

$$\sum_{j\in S} w_2^*(j)q(j|i,a) \leq -(\mu-\lambda)w_2^*(i) + b_2^* \quad \forall i \in S \text{ and } a \in A(i).$$

This inequality, together with (8.47) and (8.45), verifies Assumption 8.3. □

It should be noted that in Example 8.2 the reward function and transition rates are *unbounded*; see (8.40) and (8.41).

## 8.7 Notes

In Chaps. 7 and 8, we have studied expected AR (EAR) and pathwise AR (PAR) optimality for continuous-time MDPs with possibly unbounded transition rates and reward functions. Under suitable conditions, we have shown the existence of a solution to the AROE and the existence of a *canonical policy* which is EAR or PAR optimal. We have also shown in Example 8.1 that *in general* the pathwise and the expected AR criteria are different. Hence, it is an interesting and challenging problem to show the difference between the two criteria under some given ergodicity condition. It also *remains open* to show that a PAR-optimal stationary policy is *not* necessarily canonical.

The PAR criterion has been studied by several authors (see, for instance, Arapostathis et al. (1993) [5], Cavazos-Cadena and Fernández-Gaucherand (1995) [26], Hernández-Lerma and Lasserre (1999) [74], and their extensive bibliographies), but, to the best of our knowledge, all the existing works deal with *discrete-time* MDPs.

The main content of this chapter is from Guo and Cao (2005) [52].

On the other hand, Corollary 8.7 can also be obtained by using the uniformization technique in Remark 6.1 (Sect. 6.1), which is based on the equivalence, under some conditions, between continuous- and discrete-time MDPs.

# Chapter 9
# Advanced Optimality Criteria

In previous chapters, we have introduced conditions ensuring the existence of discounted- and average-reward optimal policies. There are situations, however, in which we wish or need to go one step further and identify optimal policies with some *additional* property. In this chapter, we discuss some of these "advanced" optimality criteria.

*The model considered in this chapter is the same as that in Chap.* 7.

## 9.1 Bias and Weakly Overtaking Optimality

The overtaking optimality criterion deals with the asymptotic maximization of the total expected reward $V_T(i, f)$ as $T \to +\infty$, in the class of deterministic stationary policies $F$. The starting point is the well-known fact that the expected average-reward optimality criterion (2.21) is very underselective, because it is only concerned with the limiting behavior of the controlled stochastic process, disregarding its behavior during any finite time interval $[0, T]$.

To account for this, we introduce the weakly overtaking optimality criterion.

**Definition 9.1** A deterministic stationary policy $\hat{f} \in F$ is called *weakly overtaking optimal* if, for all $f \in F$, $i \in S$, and $\varepsilon > 0$, there exists a positive $T_0$ (which may depend on $f$, $i$, and $\varepsilon$) such that

$$V_T(i, \hat{f}) \geq V_T(i, f) - \varepsilon \quad \text{for all } T \geq T_0.$$

Thus, the policy $\hat{f}$ "overtakes" (up to an arbitrarily small constant) any policy $f$ from some time $T_0$ onwards. Obviously, a weakly overtaking optimal policy $\hat{f} \in F$ is average optimal. If Definition 9.1 holds for $\varepsilon = 0$, then we say that $\hat{f}$ is *overtaking optimal.*

Given a deterministic stationary policy $f \in F$, under Assumptions 7.1, 7.4, and 7.5, it follows from Corollary 7.9 and Proposition 7.11 that the gain of $f$

X. Guo, O. Hernández-Lerma, *Continuous-Time Markov Decision Processes*, Stochastic Modelling and Applied Probability 62,
DOI 10.1007/978-3-642-02547-1_9, © Springer-Verlag Berlin Heidelberg 2009

is

$$g(f)=\lim_{T\to\infty}\frac{1}{T}V_T(i,f)\quad \forall i\in S.$$

This implies that

$$V_T(i,f)=g(f)T+\mathrm{o}(T)\quad \forall i\in S, \tag{9.1}$$

where $\mathrm{o}(T)$ means that $\lim_{T\to\infty}\frac{\mathrm{o}(T)}{T}=0$.

Equation (9.1) shows that every AR optimal deterministic stationary policy $f^*\in F$ weakly overtakes (in the sense of Definition 9.1) any non-average optimal $f\in F$, because $g(f^*)>g(f)$.

Therefore, to obtain a weakly overtaking optimal policy, we must determine, among the class of AR optimal deterministic stationary policies, those with the largest growth in (9.1). To this end, we must analyze the residual term $\mathrm{o}(T)$ in (9.1), which will be done using the bias of $f\in F$.

We recall that the definition of the bias $h_f$ of a policy $f\in F$ and its characterization by means of the Poisson equation have been given in (7.18) and Proposition 7.11, respectively.

**Lemma 9.2** *If Assumptions* 7.1, 7.4, *and* 7.5 *are satisfied, then*

$$\sup_{f\in F}\left\| V_T(\cdot,f)-g(f)T-h_f\right\|_w=\mathrm{O}\left(e^{-\delta T}\right)\quad \textit{as } T\to+\infty,$$

*with* $\delta>0$ *as in Assumption* 7.5, *where* $\mathrm{O}(e^{-\delta T})$ *means that* $\frac{\mathrm{O}(e^{-\delta T})}{e^{-\delta T}}$ *is bounded in* $T\geq 0$.

*Proof* Given $f\in F$, by (2.6) and (7.20),

$$\begin{aligned} E_i^f h_f\left(x(T)\right)-h_f(i)&=E_i^f\left[\int_0^T\sum_{j\in S}h_f(j)q\left(j|x(t),f\right)dt\right]\\ &=g(f)T-V_T(i,f). \end{aligned}\tag{9.2}$$

Hence,

$$V_T(i,f)=g(f)T+h_f(i)-E_i^f h_f\left(x(T)\right).$$

The uniform $w$-exponential ergodicity hypothesis (Assumption 7.5) and the fact that $\mu_f(h_f)=0$ (Proposition 7.11) yield the stated result. □

From Lemma 9.2 the interpretation of the bias is now straightforward: the total expected reward $V_T(i,f)$ of a deterministic stationary policy $f\in F$ lines up, as $T$ grows, with a straight line with slope $g(f)$ and ordinate at the origin equal to $h_f(i)$. Therefore, weakly overtaking optimal policies will be those which are AR optimal and, in addition, have the larger bias. This suggests, naturally, the following definition of *bias optimality*.

**Definition 9.3** The optimal bias function $\hat{h} \in B_w(S)$ is defined as

$$\hat{h}(i) := \sup_{f \in \mathbf{F}_{\mathrm{ao}}} h_f(i) \quad \forall i \in S,$$

with $\mathbf{F}_{\mathrm{ao}}$ the set of AR optimal stationary policies. (By (7.19), $\hat{h}$ is in $B_w(S)$.)

In other words, an AR optimal deterministic stationary policy $f \in \mathbf{F}_{\mathrm{ao}}$ is bias optimal if $h_f = \hat{h}$.

The next result shows the existence of bias optimal and weakly overtaking optimal policies.

**Theorem 9.4** *Under Assumptions* 7.1, 7.4, *and* 7.5, *the following hold*:

(a) *There exists a bias optimal policy.*
(b) *The following statements are equivalent*:
   (i) $f \in F$ *is bias optimal.*
   (ii) $f \in F$ *is weakly overtaking optimal.*
   (iii) $f \in F$ *is AR optimal, and* $\mu_f(\hat{h}) = 0$.

*Proof* (a) Suppose that $(g^*, \bar{u})$ is a solution of the AROE (by Theorem 7.8), i.e.,

$$g^* = \max_{a \in A(i)} \left\{ r(i,a) + \sum_{j \in S} \bar{u}(j) q(j|i,a) \right\} \quad \forall i \in S,$$

and, for each $i \in S$, let $A_0(i) \subseteq A(i)$ be the set of actions attaining the maximum in the AROE, that is,

$$A_0(i) := \left\{ a \in A(i) : r(i,a) + \sum_{j \in S} \bar{u}(j) q(j|i,a) = g^* \right\}.$$

It follows from Assumption 7.1(d) that $A_0(i)$ is a nonempty compact set.

Now, let $f \in F$ be an AR optimal policy. In particular (by Theorem 7.8 and (7.20)), $f$ is canonical (i.e., $f(i) \in A_0(i)$ for all $i \in S$), and

$$r(i,f) + \sum_{j \in S} \bar{u}(j) q(j|i,f) = g^* = g(f) = r(i,f) + \sum_{j \in S} h_f(j) q(j|i,f) \quad \forall i \in S.$$

This implies, as in the proof of Proposition 7.11, that $\bar{u} - h_f$ is constant; in fact,

$$\bar{u} - h_f = \mu_f(\bar{u}) - \mu_f(h_f) = \mu_f(\bar{u}).$$

Thus, $h_f = \bar{u} - \mu_f(\bar{u})$, and taking the supremum over $f \in \mathbf{F}_{\mathrm{ao}}$, we obtain

$$\hat{h} = \bar{u} + \sup_{f \in \mathbf{F}_{\mathrm{ao}}} \mu_f(-\bar{u}). \tag{9.3}$$

Observe that we may, and will, interpret $\mu_f(-\bar{u})$ as the expected average reward (or gain) of the policy $f \in F$ when the reward function is $-\bar{u}$ in lieu of $r(f)$. Hence, maximizing $h_f$ within the class of canonical policies is equivalent to solving the following expected average reward MDP (compare with the continuous-time model at the beginning of Sect. 2.2):

$$\{S, (A_0(i), i \in S), q(j|i,a), -\bar{u}(i)\}.$$

This model of a continuous-time MDP satisfies Assumptions 7.1, 7.4, and 7.5, and thus (by Theorem 7.8) there exists a bias optimal deterministic stationary policy.

(b) (i) $\Leftrightarrow$ (ii) This equivalence directly follows from Definitions 9.1 and 9.3, and Lemma 9.2.

(i) $\Rightarrow$ (iii) If $f$ is bias optimal, then $h_f = \hat{h}$. Recalling that $\mu_f(h_f) = 0$ (Proposition 7.11), $\mu_f(\hat{h}) = 0$ follows.

(iii) $\Rightarrow$ (i) If $f \in F$ is AR optimal, then $f$ is canonical, and

$$r(i,f) + \sum_{j \in S} \bar{u}(j) q(j|i,f) = g^* = g(f) = r(i,f) + \sum_{j \in S} h_f(j) q(j|i,f) \quad \forall i \in S.$$

This shows that $\bar{u}$ and $h_f$ differ by a constant. By (9.3), $\hat{h}$ and $\bar{u}$ also differ by a constant. Thus $h_f$ and $\hat{h}$ differ by a constant, but, since $\mu_f(\hat{h}) = 0 = \mu_f(h_f)$, we necessarily have $h_f = \hat{h}$, and so $f$ is bias optimal. □

Theorem 9.4 provides an existence result, but it does not show how to determine a bias optimal (or weakly overtaking optimal) policy. To fill this gap, we define the so-called *bias optimality equations*.

We say that $(g, u, v) \in \mathbb{R} \times B_w(S) \times B_w(S)$ verifies the bias optimality equations if

$$g = \max_{a \in A(i)} \left\{ r(i,a) + \sum_{j \in S} u(j) q(j|i,a) \right\} \quad \forall i \in S, \tag{9.4}$$

$$u(i) = \max_{a \in A_0(i)} \left\{ \sum_{j \in S} v(j) q(j|i,a) \right\} \quad \forall i \in S, \tag{9.5}$$

where $A_0(i)$ is the set of actions $a \in A(i)$ attaining the maximum in (9.4).

Observe that the first optimality equation (9.4) corresponds to the AROE. Also note that the maxima in (9.4) and (9.5) are attained as a consequence of the continuity–compactness requirements in Assumption 7.1(d).

**Theorem 9.5** *Suppose that Assumptions* 7.1, 7.4, *and* 7.5 *are all verified. Then the following results hold.*

(a) *There exist solutions to the bias optimality equations* (9.4)–(9.5).
(b) *If* $(g, u, v)$ *is a solution to the bias optimality equations, then* $g = g^*$ *(the optimal average-reward function) and* $u = \hat{h}$ *(the optimal bias).*

(c) *A necessary and sufficient condition for a policy $f \in F$ to be bias optimal (or weakly overtaking optimal) is that $f(i)$ attains the maximum in the bias optimality equations* (9.4)–(9.5) *for every $i \in S$.*

*Proof* (a) By (9.3) and Theorem 7.8 we see that $(g^*, \hat{h})$ satisfies (9.4). For a fixed solution $(g^*, \bar{u})$ of the AROE (9.4), it follows from (9.3) that $\hat{g} = \hat{h}(i) - \bar{u}(i)$ for all $i \in S$, where

$$\hat{g} := \sup_{f \in \mathbf{F}_{ao}} \mu_f(-\bar{u}). \tag{9.6}$$

The AROE corresponding to (9.6) is

$$\hat{g} = \max_{a \in A_0(i)} \left\{ -\bar{u}(i) + \sum_{j \in S} v(j) q(j|i,a) \right\} \quad \text{for all } i \in S \text{ and some } v \in B_w(S), \tag{9.7}$$

and this shows that $(\hat{h}, v)$ satisfies (9.5).

(b) Let $(g, u, v)$ be a solution of the bias optimality equations (i.e., (9.4) and (9.5)). Then, by Theorem 7.8 we have that $g = g^*$ and that the solution $\hat{g}$ to (9.7) is unique and, necessarily, $u = \hat{g} + \bar{u} = \hat{h}$.

(c) To prove the sufficiency condition, observe that if $f(i) \in A(i)$ attains the maximum in (9.4)–(9.5) for all $i \in S$, then $f$ is canonical, and, also,

$$\hat{h}(i) = \sum_{j \in S} v(j) q(j|i, f) \quad \forall i \in S.$$

This fact, together with Corollary 7.9, yields $\mu_f(\hat{h}) = 0$, and so it follows from Theorem 9.4 that $f$ is bias optimal. The proof of the necessity is straightforward and is omitted. □

## 9.2 Sensitive Discount Optimality

It is worth stressing that the two main optimality criteria for infinite-horizon continuous-time MDPs are the (total) expected discounted reward and the expected average reward.

When considering the discounted-reward optimality criterion, preponderance is given to the early periods of the infinite time horizon, because the discount factor $e^{-\alpha t}$ makes negligible the rewards earned at large times. On the contrary, for the average-reward criterion the rewards earned during a finite time interval, say $[0, t]$ for any fixed $t$, are negligible because $J_t(i, \pi)/t \to 0$ as $t \to +\infty$.

It is then intuitively clear that a "better" optimality criterion should be a compromise between these two extreme criteria. In particular, a neutral optimality criterion would be the total expected reward over the infinite time horizon $[0, +\infty)$, defined as

$$J^{\infty}(i, \pi) := E_i^{\pi} \left[ \int_0^{\infty} r\big(x(t), \pi_t\big)\, dt \right].$$

But, of course, unless we consider a very restrictive control model, $J^\infty(i,\pi)$ is likely to be infinite. Nevertheless, observe that (under some condition)

$$E_i^\pi\left[\int_0^\infty r\big(x(t),\pi_t\big)\,dt\right] = \lim_{T\to+\infty} E_i^\pi\left[\int_0^T r\big(x(t),\pi_t\big)\,dt\right]$$
$$= \lim_{\alpha\downarrow 0} E_i^\pi\left[\int_0^\infty e^{-\alpha t} r\big(x(t),\pi_t\big)\,dt\right],$$

or, equivalently,

$$J^\infty(i,\pi) = \lim_{T\to+\infty} V_T(i,\pi) = \lim_{\alpha\downarrow 0} V_\alpha(i,\pi).$$

Thus, it looks like there are two different approaches to overcome the fact that the total expected reward $J^\infty(i,\pi)$ is infinite: Either by analyzing the asymptotic behavior of the MDP as the time horizon grows (this has been achieved with the weakly overtaking optimality criterion in the previous section) or by letting the discount factor of a discount MDP converge to 0. In this section, we will deal with the latter approach.

For this approach, the standard tool is the so-called Laurent series expansion (as a power series in $\alpha > 0$) of the discounted reward of a policy $f \in F$, namely,

$$V_\alpha(i,f) = \sum_{k=-1}^{\infty} \alpha^k h_f^k(i) \quad \text{for small } \alpha > 0.$$

The Laurent series technique for MDPs was first used by Veinott (1969) [151] (see also Miller and Veinott (1969) [118]). It is based on the Laurent series expansion in functional analysis (see, e.g., Kato (1966) [97]) and is also related to the series expansion of the resolvent of a Markov process (see, e.g., Hernández-Lerma (1994) [70], Wang and Yang (1992) [154]). Laurent series expansions for continuous-time models are proposed by Prieto-Rumeau and Hernández-Lerma (2005a) [123], Puterman (1974) [128], Taylor (1976) [148], Sladký (1978) [144], and Veinott (1969) [151].

The usual technique to analyze the limiting behavior of the discounted reward $V_\alpha(i,f)$ when $\alpha\downarrow 0$ is to analyze the convergence rate of $V_\alpha(i,f)$ as $\alpha\downarrow 0$. To this end, we study the truncation of the Laurent series

$$\sum_{-1\le k\le n} \alpha^k h_f^k(i) \tag{9.8}$$

and then try to "asymptotically" maximize these truncations as $\alpha\downarrow 0$. This leads to the *sensitive discount optimality criteria*, which we analyze in this section.

The sensitive discount optimality criteria (defined by Veinott (1969) [151]) have interesting connections to some of the criteria analyzed so far: namely, gain and bias optimality. Also, they are characterized by means of a nested system of "bias-like" optimality equations.

When dealing with Blackwell optimality in Sect. 9.3 below, in addition to $\alpha\downarrow 0$ in (9.8), we will let $n\to\infty$.

We now give some concepts about $n$-discount optimality.

**Definition 9.6** Fix an integer $n \geq -1$.

(a) $f^* \in F$ is $n$-discount optimal (in $F$) if

$$\liminf_{\alpha \downarrow 0} \alpha^{-n}\left(V_\alpha\left(i, f^*\right) - V_\alpha(i, f)\right) \geq 0 \quad \text{for all } i \in S \text{ and } f \in F.$$

(b) $f^* \in F$ is strong $n$-discount optimal if

$$\liminf_{\alpha \downarrow 0} \alpha^{-n}\left(V_\alpha\left(i, f^*\right) - V_\alpha^*(i)\right) = 0 \quad \text{for all } i \in S.$$

The difference between $n$-discount and strong $n$-discount is obvious, and something is missing.

In our next result, we consider a deterministic stationary policy $f \in F$ and a function $u \in B_w(S)$. We will show that the expected discounted reward of $f$ when the reward rate is $u$, defined as

$$V_\alpha(i, f, u) = E_i^f\left[\int_0^\infty e^{-\alpha t} u\big(x(t)\big)\, dt\right] \quad \text{for } i \in S, \tag{9.9}$$

can be expanded as a power series in $\alpha$. This is usually known as the Laurent series expansion of $V_\alpha(i, f, u)$; see Theorem 9.7(a) below.

We define the operator $H_f : B_w(S) \to B_w(S)$ as

$$(H_f u)(i) := E_i^f\left[\int_0^\infty \left(u\big(x(t)\big) - \mu_f(u)\right) dt\right]. \tag{9.10}$$

It follows from Assumption 7.5 that $H_f u$ is in $B_w(S)$ because $\|H_f u\|_w \leq \frac{L_2}{\delta}\|u\|_w$. Also, $H_f$ can be interpreted as the "bias operator", because if $u \equiv r(f)$, with $r(f)(i) := r(i, f(i))$, then $H_f u$ is the bias of $f$; see (7.18). (Recall Remark 3.2: in our present context, "bias" and "potential" are synonymous.)

**Theorem 9.7** *Suppose that Assumptions* 7.1, 7.4, *and* 7.5 *hold, and let* $f \in F$ *and* $u \in B_w(S)$.

(a) *If* $0 < \alpha < \delta$ *(the constant* $\delta > 0$ *is taken from Assumption* 7.5), *then*

$$V_\alpha(i, f, u) = \frac{1}{\alpha}\mu_f(u) + \sum_{k=0}^{\infty}(-\alpha)^k\left(H_f^{k+1} u\right)(i) \quad \text{for all } i \in S, \tag{9.11}$$

*and the series converges in the* $w$*-norm.*

(b) *Define the* $k$*-residual of the Laurent series* (9.11) *as*

$$R_k(f, u, \alpha) := \sum_{n=k}^{\infty}(-\alpha)^n H_f^{n+1} u.$$

*Then, for all* $0 < \alpha \leq \delta' < \delta$ *and* $k = 0, 1, \ldots,$

$$\sup_{f \in F} \left\| R_k(f, u, \alpha) \right\|_w \leq (\alpha/\delta)^k \frac{L_2}{\delta - \delta'} \|u\|_w.$$

*Proof* (a) Pick arbitrary $f \in F$ and $u \in B_w(S)$. Let

$$\left(G_t^f u\right)(i) := F_t^f u(i) - \mu_f(u), \quad \text{for } i \in S \text{ and } t > 0,$$

where $F_t^f u(i) := E_i^f u(x(t))$. Notice that $F_t^f : u \mapsto F_t^f u$ is a linear operator from $B_w(S)$ into itself, and so is the "constant" mapping $\mu_f : u \mapsto \mu_f(u)$. Therefore, interpreting $G_t^f G_{t'}^f$ as a composition of operators $G_t^f$ and $G_{t'}^f$, we have

$$G_{t+t'}^f = G_t^f G_{t'}^f.$$

In fact, by the invariance of $\mu_f$ we have $\mu_f(F_{t'}^f u) = \mu_f(u)$ and also (by the Chapman–Kolmogorov equation)

$$\begin{aligned} G_t^f G_{t'}^f u &= \left(F_t^f - \mu_f\right)\left(F_{t'}^f - \mu_f\right) u \\ &= F_t^f \left(F_{t'}^f u\right) - \mu_f\left(F_{t'}^f u\right) - F_t^f \left(\mu_f(u)\right) + \mu_f\left(\mu_f(u)\right) \\ &= F_{t+t'}^f u - \mu_f(u) - \mu_f(u) + \mu_f(u) = G_{t+t'}^f u. \end{aligned}$$

The rest of the proof of (a) proceeds in two steps. First, notice that

$$\begin{aligned} V_\alpha(i, f, u) &= \int_0^\infty e^{-\alpha t} E_i^f \left[u\left(x(t)\right)\right] dt \\ &= \frac{1}{\alpha}\mu_f(u) + \int_0^\infty e^{-\alpha t} E_i^f \left[u\left(x(t)\right) - \mu_f(u)\right] dt. \end{aligned} \tag{9.12}$$

We will show that the integral in the right-hand side of (9.12) can be expressed as

$$\begin{aligned} \int_0^\infty e^{-\alpha t} E_i^f \left[u\left(x(t)\right) - \mu_f(u)\right] dt &= \int_0^\infty e^{-\alpha t} \left(G_t^f u\right)(i)\, dt \\ &= \sum_{k=0}^\infty (-\alpha)^k \int_0^\infty \frac{t^k}{k!} \left(G_t^f u\right)(i)\, dt, \end{aligned} \tag{9.13}$$

and the series converges in $w$-norm. In the second step, we show (recalling (9.10)) that

$$\int_0^\infty \frac{t^k}{k!} \left(G_t^f u\right)(i)\, dt = \left(H_f^{k+1} u\right)(i) \quad \forall i \in S \text{ and } k = 0, 1, \ldots, \tag{9.14}$$

which, together with (9.12) and (9.13), gives (a).

*Proof of* (9.13): For all $u \in B_w(S)$ and $k = 0, 1, \ldots,$ let

$$M_k^f u(i) := \int_0^\infty \frac{(-\alpha t)^k}{k!} (G_t^f u)(i)\, dt \quad \text{for } i \in S.$$

By Assumption 7.5,

$$\left\| G_t^f u \right\|_w \le L_2 e^{-\delta t} \|u\|_w. \tag{9.15}$$

Thus, $M_k^f u$ is well defined on $S$ and satisfies

$$\left\| M_k^f u \right\|_w \le \frac{\alpha^k}{\delta^{k+1}} L_2 \|u\|_w.$$

It follows from $0 < \alpha < \delta$ that

$$\sum_{k=0}^n M_k^f u = \sum_{k=0}^n \int_0^\infty \frac{(-\alpha t)^k}{k!} (G_t^f u)\, dt$$

is a Cauchy sequence in the Banach space $B_w(S)$, and, moreover, as $n \to \infty$, it converges to the left-hand side of (9.13) because by (9.15), the integrand is dominated by the integrable function $L_2 \|u\|_w w(i) e^{-(\alpha+\delta)t}$. This completes the proof of (9.13).

*Proof of* (9.14): Let $(C_k^f u)(i)$ be the left-hand side of (9.14), i.e.,

$$u \mapsto C_k^f u := \int_0^\infty \frac{t^k}{k!} (G_t^f u)\, dt.$$

We will use induction to show that

$$C_k^f u = H_f^{k+1} u \quad \forall k = 0, 1, \ldots.$$

In fact, by the definition of $C_0^f u$ and (9.10), we have $C_0^f u = H_f u$. Suppose now that $C_{k-1}^f u = H_f^k u$ for some $k \ge 1$. Then

$$\begin{aligned}
(C_k^f u) &= \int_0^\infty \left[ \int_0^t \frac{s^{k-1}}{(k-1)!} ds \right] (G_t^f u)\, dt \\
&= \int_0^\infty \frac{s^{k-1}}{(k-1)!} \left[ \int_s^\infty (G_t^f u)\, dt \right] ds \quad \text{[Fubini's theorem]} \\
&= \int_0^\infty \frac{s^{k-1}}{(k-1)!} \left[ \int_0^\infty (G_s^f G_t^f u)\, dt \right] ds \quad [\text{by } G_{s+t}^f = G_s^f G_t^f] \\
&= \int_0^\infty \frac{s^{k-1}}{(k-1)!} G_s^f (H_f u)\, ds \quad \text{[by Fubini's theorem]} \\
&= C_{k-1}^f (H_f u) = H_f^{k+1} u \quad \text{[by the induction hypothesis]}.
\end{aligned}$$

This proves (9.14).

(b) By (9.14) and (9.15),

$$\left|\left(H_f^{k+1}\right)u(i)\right| \le L_2\|u\|_w w(i)\frac{1}{\delta^{k+1}} \quad \forall k=0,1,\ldots, \tag{9.16}$$

which, together with (a) and a direct calculation, gives (b). □

Note that Theorem 9.7 gives a Laurent series expansion for a continuous-time Markov process with unbounded state space $S$. Similar Laurent series expansions for a continuous-time Markov process, but on a *compact* state space, are proposed by Puterman (1974) [128], Sladký (1978) [144], Taylor (1976) [148], and Veinott (1969) [151]. For the discrete-time counterpart, the reader is referred to Dekker and Hordijk (1988 [33], 1992 [34]).

**Definition 9.8** Let $H$ be as in (9.10). For all $f \in F$ and $k \ge 0$, define

$$h_f^k := (-1)^k H_f^{k+1} r(f).$$

This function $h_f^k$ is called the $k$-bias of $f$.

Therefore, $h_f^0$ is the bias of $f$, and for $k \ge 1$, $h_f^k$ is the bias of $f$ when the reward rate is $-h_f^{k-1}$. With this notation, the Laurent series (9.11) with $u(i) = r(i,f)$ is

$$V_\alpha(i,f) = \frac{1}{\alpha}g(f) + \sum_{k=0}^{\infty} h_f^k(i)\alpha^k \quad \text{for } i \in S \text{ and } 0 < \alpha < \delta. \tag{9.17}$$

We know that the $\mu_f$-expectation of the bias of a policy $f \in F$ is 0 (see Proposition 7.11). Thus,

$$\mu_f\left(h_f^k\right) = 0 \quad \forall k \ge 0. \tag{9.18}$$

We will next prove that the coefficients of the Laurent series (9.17) are the solution of a system of linear equations, which is a generalization of the Poisson equation.

Given $f \in F$ and $n \ge 0$, consider the system of equations, for $i \in S$,

$$g = r(i,f) + \sum_{j\in S} h^0(j)q(j|i,f), \tag{9.19}$$

$$h^0(i) = \sum_{j\in S} h^1(j)q(j|i,f), \tag{9.20}$$

$$\vdots \qquad \vdots$$

$$h^n(i) = \sum_{j\in S} h^{n+1}(j)q(j|i,f), \tag{9.21}$$

where $g$ is a constant, and $h^0, h^1, \ldots, h^{n+1}$ are in $B_w(S)$. Equations (9.19), (9.20), ..., (9.21) are referred to as the $-1, 0, \ldots, n$ Poisson equations for $f$, respectively. When $n = \infty$, these equations form an infinite system of Poisson equations.

**Proposition 9.9** *Suppose that Assumptions* 7.1, 7.4, *and* 7.5 *hold. Then, given an integer $n \geq -1$ and a policy $f \in F$, the solution to the $-1, 0, \ldots, n$ Poisson equations* (9.19)–(9.21) *for $f$ are*

$$g = g(f), \qquad h^k = h_f^k \quad \text{for } 0 \leq k \leq n$$

*and $h^{n+1} = h_f^{n+1} + z$ for some constant $z$.*

*Proof* The proof is by induction. The case $n = -1$ corresponds to the result in Proposition 7.11. Assume now that the stated result holds for some $n \geq -1$.

First of all, let us prove that $h_f^{n+1}$ and $h_f^{n+2}$ verify the $(n+1)$th Poisson equation. Consider the MDP with the reward function $-h_f^{n+1}$. The average expected reward of the policy $f$ is $\mu_f(-h_f^{n+1}) = 0$ (see (9.18)), and the bias of $f$ is $h_f^{n+2}$. Therefore, the Poisson equation for this MDP is

$$h_f^{n+1}(i) = \sum_{j \in S} h_f^{n+2}(j) q(j|i, f) \quad \forall i \in S,$$

which is precisely the $(n+1)$th Poisson equation.

Now, let us show that the solutions $(h^{n+1}, h^{n+2})$ to the $(n+1)$th Poisson equation are necessarily of the form $h^{n+1} = h_f^{n+1}$ and $h^{n+2} = h_f^{n+2} + z$, for some constant $z$. Since $h^{n+1}(i) = \sum_{j \in S} h^{n+2}(j) q(j|i, f)$ for all $i \in S$, by Corollary 7.9 we have $\mu_f(h^{n+1}) = 0$. By the induction hypothesis,

$$h^{n+1} = h_f^{n+1} + z' \quad \text{for some constant } z'.$$

Taking the $\mu_f$-expectation in this equation, together with $\mu_f(h_f^{n+1}) = 0$, shows that $z' = 0$, and so $h^{n+1} = h_f^{n+1}$. The fact that $h^{n+2} = h_f^{n+2} + z$ follows now from Proposition 7.11. □

Recall that the AROE is, roughly speaking, derived from the Poisson equation by maximizing over $f \in F$. Similarly, from the $-1, 0, \ldots, n$ Poisson equations we derive the so-called *average-reward optimality equations* (AROEs).

The $-1, 0, \ldots, n$ AROEs are respectively defined, for all $i \in S$, as

$$g = \max_{a \in A(i)} \left\{ r(i, a) + \sum_{j \in S} h^0(j) q(j|i, a) \right\}, \tag{9.22}$$

$$h^0(i) = \max_{a \in A_0(i)} \left\{ \sum_{j \in S} h^1(j) q(j|i, a) \right\}, \tag{9.23}$$

$$\vdots \quad \vdots$$

$$h^n(i) = \max_{a \in A_n(i)} \left\{ \sum_{j \in S} h^{n+1}(j) q(j|i,a) \right\}, \tag{9.24}$$

for some constant $g$ and functions $h^0, \ldots, h^{n+1} \in B_w(S)$, where $A_0(i)$ is the set of actions attaining the maximum in (9.22), and for $1 \le k \le n$, $A_k(i)$ is the set of actions $a \in A_{k-1}(i)$ attaining the maximum in the $(k-1)$th equation.

Observe that (9.22) corresponds to the AROE and that (9.22)–(9.23) are the bias optimality equations (9.4)–(9.5).

**Proposition 9.10** *Suppose that Assumptions* 7.1, 7.4, *and* 7.5 *hold. Then there exist solutions* $g, h^0, \ldots, h^{n+1}$ *to the* $-1, 0, \ldots, n$ *AROEs. Moreover,* $g, h^0, \ldots, h^n$ *are unique, and* $h^{n+1}$ *is unique up to an additive constant.*

*Proof* The proof is by induction. The proof for the cases $n = -1$ and $n = 0$ follows from Theorems 7.8 and 9.5, respectively. Suppose now that the stated result holds for the $-1, \ldots, n$ AROEs, and let us prove it for the $(n+1)$th optimality equation.

It is worth noting that, by Assumption 7.1(d), the functions that we are maximizing in (9.22)–(9.24) are continuous, and thus $A(i) \supseteq A_0(i) \supseteq \cdots \supseteq A_n(i)$ are nonempty compact sets. In particular,

$$\bigcap_{n=0}^{\infty} A_n(i) \neq \emptyset \quad \text{for all } i \in S. \tag{9.25}$$

Using the induction hypothesis, we consider the control model

$$\{S, (A_n(i), i \in S), q(j|i,a), -h^n\}.$$

Then, from (9.24) we know that its optimal gain is 0 and, moreover, the corresponding bias optimality equations are (9.24) and

$$h^{n+1}(i) = \max_{a \in A_{n+1}(i)} \left\{ \sum_{j \in S} h^{n+2}(j) q(j|i,a) \right\} \quad \text{for } i \in S,$$

where $h^{n+1}$ is unique, and $h^{n+2}$ is unique up to an additive constant (see Theorem 7.8). In particular, this shows that the nonempty compact set $A_{n+1}(i) \subseteq A_n(i)$ is well defined because it does not depend on the particular solution of the $n$th optimality equation. □

As a consequence of Propositions 9.9 and 9.10, we have the following.

**Corollary 9.11** *For every* $f \in F$, *the infinite system of Poisson equations for* $f$ *has a unique solution*

$$\left(g(f), h_f^0, h_f^1, \ldots\right).$$

*Similarly, the set of all AROEs has a unique solution.*

In the next result, we characterize $n$-discount optimal policies in $F$ as those attaining the maximum in the $-1, 0, \dots, n$ AROEs and also as those lexicographically maximizing the first $n+1$ coefficients of the Laurent series (9.17) in the class $F$. First, we need some definitions.

**Definition 9.12**

(a) Given an integer $n \geq -1$, we define $F_n$ as the set of policies $f \in F$ such that $f(i) \in A_{n+1}(i)$ for every $i \in S$. (In particular, $F_{-1} \subset F$ is the set of canonical policies.)
(b) Given vectors $u$ and $v$ in $\mathbb{R}^d$ with $1 \leq d \leq \infty$, we say that $u$ is lexicographically greater than or equal to $v$, which is denoted by $u \succeq_{lg} v$, if the first nonzero component of $u - v$ (if it exists) is positive. If $u \succeq_{lg} v$ and $u \neq v$, we write $u \succ_{lg} v$.

**Theorem 9.13** *We suppose that Assumptions* 7.1, 7.4, *and* 7.5 *hold. Given* $n \geq -1$ *and* $f \in F$, *the following statements are equivalent.*

(i) $f \in F_n$.
(ii) $f$ *is n-discount optimal in* $F$.
(iii) $(g(f), h_f^0, \dots, h_f^n) \succeq_{lg} (g(f'), h_{f'}^0, \dots, h_{f'}^n)$ *for every* $f' \in F$.

*Proof of* (i) $\Leftrightarrow$ (iii) We will make the proof by induction. For $n = -1$ and $n = 0$, the result follows from Theorems 7.8 and 9.5. Suppose now that (i) $\Leftrightarrow$ (iii) holds for some $n \geq 0$, and let us prove it for $n + 1$. □

*Proof of* (i) $\Rightarrow$ (iii) *for* $n + 1$ As a consequence of Propositions 9.9 and 9.10, if $f \in F_{n+1}$, then

$$g = g(f), \qquad h^0 = h_f^0, \qquad \dots, \qquad h^{n+1} = h_f^{n+1},$$

where $g, h^0, \dots, h^{n+1}$ are the unique solutions of the AROEs. It also follows that $f$ is bias optimal for the MDP

$$\{S, (A_n(i), i \in S), (q(j|i,a)), -h^n\}, \tag{9.26}$$

and so $h_f^{n+1} \geq h_{f'}^{n+1}$ for all $f' \in F_n$.

Let $f' \in F$. If $f' \notin F_n$, then, by the induction hypothesis,

$$(g(f), h_f^0, \dots, h_f^n) \succ_{lg} (g(f'), h_{f'}^0, \dots, h_{f'}^n),$$

and thus

$$(g(f), h_f^0, \dots, h_f^n, h_f^{n+1}) \succeq_{lg} (g(f'), h_{f'}^0, \dots, h_{f'}^n, h_{f'}^{n+1}). \tag{9.27}$$

If $f' \in F_n$, then (by Propositions 9.9 and 9.10)

$$(g(f), h_f^0, \dots, h_f^n) = (g(f'), h_{f'}^0, \dots, h_{f'}^n),$$

and $f'$ is an admissible policy for the MDP (9.26), and it has smaller bias, that is, $h_f^{n+1} \geq h_{f'}^{n+1}$, and (9.27) follows. □

*Proof of* (iii) ⇒ (i) *for* $n+1$ We suppose that $f \in F$ lexicographically maximizes $(g(f), h_f^0, \ldots, h_f^{n+1})$. By the induction hypothesis, this implies that $f \in F_n$. Given $f' \in F_n$, the terms $g(f), h_f^0, \ldots, h_f^n$ of the Laurent series expansion of $f$ and $f'$ coincide; hence, $h_f^{n+1} \geq h_{f'}^{n+1}$. Therefore, $f$ is bias optimal for the MDP (9.26), and thus it attains the maximum in the corresponding bias optimality equations, or, equivalently, $f \in F_{n+1}$. □

*Proof of* (ii) ⇔ (iii) This equivalence is a direct consequence of Definition 9.6, Theorem 9.7(b) and (9.17). □

As a consequence of Theorem 9.13, a deterministic stationary policy is AR optimal if and only if it is $(-1)$-discount optimal in $F$, and it is bias optimal (or weakly overtaking optimal) if and only if it is 0-discount optimal in $F$.

So far, we have analyzed the $n$-discount optimality criterion in the class of deterministic stationary policies $F$. Now, we turn our attention to the *strong* sensitive discount optimality criteria in Definition 9.6. Our analysis of *strong* $n$-discount optimality is, however, restricted to the cases $n = -1$ and $n = 0$.

**Theorem 9.14** *Suppose that Assumptions* 7.1, 7.4, *and* 7.5 *are satisfied. A deterministic stationary policy is AR optimal if and only if it is strong* $(-1)$*-discount optimal.*

*Proof* Suppose that a policy $f \in F$ is AR optimal. Given $\alpha > 0$, by Theorem 6.10 we can let $f_\alpha^* \in F$ be an $\alpha$-discounted reward optimal policy. We have, for small $\alpha$ (using the notation introduced in Theorem 9.7(b)),

$$V_\alpha^*(i) = \frac{1}{\alpha} g(f_\alpha^*) + R_0(f_\alpha^*, r(f_\alpha^*), \alpha)(i)$$

and

$$V_\alpha(i, f) = \frac{1}{\alpha} g(f) + R_0(f, r(f), \alpha)(i).$$

Observing that the bound on the residuals of the Laurent series in Theorem 9.7(b) is uniform on $F$, it follows that

$$\lim_{\alpha \downarrow 0} \left[\alpha\left(V_\alpha^*(i) - V_\alpha(i, f)\right) - \left(g(f_\alpha^*) - g(f)\right)\right] = 0 \quad \text{for every } i \in S.$$

But, since $g(f) \geq g(f_\alpha^*)$ and $V_\alpha^*(i) \geq V_\alpha(i, f)$, strong $(-1)$-discount optimality of $f$ follows.

The converse (i.e., a strong $(-1)$-discount optimal deterministic stationary policy is AR optimal) is a direct consequence of Theorem 9.13. □

Fix any $i_0 \in S$. As a consequence of the proof of Theorem 7.8, we see that $\lim_{\alpha\downarrow 0}\alpha V_\alpha^*(i_0) = g^*$, where $g^*$ is the optimal gain, and that $(V_\alpha^*(i) - V_\alpha^*(i_0))$ is bounded in $\alpha > 0$ (for any fixed $i \in S$). Hence, $\lim_{\alpha\downarrow 0}\alpha V_\alpha^*(i) = g^*$ *for every* $i \in S$.

**Theorem 9.15** *Suppose that Assumptions* 7.1, 7.4 *and* 7.5 *hold. A deterministic stationary policy is bias optimal if and only if it is strong* 0*-discount optimal.*

To prove this theorem, we need two preliminary lemmas. Recall that the optimal gain and the optimal bias are denoted by the constant $g^*$ and a function $\hat{h} \in B_w(S)$, respectively. We define the *discrepancy function* $\Delta_0$ (sometimes referred to as Mandl's discrepancy function) as

$$\Delta_0(i,a) := r(i,a) - g^* + \sum_{j\in S}\hat{h}(j)q(j|i,a). \tag{9.28}$$

It follows from the AROE (9.22) and (9.3) that $\Delta_0 \le 0$, and, moreover, $\Delta_0(i, f(i)) = 0$ for all $i \in S$ if and only if $f \in F$ is AR optimal, by Theorem 7.8(b).

**Lemma 9.16** *Suppose that Assumptions* 7.1, 7.4, *and* 7.5 *are satisfied. Then, for a given* $f \in F$, *a discount factor* $\alpha > 0$, *and* $i \in S$,

$$V_\alpha(i,f) = \frac{1}{\alpha}g^* + \hat{h}(i) - \alpha V_\alpha(i,f,\hat{h}) + V_\alpha(i,f,\Delta_0).$$

*Proof* Let $L\hat{h}(i,a) := \sum_{j\in S} q(j|i,a)\hat{h}(j)$ for all $i \in S$ and $a \in A(i)$. Then, by (9.28) and the Dynkin formula (C.5) in Appendix C, we obtain

$$\begin{aligned} V_\alpha(i,f,\Delta_0) &= V_\alpha(i,f) - \frac{g^*}{\alpha} + E_i^f\left[\int_0^\infty e^{-\alpha t}(L\hat{h})\big(x(t),f\big)\,dt\right] \\ &= V_\alpha(i,f) - \frac{g^*}{\alpha} - \hat{h}(i) + \alpha V_\alpha(i,f,\hat{h}), \end{aligned}$$

which implies the desired result. □

**Lemma 9.17** *Under Assumptions* 7.1, 7.4, *and* 7.5, *if* $\{f_m\}_{m\ge 1}$ *is a sequence in* $F$ *that converges pointwise to* $f \in F$ *(i.e.,* $f_m(i) \to f(i)$ *as* $m \to \infty$ *for every* $i \in S$), *then*

$$\lim_{m\to\infty} g(f_m) = g(f).$$

*Proof* Recall that $h_f^0$ and $g(f)$ denote the bias and the AR of a policy $f \in F$, respectively. Then, for any subsequence $\{g_{f_m}\}$ that converges to some constant $\hat{g}$, by Lemma 7.2 and (7.19), we know that $\{h_{f_m}^0\}$ and $\{g_{f_m}\}$ are bounded sequences in $B_w(S)$. Therefore, a diagonal argument gives the existence of $u \in B_w(S)$ such that

$h^0_{f_{m'}}$ and $g_{f_{m'}}$ converge to $u$ and $\hat{g}$, respectively, through some subsequence $\{m'\}$. Moreover, by the Poisson equation (7.20) for $f_{m'}$, we have

$$g(f_{m'}) = r(i, f_{m'}) + \sum_{j \in S} h^0_{f_{m'}}(j) q(j|i, f_{m'}) \quad \forall i \in S.$$

Hence, letting $m' \to \infty$, Proposition A.4 gives

$$\hat{g} = r(i, f) + \sum_{j \in S} u(j) q(j|i, f) \quad \forall i \in S,$$

and so $\hat{g} = g(f)$ (by Corollary 7.9). Since every convergent subsequence of $\{g(f_n)\}$ converges to the same constant $g(f)$, it follows that $\lim_{n\to\infty} g(f_n) = g(f)$, and so the proof is complete. □

Now we proceed to prove Theorem 9.15.

*Proof of Theorem 9.15* Let us prove, first of all, that a bias optimal policy is strong 0-discount optimal. The proof is by contradiction. Hence, suppose that $f \in F$ is bias optimal (i.e., $g(f) \geq g(f')$ for all $f' \in F$, and $h^0_f \geq h^0_{f'}$ for all $f' \in F_{-1} = F_{ao}$), but it is not strong 0-discount optimal. Thus, there exist some state $i \in S$, a real number $\varepsilon > 0$, and a sequence of discount factors $\alpha$ decreasing to 0 such that

$$V_\alpha(i, f) - V_\alpha\big(i, f^*_\alpha\big) + \varepsilon \leq 0, \tag{9.29}$$

where $f^*_\alpha$ is $\alpha$-discount optimal (i.e., $V^*_\alpha(i) = V_\alpha(i, f^*_\alpha)$). Using a diagonal argument, we can prove that $f^*_\alpha$ converges pointwise to some $f^* \in F$ for some (not explicit in the notation) subsequence of $\alpha$'s.

Since $g(f)$ and $h^0_f = h_f$ are the gain and the bias of $f$, respectively, it follows from Theorem 9.7 that

$$\frac{1}{\alpha}\big(g(f) - g\big(f^*_\alpha\big)\big) + h^0_f(i) - h^0_{f^*_\alpha}(i) + \frac{\varepsilon}{2} \leq 0$$

for small $\alpha > 0$. This, together with (7.19) and the AR optimality of $f$, shows that $g(f^*_\alpha) \to g(f)$, and, as a consequence of Lemma 9.17, $g(f^*) = g(f)$; hence, $f^*$ is also AR optimal.

Combining the facts that $\Delta_0(i, f(i)) = 0$ and $\Delta_0(i, f^*_\alpha(i)) \leq 0$ with Lemma 9.16, we derive from (9.29) that

$$\begin{aligned}
&\alpha V_\alpha(i, f, -\hat{h}) - \alpha V_\alpha\big(i, f^*_\alpha, -\hat{h}\big) + \varepsilon \\
&\quad = V_\alpha\big(i, f^*_\alpha, \Delta_0\big) + V_\alpha(i, f) - V_\alpha\big(i, f^*_\alpha\big) + \varepsilon \\
&\quad \leq V_\alpha(i, f) - V_\alpha\big(i, f^*_\alpha\big) + \varepsilon \leq 0.
\end{aligned}$$

Again by Theorem 9.7, but this time for the reward function $-\hat{h} \in B_w(S)$, we have

$$\mu_f(-\hat{h}) - \mu_{f^*_\alpha}(-\hat{h}) + \frac{\varepsilon}{2} \leq 0 \tag{9.30}$$

for small $\alpha > 0$. Letting $\alpha \to 0$ in (9.30) and using Lemma 9.17, we obtain

$$\mu_f(-\hat{h}) - \mu_{f^*}(-\hat{h}) + \frac{\varepsilon}{2} \leq 0. \tag{9.31}$$

From the proof of Theorem 9.4 we know that the bias optimal policies are those maximizing the gain of $-\hat{h}$ within the class of AR optimal policies. In particular,

$$\mu_f(-\hat{h}) \geq \mu_{f^*}(-\hat{h}),$$

in contradiction with (9.31). As a consequence, we have shown that a bias optimal policy is strong 0-discount optimal.

The converse (a strong 0-discount optimal policy $f \in F$ is bias optimal) directly follows from Theorem 9.13. □

## 9.3 Blackwell Optimality

Recalling the discussion at the beginning of Sect. 9.2, where we pointed out the drawbacks of both the discount and AR optimality criteria, it follows that we should look for policies being asymptotically $\alpha$-discount optimal when the discount factor $\alpha$ converges to 0.

In this direction, Blackwell (1962) [14] proposed a new definition of optimality (subsequently known in the literature as strong Blackwell optimality). We say that $f \in F$ is *strong Blackwell optimal* if there exists $\alpha > 0$ such that $f$ is $\alpha'$-discount optimal for every $\alpha'$ with $0 < \alpha' < \alpha$. The existence of such policies was proved by Blackwell (1962) for a discrete-time MDP with finite state and action spaces (see also Miller and Veinott (1969) [118] and Veinott (1969) [151]) [76].

For infinite state and action spaces, however, the concept of strong Blackwell optimality is very restrictive, and a weaker optimality criterion (known simply as Blackwell optimality) is preferred: $\pi^* \in \Pi$ is *Blackwell optimal* if, for all $i \in S$ and $\pi \in \Pi$, there exists $\alpha(i, \pi)$ such that $V_\alpha(i, \pi^*) \geq V_\alpha(i, \pi)$ for every $0 < \alpha < \alpha(i, \pi)$. This is formalized in Definition 9.18 below.

**Definition 9.18** Let $\Pi' \subseteq \Pi$ be a class of policies. A policy $\pi^* \in \Pi'$ is said to be Blackwell optimal in $\Pi'$ if, for every $\pi \in \Pi'$ and every state $i \in S$, there exists $\alpha > 0$ (depending on both $\pi$ and $i$) such that

$$V_{\alpha'}\big(i, \pi^*\big) \geq V_{\alpha'}(i, \pi) \quad \text{for } 0 < \alpha' < \alpha.$$

For every $n \geq -1$, let $F_n$ be as in Definition 9.12(a). We define $F_\infty \subseteq F$ as

$$F_\infty := \bigcap_{n \geq -1} F_n,$$

which is nonempty by (9.25).

As a direct consequence of Theorem 9.13, we obtain the following result on the existence and characterization of Blackwell optimal policies in $F$.

**Theorem 9.19** *Suppose that Assumptions* 7.1, 7.4, *and* 7.5 *are satisfied. The following statements are equivalent.*

(i) $f \in F_\infty$.
(ii) $f \in F$ *lexicographically maximizes* $(g(f), h_f^0, h_f^1, \ldots)$ *in* $F$.
(iii) $f \in F$ *is Blackwell optimal in* $F$.

Observe that, even if there might be several Blackwell optimal policies in $F$, they are, to some extent, indistinguishable because they have the same Laurent series expansion

$$\frac{1}{\alpha}g^* + \sum_{n=0}^{\infty} \alpha^n h^n \quad \text{for small } \alpha > 0$$

(recall Corollary 9.11).

Under some additional hypotheses, we can prove that a policy $f \in F_\infty$ is also Blackwell optimal in the class $\Pi$ of *all* policies. The precise statement, whose proof can be found in Prieto-Rumeau (2006) [122], is as follows.

**Theorem 9.20** *In addition to Assumptions* 7.1, 7.4, *and* 7.5, *suppose that for all* $\pi \in \Pi$, $i \in S$, *and* $\varepsilon > 0$, *there exists a finite set* $C_\varepsilon \subset S$ *(that may depend also on* $\pi$ *and* $i$*) such that*

$$\sup_{t \geq 0} \left\{ \sum_{j \notin C_\varepsilon} w(j) p_\pi(s, i, t, j) \right\} \leq \varepsilon. \tag{9.32}$$

*Then a policy* $f \in F_\infty$ *is Blackwell optimal in* $\Pi$.

If the function $w$ in Assumption 7.1 is strictly increasing, then condition (9.32) is equivalent to: for every policy $\pi \in \Pi$ and every initial state $i \in S$, the random variables $\{w(x(t))\}_{t \geq 0}$ are uniformly integrable.

For a discussion on the hypotheses of Theorem 9.20, see Prieto-Rumeau (2006) [122]. The additional uniform integrability (or tightness-like) condition (9.32) is a continuous-time version of Assumption 2.6 in Hordijk and Yushkevich (1999) [80, 81], and it seems that the extension of Blackwell optimality from $F$ to $\Pi$ cannot be achieved without conditions of this type.

Finally, it is worth noting that there are other "advanced" optimality criteria for both average- and discounted-reward problems. See Cao and Zhang (2008) [24] and Hu (1996) [87], for instance.

## 9.4 Notes

The concept of bias optimality was introduced by Veinott (1966) [150], and it has been extensively studied for discrete-time MDPs. The interested reader is referred to Hernández-Lerma and Lasserre (1999) [74], Hilgert and Hernández-Lerma (2003)

[76], Hordijk and Yushkevich (2002) [82], Lewis and Puterman (2001) [107], Lewis and Puterman (2002a, 2002b) [108, 109], and Puterman (1994) [128, 129]. However, very few papers deal with bias optimality for continuous-time MDPs. In Puterman (1974) [128], bias optimality is implicitly analyzed for controlled diffusions in compact intervals. For countable state continuous-time MDPs, see Prieto-Rumeau and Hernández-Lerma (2006) [125].

On the other hand, the overtaking optimality criterion goes back to Ramsey (1928) [132], and it is often used in economics. Here we deal with the weakly overtaking optimality, which was introduced by Gale (1967) [45] and von Weiszäcker (1965) [153]; see also Leizarowitz (1996) [105].

For the relation existing between bias and weakly overtaking optimality, the interested reader is referred to Hernández-Lerma and Lasserre (1999) [74] and Puterman (1994) [129] for discrete-time control models and to Prieto-Rumeau and Hernández-Lerma (2006) [125] for continuous-time denumerable-state MDPs; this relation is also explored for continuous-time zero-sum Markov games in Prieto-Rumeau and Hernández-Lerma (2005b) [124].

The Blackwell optimality criterion has been extensively studied for discrete-time MDPs. The case of a discrete-time MDP with a denumerable state space was studied by Dekker and Hordijk either by imposing some conditions on the deviation matrix (Dekker and Hordijk (1988) [33]) or by making some recurrence assumptions (Dekker and Hordijk (1992) [34]); see also Lasserre (1988) [102]. For discrete-time MDPs with a Borel state space, the reader is referred to Yushkevich (1994) [163] for a denumerable action space and to Yushkevich (1997) [164] for a compact action space. A thorough analysis of Blackwell optimality for discrete-time MDPs is made by Hordijk and Yushkevich (1999) [80, 81], where they impose drift and uniform geometric ergodicity conditions similar to our Assumptions 7.1, 7.4, and 7.5.

In contrast, the analysis of Blackwell optimality for continuous-time models is notably underdeveloped, and only a few papers deal with this topic; e.g., Veinott (1969) [151] for a finite state space, Puterman (1974) [128] for a compact state space, and Prieto-Rumeau and Hernández-Lerma (2005) [123] and Prieto-Rumeau (2006) [122] for a denumerable state space.

In addition to the "natural" generalizations of Blackwell optimality to more general models (from discrete-time to continuous-time, from finite to infinite state and action spaces, etc.), another challenging issue is to prove the existence of Blackwell optimal policies in a "large" class of policies. More precisely, whereas proving the existence of a Blackwell optimal policy in the class $F$ of deterministic stationary policies (see, e.g., Hordijk and Yushkevich (1999a) [80] or Prieto-Rumeau and Hernández-Lerma (2005a) [123]) is somehow "direct", the extension to the class $\Pi$ of admissible (nonstationary) policies is more complicated (see, e.g., Hordijk and Yushkevich (1999b) [81] or Prieto-Rumeau (2006) [122]).

Our main reference for this chapter is Guo, Hernández-Lerma, and Prieto-Rumeau (2006) [65] and Prieto-Rumeau and Hernández-Lerma (2006) [123]. The discrete-time version of Theorem 9.15 can be found in Hilgert and Hernández-Lerma (2003).

Finally, concerning topics for further research, from our comments in this chapter it should be clear that the advanced optimality criteria are practically unexplored for many classes of continuous-time stochastic control problems. For instance, except for Puterman (1974) [128] and Jasso-Fuentes and Hernández-Lerma (2008, 2009) [89–91], there is nothing else on sensitive discount criteria for controlled diffusions and semi-Markov decision processes. Hence, advanced optimality criteria offer a fertile source of interesting research problems.

# Chapter 10
# Variance Minimization

In Chaps. 3, 5, and 7, we have studied expected average-reward optimality. To further identify AR optimal policies, we have studied some advanced optimality criteria in Chap. 9. In this chapter, we study a variance minimization criterion, in which a decision-maker wishes to choose a policy with maximal expected average rewards and minimum variance.

*The control model in this chapter is the same as that in Chap. 7.*

## 10.1 Introduction

In previous chapters, we have introduced conditions ensuring the existence of both expected and pathwise average-reward optimal policies. Furthermore, to identify average optimal policies with some *additional* property, we have studied some "advanced" optimality criteria in Chap. 9. There are several alternative optimality criteria that improve the average optimality criterion. Among them, we can mention the variance minimization criterion. This criterion selects an average optimal policy $\pi$ for which the limiting average variance of $\int_0^T r(x(t), \pi_t)\,dt$ is minimal over a class of AR optimal policies, that is, an AR optimal policy minimizing the limiting average variance

$$\sigma^2(i,\pi) := \lim_{T\to\infty} \frac{1}{T} E_i^\pi \left( \int_0^T r\big(x(t), \pi_t\big)\,dt - V_T(i,\pi) \right)^2$$

over a class of average optimal policies.

The variance minimization criterion is particularly appealing because it determines an average optimal policy whose pathwise reward on $[0, T]$ remains "close" to its expected value $V_T(i, \pi)$ in the long run. It is also somehow related to Markowitz's portfolio problem, in which an investor chooses a portfolio or investment strategy with maximal expected reward and minimal average variance.

X. Guo, O. Hernández-Lerma, *Continuous-Time Markov Decision Processes*, Stochastic Modelling and Applied Probability 62,
DOI 10.1007/978-3-642-02547-1_10, © Springer-Verlag Berlin Heidelberg 2009

## 10.2 Preliminaries

Under some conditions, Theorem 7.8 gives the existence of AR optimal deterministic stationary policies. Hence, the class $\mathbf{F}_{\mathrm{ao}}$ of AR optimal deterministic stationary policies is not empty. When $\mathbf{F}_{\mathrm{ao}}$ has more than one element, how do we choose the "best" policy in $\mathbf{F}_{\mathrm{ao}}$? To deal with such a question, in Chap. 9, we have introduced some advanced optimality criteria. Now, we will introduce another "advanced" optimality criterion, namely, the limiting average variance.

**Definition 10.1** A deterministic stationary policy $f^* \in \mathbf{F}_{\mathrm{ao}}$ is said to be variance minimization optimal if

$$\sigma^2\bigl(i, f^*\bigr) \le \sigma^2(i, f) \quad \forall f \in \mathbf{F}_{\mathrm{ao}} \text{ and } i \in S,$$

where

$$\sigma^2(i, f) := \limsup_{T\to\infty} \frac{1}{T} E_i^f \left( \int_0^T r\bigl(x(t), f\bigr)\, dt - V_T(i, f) \right)^2$$

denotes the limiting average variance of $\int_0^T r(x(t), f)\, dt$ under $f \in F$.

Thus, a variance minimization optimal policy, in addition to being AR optimal, minimizes the limiting average variance over the class $\mathbf{F}_{\mathrm{ao}}$ of AR optimal deterministic stationary policies.

In the following sections, we will show the existence of a variance minimization optimal policy and an algorithm for computing it.

## 10.3 Computation of the Average Variance

The main goal of this section is to derive, under suitable conditions, an explicit expression for the limiting average variance in Definition 10.1.

Next, we state the assumptions imposed on the control model (2.1), which are more restrictive than Assumption 2.2. It should be noted that these assumptions are not stated in their more general form (namely, Assumptions 10.2(a) and 10.2(b) below are not strictly necessary), but in their present form they are easily verifiable in practice.

**Assumption 10.2**

(a) There exist constants $c_2 > 0$ and $b_2 \ge 0$ such that

$$\sum_{j\in S} w^2(j) q(j|i, a) \le -c_2 w^2(i) + b_2 \quad \forall (i, a) \in K,$$

with $w$ as in Assumption 7.1.

(b) The control model (2.1) is uniformly $w^2$-exponentially ergodic (recall the definition of exponential ergodicity in Assumption 7.5).

Obviously, Assumption 10.2(a) implies Assumption 6.8(c).

To state our results, we recall some notation: the invariant probability measure (i.p.m.) $\mu_f$, the expected average reward $g(f)$, and the bias $h_f$ of $f$ are given in Assumption 7.5, (7.3), and (7.18), respectively. Now define a function $\Psi_f$ by

$$\Psi_f(i) := E_i^f\left(\int_0^1 r\big(x(t), f\big)\,dt + h_f\big(x(1)\big) - h_f(i) - g(f)\right)^2 \tag{10.1}$$

for any $i \in S$.

The following result shows that the limiting average variance under any policy $f \in F$ can be transformed into the usual long-run EAR with a new reward function $\Psi_f$.

**Theorem 10.3** *Suppose that Assumptions* 7.1, 7.4, 7.5, *and* 10.2 *hold. Then*

$$\sigma^2(i, f) = \int_S \Psi_f\,d\mu_f = 2\sum_{j\in S}\big(r(j, f) - g(f)\big)h_f(j)\mu_f(j) \quad \forall i \in S \text{ and } f \in F.$$

*Proof* For any fixed $i \in S$ and $f \in F$, let

$$U_t := \int_0^t \big[r\big(x(s), f\big) - g(f)\big]\,ds \quad \text{and} \quad M_t := U_t + h_f\big(x(t)\big) - h_f(i) \tag{10.2}$$

for all $t \ge 0$.

From (10.2) and the Poisson equation (7.20) we have

$$U_t = -\int_0^t Q^f h_f\big(x(s)\big)\,ds.$$

Hence, as in the proof of (8.20), we see that $\{M_t, t \ge 0\}$ is a $P_i^f$-martingale.

The proof of this theorem now proceeds in several steps. First of all, we will show that

$$\lim_{n\to\infty}\frac{1}{n}E_i^f U_n^2 = \int_S \Psi_f\,d\mu_f. \tag{10.3}$$

To this end, by (10.2), we have

$$\begin{aligned}\frac{1}{n}E_i^f U_n^2 &= \frac{1}{n}E_i^f M_n^2 + \frac{1}{n}E_i^f\big(h_f\big(x(n)\big) - h_f(i)\big)^2 \\ &\quad - \frac{2}{n}E_i^f\big[M_n\big(h_f\big(x(n)\big) - h_f(i)\big)\big].\end{aligned} \tag{10.4}$$

Using Assumption 10.2(a), Lemma 6.3, and (7.19) with obvious notational changes, we see that

$$\begin{aligned}\sup_{t\geq 0} E_i^f\left[h_f\big(x(t)\big)-h_f(i)\right]^2 &\leq 2\sup_{t\geq 0} E_i^f h_f^2\big(x(t)\big)+2h_f(i)^2\\ &\leq \frac{2L_2M}{\delta}\left[E_i^f w^2\big(x(t)\big)+w^2(i)\right]\\ &\leq \frac{2L_2M}{\delta}\left[2w^2(i)+\frac{b_2}{c_2}\right],\end{aligned} \tag{10.5}$$

and thus

$$\lim_{t\to\infty}\frac{1}{t}E_i^f\big(h_f\big(x(t)\big)-h_f(i)\big)^2=0. \tag{10.6}$$

Again from Lemma 6.3 and Assumption 10.2(a) we have

$$\sup_n\left[E_i^f(M_{n+1}-M_n)^2\right]<\infty. \tag{10.7}$$

Now let $Y_k := M_k - M_{k-1}$ for $k \geq 1$. Then, recalling the definition of $\mathcal{F}_t$ after (8.18), from (7.20) and Dynkin's formula we have

$$\begin{aligned}E_i^f(Y_k|\mathcal{F}_{k-1}) &= E_i^f\left[-\int_{k-1}^{k} Q^f h_f\big(x(t)\big)\,dt+h_f\big(x(k)\big)-h_f\big(x(k-1)\big)|\mathcal{F}_{k-1}\right]\\ &=0.\end{aligned}$$

Consequently, for any $m > n$,

$$E_i^f[Y_mY_n]=E_i^f\left[E_i^f(Y_mY_n|\mathcal{F}_{m-1})\right]=E_i^f\left[Y_nE_i^f(Y_m|\mathcal{F}_{m-1})\right]=0.$$

Hence, since $M_n=\sum_{k=1}^n Y_k$, by (10.1) and (10.7) we have

$$\frac{1}{n}E_i^f M_n^2=\frac{1}{n}\sum_{k=1}^{n}E_i^f\Psi_f\big(x(k-1)\big)\quad\text{and}\quad E_i^f M_n^2=O(n). \tag{10.8}$$

Thus, by the Cauchy–Schwartz inequality and (10.5) we have

$$\lim_{n\to\infty}\frac{1}{n}E_i^f\left[M_n\big(h_f\big(x(n)\big)-h_f(i)\big)\right]=0. \tag{10.9}$$

It is easy to show that $\Psi_f$ is in $B_{w^2}(S)$. Moreover, by the $w^2$-ergodicity in Assumption 10.2(b), and so we obtain

$$\lim_{n\to\infty}\frac{1}{n}E_i^f M_n^2=\int_S\Psi_f\,d\mu_f=\sum_{j\in S}\Psi_f(j)\mu_f(j). \tag{10.10}$$

Combining (10.4) with (10.6)–(10.10) proves (10.3).

The next step in the proof of Theorem 10.3 is to show that

$$\lim_{t\to\infty} \frac{1}{t} E_i^f U_t^2 = \int_S \Psi_f \, d\mu_f, \tag{10.11}$$

or, equivalently (by (10.3) and (10.10)),

$$\lim_{t\to\infty} \left[ \frac{1}{t} E_i^f U_t^2 - \frac{1}{[t]} E_i^f U_{[t]}^2 \right] = 0.$$

To see this, note that

$$\left| \frac{1}{t} E_i^f U_t^2 - \frac{1}{[t]} E_i^f U_{[t]}^2 \right| \le \left( \frac{1}{[t]} - \frac{1}{t} \right) E_i^f U_{[t]}^2 + \frac{1}{t} E_i^f \left| U_t^2 - U_{[t]}^2 \right|.$$

As a consequence of (10.3), $(\frac{1}{[t]} - \frac{1}{t}) E_i^f U_{[t]}^2$ converges to zero as $t \to \infty$. On the other hand,

$$\frac{1}{t} E_i^f \left| U_t^2 - U_{[t]}^2 \right| \le \frac{1}{t} \sqrt{E_i^f (U_t - U_{[t]})^2 E_i^f (U_t + U_{[t]})^2}. \tag{10.12}$$

Also, since $|r(i, f) - g(f)|^2 \le 2[Mw^2(i) + (g(f))^2]$ by Assumption 10.2(a) and Lemma 6.3, a straightforward calculation gives

$$E_i^f w^2\bigl(x(s)\bigr) \le w^2(i) + \frac{b_2}{c_2} \quad \forall s \ge 0. \tag{10.13}$$

Hence, for each $t > 0$, by (10.2) we have

$$E_i^f (U_t - U_{[t]})^2 \le 2\left[ M^2 w^2(i) + \frac{b_2 M^2}{c_2} + \bigl(g(f)\bigr)^2 \right]$$

and

$$\left| E_i^f U_t^2 \right| \le 2t \left[ M^2 w^2(i) + \frac{b_2 M^2}{c_2} + \bigl(g(f)\bigr)^2 \right].$$

Thus, the left-hand side of (10.12) also converges to zero as $t \to \infty$. Therefore, (10.11) holds.

Moreover, by (7.20) and Dynkin's formula,

$$V_t(i, f) = g(f)t + h_f(i) - E_i^f h_f\bigl(x(t)\bigr),$$

which, together with (10.13) and $\|h_f\|_w \in B_w(S)$, implies that the limits, as $T \to \infty$, of

$$\frac{1}{T} E_i^f \left( \int_0^T r\bigl(x(s), f\bigr) \, ds - V_T(i, f) \right)^2 \quad \text{and} \quad \frac{1}{T} E_i^f \left( \int_0^T \bigl[ r\bigl(x(s), f\bigr) - g(f) \bigr] \, ds \right)^2$$

coincide. We will use this fact to obtain the last step in the proof of Theorem 10.3, namely, that the right-hand side of (10.11) satisfies

$$\int_S \Psi_f \, d\mu_f = 2\sum_{j\in S}(r(j,f)-g(f))h_f(j)\mu_f(j). \tag{10.14}$$

To this end, fix an arbitrary integer $n > 0$ and let $t_k := k/n$ for $k = 0, 1, \ldots, n$. Then, since martingale differences are orthogonal (just proved), as in the proof of (10.8), we have

$$\Psi_f(i) = \sum_{k=1}^{n} E_i^f\left[h_f\big(x(t_k)\big) - h_f\big(x(t_{k-1})\big) - \int_{t_{k-1}}^{t_k} Q^f h_f\big(x(s)\big)\,ds\right]^2. \tag{10.15}$$

Since $\mu_f$ is the unique i.p.m. of $p_f(i,t,j)$, from (10.15) and the Markov property we obtain

$$\begin{aligned}\int_S \Psi_f \, d\mu_f &= nE_{\mu_f}\left[h_f\big(x(t_1)\big) - h_f\big(x(0)\big) - \int_0^{t_1} Q^f h_f\big(x(s)\big)\,ds\right]^2\\ &= nE_{\mu_f}^f\big[h_f\big(x(t_1)\big) - h_f\big(x(0)\big)\big]^2 + nE_{\mu_f}^f\left[\int_0^{t_1} Q^f h_f\big(x(s)\big)\,ds\right]^2\\ &\quad - 2nE_{\mu_f}^f\left[\big(h_f\big(x(t_1)\big) - h_f\big(x(0)\big)\big)\int_0^{t_1} Q^f h_f\big(x(s)\big)\,ds\right]. \end{aligned}\tag{10.16}$$

In addition, since $E_{x(0)}^f h_f(x(t_1)) = h_f(x(0)) + E_{x(0)}^f[\int_0^{t_1} Q^f h_f(x(s))ds]$, we have

$$\begin{aligned} &E_{\mu_f}^f\big[h_f\big(x(t_1)\big) - h_f\big(x(0)\big)\big]^2\\ &= 2\int_S h_f^2\,d\mu_f - 2E_{\mu_f}^f\big[h_f\big(x(0)\big)E_{\mu_f}^f\big(h_f\big(x(t_1)\big)|\mathcal{F}_0\big)\big]\\ &= 2\int_S h_f^2\,d\mu_f - 2E_{\mu_f}^f\big[h_f\big(x(0)\big)E_{x(0)}^f h_f\big(x(t_1)\big)\big]\\ &= 2\int_S h_f^2\,d\mu_f - 2E_{\mu_f}^f\left[h_f\big(x(0)\big)\left[h_f\big(x(0)\big) + E_{x(0)}^f\int_0^{t_1} Q^f h_f\big(x(s)\big)\,ds\right]\right]\\ &= -2E_{\mu_f}^f\left[h_f\big(x(0)\big)E_{x(0)}^f\int_0^{t_1} Q^f h_f\big(x(s)\big)\,ds\right]. \end{aligned}\tag{10.17}$$

Moreover, by the Cauchy–Schwartz inequality we have

$$\begin{aligned} E_{\mu_f}^f\left[\int_0^{\frac{1}{n}} Q^f h_f\big(x(s)\big)\,ds\right]^2 &\le E_{\mu_f}^f\left(\frac{1}{n}\int_0^{\frac{1}{n}}\big(Q^f h_f\big(x(s)\big)\big)^2\,ds\right)\\ &= \frac{1}{n^2}\int_S \big(Q^f h_f\big)^2\,d\mu_f, \end{aligned}\tag{10.18}$$

which, together with (10.17), gives (using $t_1 := 1/n$ again)

$$\begin{aligned}
&E^f_{\mu_f}\left|\big(h_f(x(t_1)) - h_f(x(0))\big)\int_0^{t_1} Q^f h_f\big(x(s)\big)\,ds\right| \\
&\le \sqrt{E^f_{\mu_f}\big[h_f\big(x(t_1)\big) - h_f\big(x(0)\big)\big]^2}\sqrt{E^f_{\mu_f}\left[\int_0^{t_1} Q^f h_f\big(x(s)\big)\,ds\right]^2} \\
&\le \frac{1}{n}\sqrt{\left(-2E^f_{\mu_f}\left[h_f\big(x(0)\big)E^f_{x(0)}\int_0^{t_1} Q^f h_f\big(x(s)\big)\,ds\right]\right)}\sqrt{\int_S \big(Q^f h_f\big)^2\,d\mu_f} \\
&\le \frac{1}{n}2E^f_{\mu_f}\left(h_f^2\big(x(0)\big)E^f_{\mu_f}\left[\int_0^{t_1} Q^f h_f\big(x(s)\big)\,ds\right]^2\right)^{\frac{1}{4}}\sqrt{\int_S \big(Q^f h_f\big)^2\,d\mu_f} \\
&\le \frac{1}{n}2E^f_{\mu_f}\left(h_f^2\big(x(0)\big)\frac{1}{n^2}\int_S \big(Q^f h_f\big)^2\,d\mu_f\right)^{\frac{1}{4}}\sqrt{\int_S \big(Q^f h_f\big)^2\,d\mu_f} \\
&= O\big(n^{-\frac{3}{2}}\big). \qquad (10.19)
\end{aligned}$$

Combining (10.16) with (10.17)–(10.19), it follows by the Fubini theorem that

$$\begin{aligned}
\int_S \Psi_f\,d\mu_f &= -2nE^f_{\mu_f}\left[h_f\big(x(0)\big)E^f_{x(0)}\int_0^{t_1} Q^f h_f\big(x(s)\big)\,ds\right] \\
&\quad + nO\big(n^{-\frac{3}{2}}\big) + nO\big(n^{-2}\big) \\
&= 2\frac{E^f_{\mu_f}[h_f(x(0))E^f_{x(0)}\int_0^{t_1}(r(x(s),f) - g(f))\,ds]}{t_1} + o(1) \\
&= \frac{2\int_0^{t_1} E^f_{\mu_f}[h_f(x(0))E^f_{x(0)}(r(x(s),f) - g(f))\,ds]}{t_1} + o(1). \qquad (10.20)
\end{aligned}$$

On the other hand, by Lemma 6.2 and Assumption 7.1 we have

$$w(i)p_f(i,s,i) \le E^f_i w\big(x(s)\big) \le e^{-c_1 s}w(i) + \frac{b_1}{c_1}\big(1 - e^{-c_1 s}\big) \quad \forall i \in S,$$

and so $\lim_{s\downarrow 0} E^f_i w(x(s)) = w(i)$ for all $i \in S$. Then, $\lim_{s\downarrow 0}[\sum_{j\neq i} w(j)p(i,s,j)] = 0$, and thus $\lim_{s\downarrow 0}[\sum_{j\neq i} r(j,f)p(i,s,j)] = 0$. Hence, we have $\lim_{s\downarrow 0} E^f_i r(x(s), f) = r(i,f)$ for all $i \in S$. Using Assumption 10.2, the dominated convergence theorem, and Lemma 6.3, we have

$$\begin{aligned}
\lim_{s\downarrow 0} E^f_{\mu_f}\big[h_f\big(x(0)\big)E^f_{x(0)}r\big(x(s),f\big)\big] &= E^f_{\mu_f}\big[h_f\big(x(0)\big)E^f_{x(0)}r\big(x(0),f\big)\big] \\
&= \sum_{i\in S} h_f(i)r(i,f)\mu_f(i). \qquad (10.21)
\end{aligned}$$

Recall that $t_1 = \frac{1}{n}$. Letting $n \to \infty$, from (10.20) and (10.21) we obtain (10.14). This completes the proof of Theorem 10.3. □

## 10.4 Variance Minimization

In this section, we will show the existence of a variance minimization optimal policy by solving a pair of nested AROEs. But we must first impose another assumption.

**Assumption 10.4**

(a) There exist positive constants $c_3$ and $b_3$ such that $\sum_{j\in S} w^3(j)q(j|i,a) \le c_3 w^3(i) + b_3$ for all $(i,a) \in K$, with $w$ as in Assumption 10.2.
(b) The functions $r(i,a)$, $q(j|i,a)$, and $\sum_{k\in S} w^2(k)q(k|i,a)$ are all continuous in $a \in A(i)$ for any fixed $i, j \in S$.

We now state the main result in this chapter.

**Theorem 10.5** *Under Assumptions* 7.1, 7.4, 7.5, 10.2, *and* 10.4, *the following conditions hold.*

(a) *There exists a solution* $(g^*, \bar{u}, \sigma_*^2, v) \in \mathbb{R} \times B_w(S) \times \mathbb{R} \times B_{w^2}(S)$ *to the system of equations*

$$g^* = \max_{a\in A(i)} \left\{ r(i,a) + \sum_{j\in S} \bar{u}(j)q(j|i,a) \right\}, \tag{10.22}$$

$$\sigma_*^2 = \min_{a\in A^*(i)} \left\{ 2\big(r(i,a) - g^*\big)\bar{u}(i) + \sum_{j\in S} v(j)q(j|i,a) \right\} \tag{10.23}$$

*for all* $i \in S$, *where* $A^*(i)$ *is the (nonempty compact) set of actions attaining the maximum in* (10.22).
(b) *There exists a variance minimization optimal policy in* $\mathbf{F}_{ao}$, *which is also average-reward optimal, and the minimal average variance* $\min_{f\in\mathbf{F}_{ao}} \sigma^2(f)$ *equals the constant* $\sigma_*^2$ *in* (10.23).
(c) *A policy* $f \in F$ *is variance minimization optimal if and only if, for every* $i \in S$, $f(i)$ *attains the maximum and the minimum in* (10.22) *and* (10.23), *respectively.*

*Proof* (a) The existence of a solution to the AROE (10.22) follows from Theorem 7.8(a). Suppose now that $f$ is in $\mathbf{F}_{ao}$. Then Theorem 7.8(b) shows that $f$ is canonical, and so the gain and the bias of $f$ satisfy that $g(f) = g^*$ and $h_f = \bar{u} - \mu_f(\bar{u})$ (recalling the formula above (9.3)). Thus, since $\sum_{j\in S} r(j,f)\mu_f(j) = g(f) = g^*$, from Theorem 10.3 we obtain

$$\sigma^2(i,f) = 2\sum_{j\in S} \big(r(j,f) - g(f)\big)h_f(j)\mu_f(j) = 2\sum_{j\in S} \big(r(j,f) - g^*\big)\bar{u}(j)\mu_f(j)$$

for all $i \in S$. This shows that $\sigma^2(f)$ is the expected average reward of the policy $f$ when the reward function is

$$r'(i,a) := 2\bigl(r(i,a) - g^*\bigr)\bar{u}(i) \quad \forall i \in S \text{ and } a \in A^*(i).$$

By Assumption 7.1 and (7.19), there exists a constant $M' > 0$ such that $|r'(i,a)| \le M' w^2(i)$ for all $i \in S$ and $a \in A^*(i)$. Hence, the control model with state space $S$, compact action sets $A^*(i)$ for $i \in S$, transition rates $q(j|i,a)$, and the reward function given by $(-r'(i,a))$ above verifies the hypothesis in Theorem 7.8 when the Lyapunov function $w$ is replaced with $w^2$. This yields that, under Assumptions 10.2 and 10.4, there exist solutions to the corresponding AROE (10.23).

The proof of (b) and (c) is now straightforward (by Theorem 7.8). □

## 10.5 Examples

In this section, we will apply our results to a controlled birth-and-death system.

*Example 10.1* (On average variance minimization) Consider a controlled birth-and-death system in which the state variable denotes the population size at any time $t \ge 0$. There are *fixed* birth and death rates denoted by *positive* constants $\lambda$ and $\mu$, respectively, as well as a *nonnegative* emigration rate that is assumed to be controlled by a decision-maker and represented by actions $a$. When the system is at state $i \ge 1$, the decision-maker takes an action $a$ from a *compact* set $A(i)$ of available actions, which increases/decreases the emigration rate and may incur a cost $c(i,a)$. Moreover, suppose that the benefit earned from each individual in the population is represented by $p > 0$. Then the *net* income in this system is $r(i,a) := pi - c(i,a)$ for all $i \in S := \{0, 1, \ldots\}$ and $a \in A(i)$. On the other hand, when the system is empty (i.e., $i = 0$), we may assume that the decision-maker hopes to increase the immigration with a rate $a$ in $[\lambda_1, \lambda_2]$, where the constants $\lambda_2 > \lambda_1 > 0$ are fixed.

Then the corresponding transition rates $q(j|i,a)$ and the reward function $r(i,a)$ are as follows. For $i = 0$ and each $a \in A(0) = [\lambda_1, \lambda_2]$, we have

$$q(1|0,a) = -q(0|0,a) := a > 0, \qquad q(j|0,a) = 0 \quad \text{for } j \ge 2.$$

For $i \ge 1$ and each $a \in A(i)$,

$$q(j|i,a) := \begin{cases} \mu i + a & \text{if } j = i - 1, \\ -(\mu + \lambda)i - a & \text{if } j = i, \\ \lambda i & \text{if } j = i + 1, \\ 0 & \text{otherwise;} \end{cases} \tag{10.24}$$

and

$$r(i,a) := pi - c(i,a) \quad \text{for all } i \in S \text{ and } a \in A(i). \tag{10.25}$$

To ensure the existence of a variance minimization optimal policy, we consider the following conditions:

$\mathbf{G_1}$. $A(i) = [\mu_1, \mu_2]$ for all $i \geq 1$, with $\mu_2 > \mu_1 \geq \frac{2\lambda}{3}$ and $\mu - \lambda > 0$.
$\mathbf{G_2}$. The function $c(i, \cdot)$ is continuous on $A(i)$ for each fixed $i \in S$.
$\mathbf{G_3}$. $|c(i, a)| \leq L'(i+1)$ for all $(i, a) \in K$ and some constant $L' > 0$.

**Proposition 10.6** *Under Conditions* $\mathbf{G_1}$, $\mathbf{G_2}$, *and* $\mathbf{G_3}$, *the above controlled birth-and-death system satisfies Assumptions* 7.1, 7.4, 7.5, 10.2, *and* 10.4. *Therefore (by Theorem* 10.5), *there exists a variance minimization optimal policy which can be computed (or at least approximated) by Policy Iteration Algorithm* 7.1.

*Proof* Let $w(i) := i + 1$ for all $i \in S$. Then, for each $a \in A(i)$ with $i \geq 1$, by $\mathbf{G_1}$ and (10.24) we have

$$\sum_{j \in S} w(j) q(j|i, a) = (\lambda - \mu)(i + 1) + \mu - \lambda - a$$

$$\leq -\frac{1}{2}(\mu - \lambda) w(i); \tag{10.26}$$

$$\sum_{j \in S} w^2(j) q(j|i, a) = -2\mu i^2 + 2\lambda i^2 - \mu i + 3\lambda i - 2ia - a$$

$$\leq -\frac{1}{2}(\mu - \lambda)(i + 1)^2 + (2\lambda - 3a)i$$

$$\leq -\frac{1}{2}(\mu - \lambda) w^2(i); \tag{10.27}$$

$$\sum_{j \in S} w^3(j) q(j|i, a) = -3\mu i^3 + 2\lambda i^3 - 3\mu i^2 + 2\lambda i^2 - \mu i + 3\lambda i - (3i + 1)a$$

$$\leq -(\mu - \lambda) w^3(i) + 3\mu. \tag{10.28}$$

In particular, for $i = 0$ and each $a \in A(0) = [\lambda_1, \lambda_2]$, we have

$$\sum_{j \in S} w(j) q(j|0, a) = a \leq -(\mu - \lambda) w(0) + \lambda_2 + \mu; \tag{10.29}$$

$$\sum_{j \in S} w^2(j) q(j|0, a) = 3a \leq -(\mu - \lambda) w^2(0) + 3\lambda_2 + \mu; \tag{10.30}$$

$$\sum_{j \in S} w^3(j) q(j|0, a) = 7a \leq -(\mu - \lambda) w^3(0) + 7\lambda_2 + \mu. \tag{10.31}$$

By inequalities (10.26) and (10.29) we see that Assumption 7.1(a) holds. By (10.24) we obtain Assumption 7.1(b). Also, by $\mathbf{G_3}$ and (10.25), we have $|r(i, a)| \leq pi + L'(i + 1) \leq (p + L') w(i)$ for all $i \in S$ and $a \in A(i)$, which verifies Assumption 7.1(c). Moreover, by (10.24), (10.27), and (10.30) and the model's description, we see that Assumption 6.8 holds. Hence, Assumption 7.1 is verified. On the other hand, by the model's description we see that Assumption 7.4 is also true. In addition, Assumption 7.5 follows from Proposition 7.6 and inequalities (10.26) and (10.29) together with (10.24). Thus, by inequalities (10.27) and (10.30) we obtain Assumption 10.2(a). Moreover, Assumption 10.2(b) follows from Proposition 7.6 and inequalities (10.27) and (10.30) together with (10.24). Hence, Assumption 10.2 holds.

Finally, it only remains to verify Assumption 10.4. This is straightforward, however, because Assumption 10.4(b) follows from the description of the model, whereas by the above verification of Assumption 10.2 and inequalities (10.28) and (10.31) we see that Assumption 10.4(a) is indeed true. □

## 10.6 Notes

The variance minimization problem has been studied by several authors; see, for instance, Prieto-Rumeau and Hernández-Lerma (2007) [127] for the continuous-time case, Hernández-Lerma and Lasserre (1999) [74] and Hernández-Lerma, Vega-Amaya, and Carrasco (1999) [75] for discrete-time Markov control processes, Guo (1999) [48] and the references therein for nonstationary discrete-time Markov decision processes. The main results in this chapter are from Prieto-Rumeau and Hernández-Lerma (2007) [127]. The proof of Theorem 10.3 and Example 10.1 in this chapter are both new.

# Chapter 11
# Constrained Optimality for Discount Criteria

In previous chapters, we have studied discount, average, and advanced optimality criteria, and in each of these cases, we have shown the existence of an optimal stationary policy. However, in real-world situations, there are always some constraints on the policies that can be used, and so it is desirable to look for an optimal policy over a class of policies with some constraints. For example, we might wish to maximize expected rewards over a class of policies subject to constraints on some associated costs. Such a constrained optimality problem will be considered in this chapter and the next one. In this chapter, we consider discount optimality criteria, and the case of average criteria is dealt with in the following chapter.

## 11.1 The Model with a Constraint

The model of a constrained continuous-time MDP is of the form

$$\{S, A, (A(i)|i \in S), q(j|i,a), r(i,a), c(i,a), \beta, \nu\}, \tag{11.1}$$

where $S, A, A(i), q(j|i,a)$ are the same as in (2.1), while $r(i,a)$ and $c(i,a)$ are, respectively, given reward and cost functions on $K$ as in (2.2). These functions are assumed to be measurable on $A(i)$ for each fixed $i \in S$. Finally, $\beta$ in (11.1) is a given *constraint* constant, and $\nu$ denotes a given initial distribution on $S$.

Fix a discount factor $\alpha > 0$. For each $\pi \in \Pi$, we may define the expected discounted reward (recalling (2.19))

$$V_\alpha(\nu, \pi) := E_\nu^\pi \left[ \int_0^\infty e^{-\alpha t} r\big(x(t), \pi_t\big)\, dt \right] \tag{11.2}$$

and the discounted cost (recalling (4.3))

$$J_\alpha(\nu, \pi) := E_\nu^\pi \left[ \int_0^\infty e^{-\alpha t} c\big(x(t), \pi_t\big)\, dt \right]. \tag{11.3}$$

X. Guo, O. Hernández-Lerma, *Continuous-Time Markov Decision Processes*, Stochastic Modelling and Applied Probability 62,
DOI 10.1007/978-3-642-02547-1_11, © Springer-Verlag Berlin Heidelberg 2009

For (11.2) and (11.3) to be well defined and finite, in addition to Assumption 2.2, we suppose the following.

**Assumption 11.1**

(a) Assumptions 2.2, 6.4, and 6.8 hold.
(b) $c(i,a)$ is continuous on $A(i)$ for each fixed $i \in S$ and satisfies

$$0 \le c(i,a) \le Mw(i) \quad \forall (i,a) \in K \tag{11.4}$$

with $w(i)$ as in Assumption 2.2 and $M$ as in Assumption 6.4(a).

Assumption 11.1(b) trivially holds if the cost function is nonnegative and bounded on $K$.

The constrained problem is to maximize the reward $V_\alpha(\nu,\pi)$ over the set of policies that satisfy the constraint $J_\alpha(\nu,\pi) \le \beta$. For the given constraint constant $\beta$ in (11.1), let

$$U := \{\pi \in \Pi : J_\alpha(\nu,\pi) \le \beta\}$$

be the set of "constrained" policies.

**Definition 11.2** A policy $\pi^* \in U$ is said to be *discount constrained-optimal* if $\pi^*$ maximizes the discounted reward in (11.2) over all $\pi$ in $U$, that is,

$$V_\alpha(\nu,\pi^*) = \sup_{\pi \in U} V_\alpha(\nu,\pi).$$

To ensure the existence of a discount constrained-optimal policy, we consider the following conditions.

**Assumption 11.3**

(a) $\sum_{j\in S} w(j)\nu(j) < \infty$ with $\nu$ as in (11.1) and $w(i)$ as in (11.4).
(b) The set $U_0 := \{\pi \in \Pi : J_\alpha(\nu,\pi) < \beta\} \neq \emptyset$.

Assumption 11.3(a) is a condition on the "tails" of the initial distribution $\nu$, whereas Assumption 11.3(b) is a Slater-like hypothesis, typical for constrained problems (see, for instance, Beutler and Ross (1985) [13], Guo (2000) [49], and Sennott (1991) [140]).

Now we state the main result in this chapter.

**Theorem 11.4** *Suppose that Assumptions* 11.1 *and* 11.3 *hold. Then there exists a discount constrained-optimal policy that is either a stationary policy or a randomized stationary policy that randomizes between two stationary policies which differ in at most one state; that is, there exist two stationary policies* $f^1, f^2$, *a state* $i^* \in S$, *and a number $p \in [0, 1]$ such that $f^1(i) = f^2(i)$ for all $i \neq i^*$, and, in addition, the randomized stationary policy $\pi^p(\cdot|i)$ is discount constrained-optimal, where*

$$\pi^p(a|i) = \begin{cases} p & \text{for } a = f^1(i^*) \text{ when } i = i^*, \\ 1-p & \text{for } a = f^2(i^*) \text{ when } i = i^*, \\ 1 & a = f^1(i) \text{ when } i \neq i^*. \end{cases}$$

In other words, when the state is $i = i^*$, the randomized stationary policy $\pi^p(\cdot|i^*)$ takes the action $f^1(i^*)$ with probability $p$ and the action $f^2(i^*)$ with probability $1 - p$; when the state is in $i \neq i^*$, $\pi^p(\cdot|i)$ takes the action $f^1(i)$ with probability 1.

Theorem 11.4 is proved in Sect. 11.3 after some technical preliminaries in Sect. 11.2. An example in Sect. 11.4 illustrates Theorem 11.4.

## 11.2 Preliminaries

In this section, we present some results needed to prove Theorem 11.4. First, we recall some notation: If the initial distribution $\nu$ is concentrated at some state $i \in S$, we write $E_\nu^\pi$, $V_\alpha(\nu, \pi)$, and $J_\alpha(\nu, \pi)$ as $E_i^\pi$, $V_\alpha(i, \pi)$, and $J_\alpha(i, \pi)$, respectively.

**Lemma 11.5** *Suppose that Assumption* 11.1 *holds. Then for each $\pi \in \Pi_s$, $V_\alpha(\cdot, \pi)$ is the unique solution in $B_w(S)$ to the equation*

$$\alpha u(i) = r(i, \pi) + \sum_{j \in S} u(j) q(j|i, \pi) \quad \forall i \in S. \tag{11.5}$$

*Proof* Obviously, this lemma follows from Theorem 6.9(c). □

Note that $F$ can be written as the product space $F = \prod_{i \in S} A(i)$. Hence, by Assumption 11.1(a) and Tychonoff's theorem (see Proposition A.6), $F$ is a compact metric space.

**Lemma 11.6** *Suppose that Assumptions* 11.1 *and* 11.3(a) *hold. Then the functions $V_\alpha(\nu, f)$ and $J_\alpha(\nu, f)$ are continuous in $f \in F$.*

*Proof* We only prove the continuity of $V_\alpha(\nu, f)$ in $f \in F$ because the other case is similar. Let $f_n \to f$ as $n \to \infty$, and choose any subsequence $\{V_\alpha(\cdot, f_{n_k})\}$ of $\{V_\alpha(\cdot, f_n)\}$ converging to some function $v$ on $S$. Then

$$\lim_{k \to \infty} V_\alpha(i, f_{n_k}) =: \bar{v}(i) \quad \text{and} \quad \lim_{k \to \infty} f_{n_k}(i) = f(i) \quad \forall i \in S. \tag{11.6}$$

By Theorem 6.5(a),

$$|\bar{v}(i)| \leq \frac{b_0 M}{\alpha(\alpha - c_0)} + \frac{M}{\alpha - c_0} w(i) \quad \forall i \in S, \tag{11.7}$$

with $c_0$ and $b_0$ as in Assumption 2.2.

On the other hand, for all $i \in S$ and $k \geq 1$, by Lemma 11.5 we have

$$\alpha V_\alpha(i, f_{n_k}) = r(i, f_{n_k}) + \sum_{j \in S} V_\alpha(j, f_{n_k}) q(j|i, f_{n_k}),$$

which, equivalently, can be expressed as

$$V_\alpha(i, f_{n_k}) = \frac{r(i, f_{n_k})}{\alpha - q(i|i, f_{n_k})} + \frac{1}{\alpha - q(i|i, f_{n_k})} \sum_{j \neq i} V_\alpha(j, f_{n_k}) q(j|i, f_{n_k}). \quad (11.8)$$

Then, under Assumption 11.1, from (11.6)–(11.8) and Proposition A.4 we get

$$\bar{v}(i) = \frac{r(i, f)}{\alpha - q(i|i, f)} + \frac{1}{\alpha - q(i|i, f)} \sum_{j \neq i} \bar{v}(j) q(j|i, f),$$

that is,

$$\alpha \bar{v}(i) = r(i, f) + \sum_{j \in S} \bar{v}(j) q(j|i, f) \quad \forall i \in S.$$

Hence, Lemma 11.5 yields

$$\bar{v}(i) = V_\alpha(i, f) \quad \forall i \in S. \quad (11.9)$$

Thus, as the above subsequence $\{V_\alpha(\cdot, f_{n_k})\}$ was arbitrarily chosen and (by (11.9) and (11.6)) all such subsequences have the same limit $V_\alpha(\cdot, f)$, we have

$$\lim_{n \to \infty} V_\alpha(i, f_n) = V_\alpha(i, f) \quad \forall i \in S.$$

Therefore, from Assumption 11.3(a) and the dominated convergence theorem we obtain

$$\lim_{n \to \infty} V_\alpha(\nu, f_n) = \sum_{j \in S} \Big[\lim_{n \to \infty} V_\alpha(j, f_n)\Big] \nu(j) = \sum_{j \in S} V_\alpha(j, f) \nu(j) = V_\alpha(\nu, f),$$

which gives the desired conclusion, $V_\alpha(\nu, f_n) \to V_\alpha(\nu, f)$. □

Now, by Definition 11.2, let us express the constrained control problem as an optimization problem of the form

$$\begin{aligned} &\text{maximize} \quad V_\alpha(\nu, \pi) \\ &\text{subject to:} \quad J_\alpha(\nu, \pi) \leq \beta, \quad \pi \in U. \end{aligned}$$

To analyze this problem, we introduce a Lagrange multiplier $\gamma \geq 0$ as follows. For all $i \in S$ and $a \in A(i)$, let

$$b^\gamma(i, a) := r(i, a) - \gamma c(i, a) \quad (11.10)$$

be a *new* reward function on $K$ (depending on $\gamma$).

Furthermore, for each policy $\pi = (\pi_t, t \geq 0) \in \Pi$ and each $i \in S$, let

$$b^\gamma(i,\pi_t) := \int_{A(i)} b^\gamma(i,a)\pi_t(da|i),$$

$$V_b^\gamma(i,\pi) := E_i^\pi\left[\int_0^\infty e^{-\alpha t} b^\gamma\big(x(t),\pi_t\big)\,dt\right], \tag{11.11}$$

$$V_b^\gamma(\nu,\pi) := \sum_{j\in S} V_b^\gamma(j,\pi)\nu(j),$$

$$V_b^{*\gamma}(i) := \sup_{\pi\in\Pi} V_b^\gamma(i,\pi), \qquad V_b^{*\gamma}(\nu) := \sup_{\pi\in\Pi} V_b^\gamma(\nu,\pi). \tag{11.12}$$

Under Assumption 11.1, by Theorem 6.10(b) we have

$$\alpha V_b^{*\gamma}(i) = \sup_{a\in A(i)}\left\{b^\gamma(i,a) + \sum_{j\in S} V_b^{*\gamma}(j)q(j|i,a)\right\} \quad \forall i \in S, \tag{11.13}$$

and the maximum in (11.13) is realized by some $a \in A(i)$ for each $i \in S$. In other words, the set

$$F_\gamma^* := \left\{f \in F \mid \alpha V_b^{*\gamma}(i) = b^\gamma(i,f) + \sum_{j\in S} V_b^{*\gamma}(j)q(j|i,f)\ \forall i \in S\right\} \tag{11.14}$$

is *nonempty*. This implies (by Lemma 11.5) that $V_b^{*\gamma}(i) = V_b^\gamma(i,f)$ for all $i \in S$ if $f$ is in $F_\gamma^*$, and, therefore, the set

$$\Pi_s^\gamma := \left\{\pi \in \Pi_s \mid V_b^\gamma(i,\pi) = V_b^{*\gamma}(i)\ \forall i \in S\right\}$$

is *nonempty*. The following lemma further shows that $\Pi_s^\gamma$ is *convex*.

**Lemma 11.7** *Under Assumptions* 11.1 *and* 11.3(a), *the set* $\Pi_s^\gamma$ *is convex. That is, if* $\pi_1$ *and* $\pi_2$ *are in* $\Pi_s^\gamma$ *and, for any* $p \in [0,1]$,

$$\pi^p(\cdot|i) := p\pi_1(\cdot|i) + (1-p)\pi_2(\cdot|i) \quad \forall i \in S, \tag{11.15}$$

*then the policy* $\pi^p$ *is also in* $\Pi_s^\gamma$.

*Proof* For each $\pi \in \Pi_s$, let

$$b^\gamma(i,\pi) := \int_{A(i)} b^\gamma(i,a)\pi(da|i). \tag{11.16}$$

Then, by Lemma 11.5 and the definition of $\Pi_s^\gamma$, we have

$$\alpha V_b^{*\gamma}(i) = b^\gamma(i,\pi_n) + \sum_{j\in S} V_b^{*\gamma}(j)q(j|i,\pi_n) \quad \forall i \in S \text{ and } n = 1,2,$$

which, together with (11.15) and (11.16), gives that

$$\alpha V_b^{*\gamma}(i) = b^\gamma(i, \pi^p) + \sum_{j\in S} V_b^{*\gamma}(j) q(j|i, \pi^p) \quad \forall i \in S. \tag{11.17}$$

Therefore, by Lemma 11.5 and (11.17) we have that $V_b^\gamma(i, \pi^p) = V_b^{*\gamma}(i)$ for all $i \in S$, and the lemma follows. □

*Notation* For each $\gamma \geq 0$, we take an arbitrary but *fixed* policy $f^\gamma \in F_\gamma^*$ and denote $V_\alpha(\nu, f^\gamma)$, $J_\alpha(\nu, f^\gamma)$, and $V_b^\gamma(\nu, f^\gamma)$ by $V_\alpha(\gamma)$, $J_\alpha(\gamma)$, and $V_b(\gamma)$, respectively. By Lemma 11.5 we have that $V_b^\gamma(\nu, f) = V_b^{*\gamma}(\nu)$ for all $f \in F_\gamma^*$. Hence, $V_b(\gamma) = V_b^\gamma(\nu, f^\gamma) = V_b^{*\gamma}(\nu)$.

**Lemma 11.8** *If Assumptions* 11.1 *and* 11.3(a) *hold, then* $V_\alpha(\gamma)$, $J_\alpha(\gamma)$, *and* $V_b(\gamma)$ *are nonincreasing in* $\gamma \in [0, \infty)$.

*Proof* By (11.2), (11.3), and (11.10)–(11.11), for each $\pi \in \Pi$,

$$V_b^\gamma(\nu, \pi) = V_\alpha(\nu, \pi) - \gamma J_\alpha(\nu, \pi) \quad \forall \gamma \geq 0.$$

Moreover, noting that $V_b(\gamma) = V_b^\gamma(\nu, f^\gamma) = V_b^{*\gamma}(\nu)$ for all $\gamma \geq 0$ and $f^\gamma \in F_\gamma^*$, we have, for any $h > 0$,

$$\begin{aligned}
-hJ_\alpha(\gamma) &= V_b^{\gamma+h}(\nu, f^\gamma) - V_b(\gamma) \\
&\leq V_b(\gamma + h) - V_b(\gamma) \\
&\leq V_b(f^{\gamma+h}) - V_b^\gamma(\nu, f^{\gamma+h}) = -hJ_\alpha(\gamma + h).
\end{aligned}$$

Thus, from the last two relations we get

$$-hJ_\alpha(\gamma) \leq V_b(\gamma + h) - V_b(\gamma) \leq -hJ_\alpha(\gamma + h).$$

Hence, as the cost function $c(i, a)$ is nonnegative, it follows that $J_\alpha(\gamma)$ and $V_b(\gamma)$ are both nonincreasing in $\gamma$.

Now, if $V_\alpha(\gamma)$ is not nonincreasing in $\gamma \geq 0$, then there exist constants $\gamma_0 \geq 0$ and $h_0 > 0$ such that $V_\alpha(\gamma_0) < V_\alpha(\gamma_0 + h_0)$. Thus, since $J_\alpha(\gamma_0) \geq J_\alpha(\gamma_0 + h_0)$ (just proved), we have

$$V_\alpha(\gamma_0) - \gamma_0 J_\alpha(\gamma_0) < V_\alpha(\gamma_0 + h_0) - \gamma_0 J_\alpha(\gamma_0 + h_0),$$

that is, $V_b(\gamma_0) < V_b(\gamma_0 + h_0)$, which contradicts that $V_b(\gamma)$ is nonincreasing. Hence, $V_\alpha(\gamma)$ is nonincreasing. □

**Lemma 11.9** *Suppose that Assumptions* 11.1 *and* 11.3(a) *hold. Then* $V_b(\gamma)$ *and* $V_b^{*\gamma}(i)$ *are continuous in* $\gamma \in [0, \infty)$ *for each fixed* $i \in S$.

*Proof* As $V_b^\gamma(i,\pi) = V_\alpha(i,\pi) - \gamma J_\alpha(i,\pi)$, the function $V_b^\gamma(i,\pi)$ is convex in $\gamma$ for any fixed $\pi \in \Pi$ and $i \in S$, that is, for all $\gamma_1, \gamma_2 \geq 0$ and $p \in [0,1]$,

$$V_b^{(p\gamma_1+(1-p)\gamma_2)}(i,\pi) \leq p V_b^{\gamma_1}(i,\pi) + (1-p) V_b^{\gamma_2}(i,\pi).$$

Hence, by (11.12), $V_b^{*\gamma}(i)$ is a convex function of $\gamma$ (for each fixed $i \in S$) and, therefore, is continuous in $\gamma \in (0,\infty)$ (by Theorem 6.2 in Hiriart-Urruty and Lemaréchal (2001) [78, p. 15]). On the other hand, for any sequence $\gamma_k \in (0,\infty)$ such that $\gamma_k \to 0$ as $k \to \infty$, by Lemma 11.8 and Theorem 6.5(a) we have

$$\begin{aligned} V_b^{*0}(i) \geq V_b^{*\gamma_k}(i) &= \sup_{\pi\in\Pi}\bigl\{V_\alpha(i,\pi) - \gamma_k J_\alpha(i,\pi)\bigr\} \\ &\geq \sup_{\pi\in\Pi}\bigl\{V_\alpha(i,\pi)\bigr\} - \gamma_k\left[\frac{b_0 M}{\alpha(\alpha-c_0)} + \frac{M}{\alpha-c_0}w(i)\right] \\ &= V_b^{*0}(i) - \gamma_k\left[\frac{b_0 M}{\alpha(\alpha-c_0)} + \frac{M}{\alpha-c_0}w(i)\right]. \end{aligned} \tag{11.18}$$

If $k \to \infty$, from (11.18) we have that $\lim_{k\to\infty} V_b^{*\gamma_k}(i) = V_b^{*0}(i)$. Thus, $V_b^{*\gamma}(i)$ is continuous on $[0,\infty)$ for each $i \in S$. Finally, as $V_b(\gamma) = V_b^{*\gamma}(\nu) = \sum_{j\in S} V_b^{*\gamma}(j) \times \nu(j)$, using dominated convergence (which can be obtained from Assumption 11.3(a)), we get that $V_b^{*\gamma}(\nu)$ is continuous in $\gamma \in [0,\infty)$. □

**Lemma 11.10** *Suppose that Assumptions* 11.1 *and* 11.3(a) *hold, and let* $F_\gamma^*$ *be as in* (11.14). *If* $\lim_{k\to\infty}\gamma_k = \gamma$ *and* $f^{\gamma_k} \in F_{\gamma_k}^*$ *is such that* $\lim_{k\to\infty} f^{\gamma_k} = f$, *then* $f \in F_\gamma^*$.

*Proof* As $f^{\gamma_k} \in F_{\gamma_k}^*$, by (11.14) we have

$$\alpha V_b^{*\gamma_k}(i) = b^{\gamma_k}\bigl(i, f^{\gamma_k}\bigr) + \sum_{j\in S} V_b^{*\gamma_k}(j) q\bigl(j|i, f^{\gamma_k}\bigr) \quad \forall i \in S.$$

Letting $k \to \infty$, Lemma 11.9, together with Proposition A.4, yields

$$\alpha V_b^{*\gamma}(i) = b^\gamma(i,f) + \sum_{j\in S} V_b^{*\gamma}(j) q(j|i,f) \quad \forall i \in S,$$

which implies that $f \in F_\gamma^*$. □

Under Assumptions 11.1 and 11.3(a) it follows from Lemma 11.8 that the nonnegative constant

$$\tilde{\gamma} := \inf\bigl\{\gamma : J_\alpha(\gamma) \leq \beta\bigr\} \tag{11.19}$$

is well defined.

**Lemma 11.11** *Suppose that Assumptions* 11.1 *and* 11.3 *hold. Then the constant* $\tilde{\gamma}$ *in* (11.19) *is finite; that is,* $\tilde{\gamma}$ *is in* $[0,\infty)$.

*Proof* Suppose that $\tilde{\gamma} = \infty$. By Assumption 11.3(b), there exists a policy $\pi' \in \Pi$ such that $J_\alpha(\nu, \pi') < \beta$. Let $d := \beta - J_\alpha(\nu, \pi') > 0$. Then, for any $\gamma > 0$, we have

$$V_b^\gamma(\nu, \pi') = V_\alpha(\nu, \pi') - \gamma J_\alpha(\nu, \pi') = V_\alpha(\nu, \pi') - \gamma(\beta - d). \tag{11.20}$$

As $\tilde{\gamma} = \infty$, by (11.19) and Lemma 11.8 we obtain that $J_\alpha(\gamma) > \beta$ for all $\gamma > 0$. Therefore, $V_b(\gamma) = V_\alpha(\gamma) - \gamma J_\alpha(\gamma) < V_\alpha(\gamma) - \gamma\beta$. Since $V_b(\gamma) = V_b^{*\gamma}(\nu) \geq V_b^\gamma(\nu, \pi')$, from (11.20) we have

$$V_\alpha(\gamma) - \gamma\beta > V_b(\gamma) \geq V_b^\gamma(\nu, \pi') = V_\alpha(\nu, \pi') - \gamma(\beta - d) \quad \forall \gamma > 0,$$

which gives

$$V_\alpha(\gamma) > V_\alpha(\nu, \pi') + \gamma d \quad \forall \gamma > 0. \tag{11.21}$$

On the other hand, by Theorem 6.5 and Assumption 11.3(a) we have

$$\max\{|V_\alpha(\nu, \pi')|, |V_\alpha(\gamma)|\} \leq \frac{M}{\alpha - c_0}\left[\frac{b_0}{\alpha} + \sum_{j \in S} w(j)\nu(j)\right] =: \tilde{M} < \infty \tag{11.22}$$

for all $\gamma > 0$. The latter inequality, together with (11.21), gives that

$$2\tilde{M} > \gamma d \quad \forall \gamma > 0,$$

which is clearly a contradiction; take $\gamma = \frac{3}{d}\tilde{M} > 0$, for instance. Hence, $\tilde{\gamma}$ is finite. □

## 11.3 Proof of Theorem 11.4

In this section, we prove Theorem 11.4 using the Lagrange approach and the following lemma.

**Lemma 11.12** *If there exist $\gamma_0 \geq 0$ and $\pi^* \in U$ such that*

$$J_\alpha(\nu, \pi^*) = \beta \quad \text{and} \quad V_b^{\gamma_0}(\nu, \pi^*) = V_b^{*\gamma_0}(\nu),$$

*then the policy $\pi^*$ is discount constrained-optimal.*

*Proof* For any $\pi \in U$, since $V_b^{\gamma_0}(\nu, \pi^*) = V_b^{*\gamma_0}(\nu) \geq V_b^{\gamma_0}(\nu, \pi)$, we have

$$V_\alpha(\nu, \pi^*) - \gamma_0 J_\alpha(\nu, \pi^*) \geq V_\alpha(\nu, \pi) - \gamma_0 J_\alpha(\nu, \pi). \tag{11.23}$$

As $J_\alpha(\nu, \pi^*) = \beta$ and $J_\alpha(\nu, \pi) \leq \beta$ (because $\pi \in U$), from (11.23) we get

$$V_\alpha(\nu, \pi^*) \geq V_\alpha(\nu, \pi) + \gamma_0(\beta - J_\alpha(\nu, \pi)) \geq V_\alpha(\nu, \pi) \quad \forall \pi \in U,$$

which means that $\pi^*$ is discount constrained-optimal. □

*Proof of Theorem 11.4* By Lemma 11.11, the constant $\tilde{\gamma}$ is in $[0,\infty)$. Thus, we shall consider two cases, $\tilde{\gamma}=0$ and $\tilde{\gamma}>0$.

**The case** $\tilde{\gamma}=0$: By (11.19), there exists a sequence $f^{\gamma_k}\in F^*_{\gamma_k}$ such that $\gamma_k\downarrow 0$ as $k\to\infty$. Because $F$ is compact, without loss of generality we may assume that $f^{\gamma_k}\to\tilde{f}\in F$. Thus, by Lemma 11.7, we have $J_\alpha(\nu,f^{\gamma_k})\le\beta$ for all $k\ge 1$, and then it follows from Lemma 11.6 that $\tilde{f}\in U$. Moreover, for each $\pi\in U$, we have that $V_b(\gamma_k)=V_b^{\gamma_k}(\nu,f^{\gamma_k})\ge V_b^{\gamma_k}(\nu,\pi)$. Hence, by (11.22),

$$V_\alpha\big(\nu,f^{\gamma_k}\big)-V_\alpha(\nu,\pi)\ge\gamma_k\big[J_\alpha\big(\nu,f^{\gamma_k}\big)-J_\alpha(\nu,\pi)\big]\ge-2\gamma_k\tilde{M}. \tag{11.24}$$

Letting $k\to\infty$ in (11.24), by Lemma 11.6 we have

$$V_\alpha(\nu,\tilde{f})-V_\alpha(\nu,\pi)\ge 0\quad\forall\pi\in U,$$

which means that $\tilde{f}$ is a discount constrained-optimal stationary policy.

**The case** $\tilde{\gamma}\in(0,\infty)$: First, if there exists some $\gamma'\in(0,\infty)$ satisfying $J_\alpha(\gamma')=\beta$, then there exists an associated $f^{\gamma'}\in F^*_{\gamma'}$ such that $J_\alpha(\gamma')=J_\alpha(\nu,f^{\gamma'})=\beta$ and $V_b^{*\gamma'}(\nu)=V_b^{\gamma'}(\nu,f^{\gamma'})$. Thus, by Lemma 11.12, $f^{\gamma'}$ is a discount constrained-optimal stationary policy.

Now suppose that $J_\alpha(\gamma)\ne\beta$ for all $\gamma\in(0,\infty)$. Then, as $\tilde{\gamma}$ is in $(0,\infty)$, there exist two sequences of positive numbers $\{\gamma_k\}$ and $\{\delta_k\}$ such that $\gamma_k\uparrow\tilde{\gamma}$ and $\delta_k\downarrow\tilde{\gamma}$. Since $F$ is compact, we may take $f^{\gamma_k}\in F^*_{\gamma_k}$ and $f^{\delta_k}\in F^*_{\delta_k}$ such that $f^{\gamma_k}\to f^1\in F$ and $f^{\delta_k}\to f^2\in F$. By Lemma 11.10 we have that $f^1,f^2\in F^*_{\tilde{\gamma}}$. By Lemmas 11.6 and 11.8 we obtain that $J_\alpha(\nu,f^1)\ge\beta$ and $J_\alpha(\nu,f^2)\le\beta$. If $J_\alpha(\nu,f^1)$ (or $J_\alpha(\nu,f^2)$) $=\beta$, by Lemma 11.12 we have that $f^1$ (or $f^2$) is a discount constrained-optimal stationary policy, and the theorem follows. Hence, to complete the proof, we shall consider the following case:

$$J_\alpha\big(\nu,f^1\big)>\beta\quad\text{and}\quad J_\alpha\big(\nu,f^2\big)<\beta. \tag{11.25}$$

Now using $f^1$ and $f^2$, we construct a sequence of stationary policies $\{f_n\}$ as follows. For all $n\ge 1$ and $i\in S$, let

$$f_n(i):=\begin{cases}f^1(i), & i<n,\\ f^2(i), & i\ge n,\end{cases}$$

where, without loss of generality, the denumerable state space $S$ is supposed to be the set $\{1,2,\dots\}$. Obviously, $f_1=f^2$ and $\lim_{n\to\infty}f_n=f^1$. Hence, by Lemma 11.6, $\lim_{n\to\infty}J_\alpha(\nu,f_n)=J_\alpha(\nu,f^1)$. Since $f^1,f^2\in F^*_{\tilde{\gamma}}$ (just mentioned), by (11.14) we see that $f_n\in F^*_{\tilde{\gamma}}$ for all $n\ge 1$. As $f_1=f^2$, by (11.25) we have $J_\alpha(\nu,f_1)<\beta$. If there exists $n^*$ such that $J_\alpha(\nu,f_{n^*})=\beta$, then by Lemma 11.12 and $f_{n^*}\in F^*_{\tilde{\gamma}}$, $f_{n^*}$ is a discount constrained-optimal stationary policy. Thus, in the remainder of this section, we may assume that $J_\alpha(\nu,f_n)\ne\beta$ for all $n\ge 1$. If $J_\alpha(\nu,f_n)<\beta$ for all $n\ge 1$, then $\lim_{n\to\infty}J_\alpha(\nu,f_n)=J_\alpha(\nu,f^1)\le\beta$, which is a contradiction to (11.25). Thus, there exists some $n \geq 1$ such that $J_\alpha(\nu, f_n) > \beta$, which, together with $J_\alpha(\nu, f_1) < \beta$, gives the existence of some $\tilde{n}$ such that

$$J_\alpha(\nu, f_{\tilde{n}}) < \beta \quad \text{and} \quad J_\alpha(\nu, f_{\tilde{n}+1}) > \beta.$$

Obviously, the stationary policies $f_{\tilde{n}}$ and $f_{\tilde{n}+1}$ differ in at most the state $\tilde{n}$.

For any $p \in [0, 1]$, using the stationary policies $f_{\tilde{n}}$ and $f_{\tilde{n}+1}$, we construct a randomized stationary policy $\pi^p$ as follows. For each $i \in S$,

$$\pi^p(a|i) = \begin{cases} p & \text{if } a = f_{\tilde{n}}(\tilde{n}) \text{ when } i = \tilde{n}, \\ 1-p & \text{if } a = f_{\tilde{n}+1}(\tilde{n}) \text{ when } i = \tilde{n}, \\ 1 & \text{if } a = f_{\tilde{n}}(i) \text{ when } i \neq \tilde{n}. \end{cases} \tag{11.26}$$

Since $f_{\tilde{n}}, f_{\tilde{n}+1} \in F^*_{\tilde{\gamma}} \subseteq \Pi_s^{\tilde{\gamma}}$, by Lemma 11.7 we have $V_b^{\tilde{\gamma}}(\nu, \pi^p) = V_b^{*\tilde{\gamma}}(\nu)$ for all $p \in [0, 1]$. We also have that $J_\alpha(\nu, \pi^p)$ is continuous in $p \in [0, 1]$. Indeed, for any $p \in [0, 1]$ and any sequence $\{p_m\}$ in $[0, 1]$ such that $\lim_{m\to\infty} p_m = p$, by Lemma 11.5, we have

$$\alpha J_\alpha\big(i, \pi^{p_m}\big) = c\big(i, \pi^{p_m}\big) + \sum_{j\in S} J_\alpha\big(j, \pi^{p_m}\big) q\big(j|i, \pi^{p_m}\big) \quad \forall i \in S. \tag{11.27}$$

Hence, as in the proof of Lemma 11.6, from (11.26) and (11.27) we can obtain

$$\lim_{m\to\infty} J_\alpha\big(\nu, \pi^{p_m}\big) = J_\alpha\big(\nu, \pi^p\big),$$

and so $J_\alpha(\nu, \pi^p)$ is continuous in $p \in [0, 1]$.

Finally, let $p_0 = 0$ and $p_1 = 1$. Then $J_\alpha(\nu, \pi^{p_0}) = J_\alpha(\nu, f_{\tilde{n}+1}) > \beta$ and $J_\alpha(\nu, \pi^{p_1}) = J_\alpha(\nu, f_{\tilde{n}}) < \beta$. Therefore, by the continuity of $J_\alpha(\nu, \pi^p)$ in $p \in [0, 1]$, there a exists $p^* \in (0, 1)$ such that $J_\alpha(\nu, \pi^{p^*}) = \beta$. Since $V_b^{\tilde{\gamma}}(\nu, \pi^{p^*}) = V_b^{*\tilde{\gamma}}(\nu)$, by Lemma 11.12 we have that $\pi^{p^*}$ is a discount constrained-optimal randomized stationary policy, which randomizes between the two stationary policies $f_{\tilde{n}}$ and $f_{\tilde{n}+1}$ that differ in at most the state $\tilde{n}$. □

## 11.4 An Example

In this section, we use an example to illustrate Theorem 11.4.

*Example 11.1* Consider a controlled birth-and-death system in which the state variable denotes the population size at each time $t \geq 0$. There is a "natural" birth rate, say $\lambda$ ($\geq 0$), and a death parameter $a$ controlled by a decision-maker. (This example can be seen, in particular, as the control of service in a queueing system.) Thus, when the state of the system is $i \geq 1$, the decision-maker takes an action $a$ from a given set $A(i) := [\mu_1, \mu_2]$; when the system is empty (i.e., $i = 0$), we suppose that there is an immigration rate $a \in A(0) := [\lambda_1^*, \lambda_2^*]$ (for given constants

$0 < \lambda_1^* < \lambda_2^*$), which is assumed to also be controlled by the decision-maker. When the decision-maker takes an action $a$, a cost denoted by $\tilde{c}(a) \geq 0$ per unit time is produced, where $\mu_2 > \mu_1 > 0$ are two given constants. In addition, suppose that $p > 0$ is a fixed fee per individual. Then the net reward gained by the decision-maker is $r(i,a) := pi - \tilde{c}(a)$ for each unit of time during which the system remains in the state $i \in S := \{0, 1, \ldots\}$. Moreover, the decision-maker wishes to keep the associated cost bounded above by $\beta > 0$.

We formulate this system as a constrained continuous-time MDP. The corresponding transition rates and reward/cost functions are as follows: For $i = 0$ and $a \in A(0)$,

$$q(1|0,a) = -q(0|0,a) := a, \quad \text{and} \quad q(j|0,a) = 0 \quad \text{for } j \geq 2,$$

and, for $i \geq 1$,

$$q(j|i,a) = \begin{cases} a & \text{if } j = i-1, \\ -a - \lambda i & \text{if } j = i, \\ \lambda i & \text{if } j = i+1, \\ 0 & \text{otherwise,} \end{cases} \tag{11.28}$$

$$r(i,a) = pi - \tilde{c}(a), \quad c(i,a) = \tilde{c}(a) \quad \text{for all } (i,a) \in K, \tag{11.29}$$

with $K$ as in (2.1).

For a given discount factor $\alpha > 0$, the aim here is to find conditions under which there exists a stationary policy achieving the maximum $\alpha$-discounted rewards, and at the same time, the associated $\alpha$-discounted cost does not exceed the cost $\beta$. To this end, we assume that

**H$_1$**. The discount factor $\alpha$ verifies that $\alpha > \lambda$.
**H$_2$**. $\tilde{c}(a)$ is continuous in $a$.
**H$_3$**. The initial distribution $\nu$ is such that $\sum_{i \in S} i\nu(i) < \infty$.
**H$_4$**. $\beta > \max\{\alpha^{-1}\tilde{c}(\mu_1), \alpha^{-1}\tilde{c}(\lambda_1^*)\}$.

Under these conditions, we obtain the following.

**Proposition 11.13** *Under conditions* **H$_1$**, **H$_2$**, **H$_3$**, *and* **H$_4$**, *the above birth-and-death system satisfies Assumptions* 11.1 *and* 11.3. *Therefore (by Theorem* 11.4), *there exists a discount constrained-optimal policy.*

*Proof* Let $w(i) := 1 + i$ and $w'(i) := (1+i)^2$ for all $i \in S$, and let $\|\tilde{c}\| := \sup_{a \in A} |\tilde{c}(a)|$. By **H$_2$**, we have $\|\tilde{c}\| < \infty$. Furthermore, for each $(i,a) \in K$,

$$\text{when } i \geq 1\text{:} \quad \sum_{j \in S} w(j)q(j|i,a) = \lambda i - a \leq \lambda w(i) + \mu_2, \tag{11.30}$$

$$\text{when } i = 0\text{:} \quad \sum_{j \in S} w(j)q(j|0,a) \leq \lambda w(0) + \lambda_2^*, \tag{11.31}$$

$$\text{for } i \geq 0: \quad \sum_{j \in S} w'(j) q(j|i,a) \leq 2(\lambda + \lambda_2^*) w'(i). \tag{11.32}$$

Then, from (11.28), (11.30), and (11.31) we see that Assumption 2.2 holds. Moreover, Assumption 6.4 follows from (11.28), (11.30)–(11.31), and $\mathbf{H_1}$. Also, from (11.28), (11.32), and $\mathbf{H_2}$ we see that Assumption 6.8 is true, and thus Assumption 11.1(a) is verified. Similarly, Assumption 11.1(b) follows from $\mathbf{H_2}$, (11.29), and the model's description. Hence, Assumption 11.1 is satisfied. Finally, it remains to verify Assumption 11.3(b), because Assumption 11.3(a) trivially follows from $\mathbf{H_3}$ and $w(i) = i + 1$. Let $f$ be the stationary policy that always chooses the death parameter $\mu_1$ at $i \geq 1$ and $\lambda_1^*$ at $i = 0$ (i.e., $f(i) := \mu_1$ for all $i \geq 1$, and $f(0) := \lambda_1^*$). By (11.29) and (11.3), the associated $\alpha$-discounted cost $J_\alpha(\nu, f)$ does not exceed $\max\{\alpha^{-1}\tilde{c}(\mu_1), \alpha^{-1}\tilde{c}(\lambda_1^*)\}$, which, together with $\mathbf{H_4}$, yields Assumption 11.3(b). □

## 11.5 Notes

Constrained Markov decision processes form an important class of stochastic control problems with applications in many areas. See, for instance, Beutler and Ross (1985) [13], Feinberg and Shwartz (1996) [40], Guo (2000) [49], Hernández-Lerma and González-Hernández (2000) [71], Hordijk and Kallenberg (1984) [79], Kurano, Nakagami and Huang (2002) [101], Piunovskiy and Mao (2000) [121], Sennott (1991) [140], Tanaka (1991) [147], as well as the books by Altman (1999) [3], Hou and Guo (1998) [84], Kallenberg (1983) [96], Piunovskiy (1997) [119], Puterman (1994) [129], and their extensive references. The main results in this chapter are from Guo and Hernández-Lerma (2003e) [57].

# Chapter 12
# Constrained Optimality for Average Criteria

In this chapter, we extend the main results in Chap. 11 to the average-reward case. Since the approach in this chapter is very similar to that in Chap. 11, some arguments are essentially a repetition of those in that chapter. However, for completeness and ease of reference, we sketch here the main arguments.

## 12.1 Average Optimality with a Constraint

The constrained continuous-time MDP model we are concerned with is the same as in (11.1), i.e.,

$$\{S, A, (A(i)|i \in S), q(j|i,a), r(i,a), c(i,a), \beta, \nu\}. \tag{12.1}$$

However, we here consider the average criteria, which are defined as follows:

Under Assumptions 7.1 and 11.1(b), for each $\pi \in \Pi$, let

$$\bar{V}(\nu, \pi) := \liminf_{T\to\infty} \frac{1}{T} E_\nu^\pi \left[ \int_0^T r(x(t), \pi_t)\, dt \right], \tag{12.2}$$

$$J_c(\nu, \pi) := \limsup_{T\to\infty} \frac{1}{T} E_\nu^\pi \left[ \int_0^T c(x(t), \pi_t)\, dt \right]. \tag{12.3}$$

For the constraint constant $\beta$ in (12.1), let

$$U := \{\pi \in \Pi, J_c(\nu, \pi) \le \beta\}$$

be the set of "constrained" policies.

**Definition 12.1** A policy $\pi^* \in U$ is said to be *average constrained-optimal* if $\pi^*$ maximizes the expected average reward in (12.2) over all $\pi$ in $U$, that is,

$$\bar{V}(\nu, \pi^*) = \sup_{\pi \in U} \bar{V}(\nu, \pi).$$

X. Guo, O. Hernández-Lerma, *Continuous-Time Markov Decision Processes*, Stochastic Modelling and Applied Probability 62, DOI 10.1007/978-3-642-02547-1_12, © Springer-Verlag Berlin Heidelberg 2009

To show the existence of an average constrained-optimal policy, we impose the following conditions.

**Assumption 12.2**

(a) Assumption 7.1 is satisfied.
(b) Assumptions 7.4 and 7.5 hold for all $f := \pi \in \Pi_s$.
(c) $c(i,a)$ is continuous in on $A(i)$ for each fixed $i \in S$, and $0 \le c(i,a) \le Mw(i)$ for all $i \in S$ and $a \in A(i)$, with $w(i)$ as in Assumption 2.2 and $M$ as in Assumption 7.1(c).

**Assumption 12.3** The set $U_0 := \{\pi \in \Pi : J_c(\nu,\pi) < \beta\}$ is nonempty.

The main result in this chapter is as follows.

**Theorem 12.4** *Suppose that Assumptions* 12.2 *and* 12.3 *hold. Then there exists an average constrained-optimal policy that may be either a stationary policy or a randomized stationary policy which differ in at most one state; that is, there exist two stationary policies* $f^1, f^2$, *a state* $i^* \in S$, *and a number* $p \in [0,1]$ *such that* $f^1(i) = f^2(i)$ *for all* $i \neq i^*$ *and, in addition, the randomized stationary policy* $\pi^p(\cdot|i)$ *given by*

$$\pi^p(a|i) = \begin{cases} p & \text{if } a = f^1(i^*), i = i^*, \\ 1-p & \text{if } a = f^2(i^*), i = i^*, \\ 1 & \text{if } a = f^1(i), i \neq i^*, \end{cases} \tag{12.4}$$

*is average constrained-optimal.*

Theorem 12.4 will be proved in Sect. 12.3.

## 12.2 Preliminaries

In this section we present some results needed to prove Theorem 12.4.

**Lemma 12.5** *Suppose that Assumption* 12.2 *holds. Then the functions* $\bar{V}(\nu, f)$ *and* $J_c(\nu, f)$ *are continuous in* $f \in F$.

*Proof* It suffices to prove the continuity of $\bar{V}(\nu, f)$ on $F$ because, obviously, the other case is similar. In fact, under Assumption 12.2, by (7.3) and (12.2) we have $g(f) = \bar{V}(\nu, f)$ for all $f \in F$. Thus, this lemma follows from Lemma 9.17. □

To analyze the constrained control problem in Definition 12.1, we introduce a Lagrange multiplier $\gamma \ge 0$ as follows.

With $b^\gamma(i,a)$ as in (11.10), for any policy $\pi=(\pi_t)\in\Pi$ and initial distribution $\nu$ on $S$, let

$$J_b^\gamma(\nu,\pi):=\liminf_{T\to\infty}\frac{1}{T}E_\nu^\pi\left[\int_0^T b^\gamma\big(x(t),\pi_t\big)\,dt\right], \tag{12.5}$$

which is expressed as

$$J_b^\gamma(i,\pi):=\liminf_{T\to\infty}\frac{1}{T}E_i^\pi\left[\int_0^T b^\gamma\big(x(t),\pi_t\big)\,dt\right]$$

if $\nu$ is concentrated at the state $i$. Now define the associated optimal value functions

$$J_b^{*\gamma}(i):=\sup_{\pi\in\Pi}J_b^\gamma(i,\pi)\quad\text{and}\quad J_b^{*\gamma}(\nu):=\sup_{\pi\in\Pi}J_b^\gamma(\nu,\pi).$$

Under Assumption 12.2, all the conclusions in Chap. 7 remain true when we replace $r(i,a)$ with $c(i,a)$ or $b^\gamma(i,a)$. Thus, by Theorem 7.8 we know that $J_b^{*\gamma}(i)$ is a constant, denoted by $g_b^{*\gamma}$, and, moreover,

$$g_b^{*\gamma}:=J_b^{*\gamma}(i)=J_b^{*\gamma}(\nu)=\sup_{\pi\in\Pi_s}J_b^\gamma(\nu,\pi)\quad\forall i\in S. \tag{12.6}$$

Furthermore, there exists a function $u_b^\gamma\in B_w(S)$ (depending on $b^\gamma$) such that

$$g_b^{*\gamma}=\sup_{a\in A(i)}\left\{b^\gamma(i,a)+\sum_{j\in S}u_b^\gamma(j)q(j|i,a)\right\}\quad\forall i\in S, \tag{12.7}$$

and the maximum in (12.7) can be actually attained, that is, the set

$$A_\gamma^*(i):=\left\{a\in A(i)\mid g_b^{*\gamma}=b^\gamma(i,a)+\sum_{j\in S}u_b^\gamma(j)q(j|i,a)\ \forall i\in S\right\} \tag{12.8}$$

is nonempty. Hence, the sets

$$\bar F_\gamma^*:=\left\{f\in F: f(i)\in A_\gamma^*(i)\ \forall i\in S\right\} \tag{12.9}$$

and

$$\bar\Pi_s^\gamma:=\left\{\pi\in\Pi_s:\pi\big(A_\gamma^*(i)|i\big)=1\ \forall i\in S\right\}$$

are nonempty. The following lemma further shows that $\bar\Pi_s^\gamma$ is convex.

**Lemma 12.6** *Under Assumption* 12.2, *the set* $\bar\Pi_s^\gamma$ *is convex.*

*Proof* For all $\pi_1,\pi_2\in\bar\Pi_s^\gamma$ and $p\in[0,1]$, let

$$\pi^p(\cdot|i):=p\pi_1(\cdot|i)+(1-p)\pi_2(\cdot|i)\quad\forall i\in S.$$

Then

$$\pi^p\big(A_\gamma^*(i)|i\big) = p\pi_1\big(A_\gamma^*(i)|i\big) + (1-p)\pi_2\big(A_\gamma^*(i)|i\big) = 1,$$

and so $\Pi_s^\gamma$ is convex. □

As in Chap. 11, we introduce some convenient notation: For each $\gamma \ge 0$, we take an arbitrary but fixed policy $f^\gamma \in \bar{F}_\gamma^*$ and write $\bar{V}(\nu, f^\gamma)$, $J_c(\nu, f^\gamma)$, and $J_b^\gamma(\nu, f^\gamma)$ as $\bar{V}(\gamma)$, $J_c(\gamma)$, and $J_b(\gamma)$, respectively. Then, by Theorem 7.8 we have that $J_b^\gamma(i, f^\gamma) = g_b^{*\gamma}$ for all $i \in S$ and $f^\gamma \in \bar{F}_\gamma^*$. Hence, $J_b(\gamma) = J_b^\gamma(\nu, f^\gamma) = g_b^{*\gamma}$.

**Lemma 12.7.** *Suppose that Assumption* 12.2 *holds. Then* $\bar{V}(\gamma)$, $J_c(\gamma)$, *and* $J_b(\gamma)$ *are nonincreasing in* $\gamma \in [0, \infty]$.

*Proof* This proof is similar to that of Lemma 11.8 and thus it is omitted. □

**Lemma 12.8** *Suppose that Assumption* 12.2 *holds. Then* $J_b(\gamma) = g_b^{*\gamma}$ *is continuous in* $\gamma \in [0, \infty)$.

*Proof* Since it follows from Assumption 12.2 and (7.2) that $J_b^\gamma(\nu, \pi) = \bar{V}(\nu, \pi) - \gamma J_c(\nu, \pi)$ for all $\pi \in \Pi_s$, for all $\gamma_1, \gamma_2 \ge 0$ and $p \in [0, 1]$, by (12.6) we have

$$\begin{aligned}
&g_b^{*(p\gamma_1+(1-p)\gamma_2)} \\
&\quad = \sup_{\pi\in\Pi_s} \big\{\bar{V}(\nu, \pi) - \big(p\gamma_1 + (1-p)\gamma_2\big) J_c(\nu, \pi)\big\} \\
&\quad = \sup_{\pi\in\Pi_s} \big\{p\big[\bar{V}(\nu, \pi) - \gamma_1 J_c(\nu, \pi)\big] + (1-p)\big[\bar{V}(\nu, \pi) - \gamma_2 J_c(\nu, \pi)\big]\big\} \\
&\quad \le p g_b^{*\gamma_1} + (1-p) g_b^{*\gamma_2},
\end{aligned}$$

that is, $g_b^{*\gamma}$ is a convex function of $\gamma$, and, therefore, it is continuous in $\gamma \in (0, \infty)$.

On the other hand, for any sequence $\gamma_k \in (0, \infty)$ such that $\lim_{k\to\infty} \gamma_k = 0$, by Lemmas 12.7 and 7.2 we have

$$g_b^{*0} \ge g_b^{*\gamma_k} = \sup_{\pi\in\Pi_s} \big\{\bar{V}(\nu, \pi) - \gamma_k J_c(\nu, \pi)\big\} \ge g_b^{*0} - \frac{2b_1 M}{c_1}\gamma_k.$$

Hence, letting $k \to \infty$ gives $\lim_{k\to\infty} g_b^{*\gamma_k} = g_b^{*0}$. Thus, $g_b^{*\gamma}$ is continuous on $[0, \infty)$. □

**Lemma 12.9** *Suppose that Assumption* 12.2 *holds. If* $\lim_{k\to\infty} \gamma_k = \gamma$ *and* $f^{\gamma_k} \in \bar{F}_{\gamma_k}^*$ *is such that* $\lim_{k\to\infty} f^{\gamma_k} = f$, *then* $f$ *belongs to* $\bar{F}_\gamma^*$.

*Proof* As $f^{\gamma_k} \in \bar{F}_{\gamma_k}^*$, by (12.9) and (12.8) we have

$$g_b^{*\gamma} = b^{\gamma_k}\big(i, f^{\gamma_k}\big) + \sum_{j\in S} u_b^{\gamma_k}(j) q\big(j|i, f^{\gamma_k}\big) \quad \forall i \in S. \tag{12.10}$$

Moreover, since $\lim_{k\to\infty}\gamma_k=\gamma$, there exists a constant $B$ such that $\sup_{k\geq 1}(|\gamma_k|+|\gamma|)\leq B$. Thus, it follows from Assumption 12.2 that $|b^{\gamma'}(i,a)|\leq(1+B)Mw(i)$ for all $(i,a)\in K$, and $\gamma'=\gamma_k,\gamma$. Hence, by Lemma 7.7 and the proof of Theorem 7.8, $\{u_b^{\gamma_k}\}$ is a sequence in the compact metric space $\prod_{i\in S}[-\bar{h}(i),\bar{h}(i)]$ with $\bar{h}(i):=\frac{2L_2(1+B)M}{\delta}[1+w(i_0)]w(i)$ and $i_0$ an arbitrary but fixed state. Then, the Tychonoff Theorem (see Propositions A.6 and A.7) gives the existence of both a subsequence $\{\gamma_{k_m}\}$ of $\{\gamma_k\}$ and a function $\bar{v}$ on $S$ such that $\lim_{m\to\infty}u_b^{\gamma_{k_m}}(i)=\bar{v}(i)$ for all $i\in S$. Moreover, by Lemma 12.8 we have

$$\lim_{m\to\infty} g_b^{*\gamma_{k_m}}=g_b^{*\gamma}.$$

On the other hand, for all $m\geq 1$ and $i\in S$, by the conservativeness property (2.3) and (12.8) we have (recall (6.11))

$$\frac{g_b^{*\gamma_{k_m}}}{m(i)}+u_b^{\gamma_{k_m}}(i)=\frac{b^{\gamma_{k_m}}(i,f^{\gamma_{k_m}})}{m(i)}+\sum_{j\in S}u_b^{\gamma_{k_m}}(j)\left[\frac{q(j|i,f^{\gamma_{k_m}})}{m(i)}+\delta_{ij}\right]\quad\forall i\in S. \tag{12.11}$$

Letting $m\to\infty$, Proposition A.3, together with (12.11), gives

$$\frac{g_b^{*\gamma}}{m(i)}+\bar{v}(i)=\frac{b^{\gamma}(i,f)}{m(i)}+\sum_{j\in S}\bar{v}(j)\left[\frac{q(j|i,f)}{m(i)}+\delta_{ij}\right]\quad\forall i\in S,$$

and so

$$g_b^{*\gamma}=b^{\gamma}(i,f)+\sum_{j\in S}\bar{v}(j)q(j|i,f)\quad\forall i\in S. \tag{12.12}$$

By Proposition 7.3 and (12.12) we have

$$J_b^{\gamma}(\nu,f)=g_b^{*\gamma},$$

which means that $f$ is AR optimal (with respect to the new reward $b^{\gamma}(i,a)$). Then Theorem 7.8(b) (with $r(i,a)$ being replaced by $b^{\gamma}(i,a)$) implies that $f$ realizes the maximum of (12.7), that is, $f\in\bar{F}_{\gamma}^{*}$. □

Under Assumption 12.2, it follows from Lemma 12.7 that the nonnegative constant

$$\bar{\gamma}:=\inf\{\gamma\geq 0: J_c(\gamma)\leq\beta\} \tag{12.13}$$

is well defined.

**Lemma 12.10** *Suppose that Assumptions* 12.2 *and* 12.3 *hold. Then the constant* $\bar{\gamma}$ *in* (12.13) *is finite, that is,* $\bar{\gamma}$ *is in* $[0,\infty)$.

*Proof* By Theorem 7.8, it follows that $\inf_{\pi\in\Pi}J_c(\nu,\pi)=\inf_{f\in F}J_c(\nu,f)$. Thus, Assumption 12.3 gives that existence of $f\in F$ such that $J_c(\nu,f)<\beta$. Then the proof is completed as in the proof of Lemma 11.11. □

## 12.3 Proof of Theorem 12.4

In this section, we prove Theorem 12.4 using the following lemma.

**Lemma 12.11** *Under Assumptions* 12.2 *and* 12.3, *if there exist* $\gamma_0 \ge 0$ *and* $\pi^* \in \Pi_s$ *such that*

$$J_c(\nu, \pi^*) = \beta \quad \text{and} \quad J_b^{\gamma_0}(\nu, \pi^*) = g_b^{*\gamma_0}, \tag{12.14}$$

*then the policy* $\pi^*$ *is average constrained-optimal.*

*Proof* For any $\pi \in U$, since $b^\gamma(i,a) = r(i,a) - \gamma c(i,a)$, by (12.5) and Proposition A.1(d) we have

$$J_b^\gamma(\nu, \pi) \ge \bar{V}(\nu, \pi) - \gamma J_c(\nu, \pi) \quad \forall \gamma \ge 0. \tag{12.15}$$

Also, since $\pi^* \in \Pi_s$, by (7.2) and Assumption 12.2 we have $\bar{V}(\nu, \pi^*) - \gamma_0 J_c(\nu, \pi^*) = J_b^{\gamma_0}(\nu, \pi^*) = g_b^{*\gamma_0} \ge J_b^{\gamma_0}(\nu, \pi)$. Hence, from (12.15) we have

$$\bar{V}(\nu, \pi^*) - \gamma_0 J_c(\nu, \pi^*) \ge J_b^{\gamma_0}(\nu, \pi) \ge \bar{V}(\nu, \pi) - \gamma_0 J_c(\nu, \pi),$$

which, together with $J_c(\nu, \pi^*) = \beta$ and $J_c(\nu, \pi) \le \beta$ (because $\pi \in U$), implies

$$\bar{V}(\nu, \pi^*) \ge \bar{V}(\nu, \pi) + \gamma_0(\beta - J_c(\nu, \pi)) \ge \bar{V}(\nu, \pi) \quad \forall \pi \in U.$$

This means that $\pi^*$ is average constrained-optimal. □

*Proof of Theorem 12.4* By Lemma 12.10, the constant $\bar{\gamma}$ is in $[0, \infty)$. Thus, we may consider two cases, $\bar{\gamma} = 0$ and $\bar{\gamma} > 0$. For the case $\bar{\gamma} = 0$, as in the proof of Theorem 11.4, using Lemmas 7.2 and 12.11, we obtain Theorem 12.4. Also, for the other case $\bar{\gamma} \in (0, \infty)$, following the proof of Theorem 11.4, but now using Lemmas 12.7–12.11 and Proposition 7.11 with $f$ replaced by $\pi$ in $\Pi_s$, we again obtain Theorem 12.4. (The details are omitted but can be found in Zhang and Guo (2008) [167].) □

## 12.4 An Example

In this section, we illustrate our conditions with an example.

*Example 12.1* Consider a queueing system in which the state variable $i \in S := \{0, 1, \ldots\}$ denotes the total number of customers (in service and waiting in the queue) at any time $t \ge 0$. There are "natural" arrival and service rates, say $\hat{\lambda}(> 0)$ and $\hat{\mu}(> 0)$, respectively. When the state of the system is $i \ge 1$ and a decision-maker takes an action (namely, a service rate) $a$ from a given set $A(i) \equiv [\mu_1, \mu_2]$ $(\mu_2 > \mu_1 \ge 0)$, a cost denoted by $\hat{c}(a) \ge 0$ per unit time is produced. In addition, suppose that $\hat{p}(a) > 0$ is a fixed fee per customer, so that the net reward for the decision-maker is $r(i,a) := \hat{p}(a)i - \hat{c}(a)$ for each unit of time that the system remains in the state $i \geq 1$. When $i = 0$ (i.e., there is no any customer in the queueing system, and thus the decision-maker needs no service), the transition rates of this system are assumed to be given as (12.16) below, and let $r(0,a) \equiv 0$. On the other hand, suppose that the decision-maker wishes to keep the service cost bounded above by a constant $\beta (> 0)$.

We formulate this controlled queueing system as a constrained continuous-time MDP. The state space is $S = \{0, 1, \ldots\}$, and the corresponding transition rates and reward/cost functions are as follows: For $i = 0$ and $a \in A(0) := \{\hat{\lambda}\}$,

$$q(1|0,a) = -q(0|0,a) := \hat{\lambda}, \qquad q(j|0,a) = 0 \quad \text{for } j \geq 2, \tag{12.16}$$

and, for $i \geq 1$ and $a \in A(i)$,

$$q(j|i,a) := \begin{cases} \hat{\mu}i + a & \text{if } j = i-1, \\ -(\hat{\mu} + \hat{\lambda})i - a & \text{if } j = i, \\ \hat{\lambda}i & \text{if } j = i+1, \\ 0 & \text{otherwise,} \end{cases} \tag{12.17}$$

$$r(0,\hat{\lambda}) = 0, \qquad r(i,a) = \hat{p}(a)i - \hat{c}(a), \quad c(i,a) = \hat{c}(a)$$
$$\forall i \geq 1 \text{ and } a \in A(i). \tag{12.18}$$

The aim now is to find conditions under which there exists a stationary policy achieving the maximum expected average reward, and at the same time, the associated expected average cost does not exceed the cost $\beta$. To do this, we consider the following conditions.

**I$_1$**. $\hat{\mu} > \hat{\lambda} > 0$.
**I$_2$**. $\hat{c}(a)$ and $\hat{p}(a)$ are continuous in $a \in [\mu_1, \mu_2]$.
**I$_3$**. $\beta > \hat{c}(\mu_1)$.

Under these conditions, we obtain the following.

**Proposition 12.12** *Under conditions* **I$_1$**, **I$_2$**, *and* **I$_3$**, *the above queueing system satisfies Assumptions* 12.2 *and* 12.3. *Therefore (by Theorem* 12.4), *there exists an average constrained-optimal policy.*

*Proof* Let $w(i) := i + 1$ for all $i \in S$, and $\|\hat{c}\| := \sup_{a \in A} |\hat{c}(a)|$. Then, for each $(i,a) \in K$,

$$\text{when } i \geq 1: \quad \sum_{j \in S} w(j) q(j|i,a) \leq -(\hat{\mu} - \hat{\lambda}) w(i) + \mu_2 + \hat{\mu}, \tag{12.19}$$

$$\text{when } i = 0: \quad \sum_{j \in S} w(j) q(j|0,a) = \hat{\lambda} \leq -(\hat{\mu} - \hat{\lambda}) w(0) + \hat{\mu}. \tag{12.20}$$

Hence, $\mathbf{I}_1$ and $\mathbf{I}_2$, together with (12.17)–(10.28), give Assumptions 7.1(a)–(c). Also, as in the verification of Proposition 7.16, we see that Assumptions 6.8, 7.4, and 7.5 are all true, and so Assumption 12.2 follows.

Finally, it only remains to verify Assumption 12.3. Let $f \equiv \mu_1$. By (12.18), the associated average cost $J_c(\nu, f)$ equals $\hat{c}(\mu_1)$, which, together with $\mathbf{I}_3$, yields Assumption 12.3. □

## 12.5 Notes

In the last two Chaps. 11 and 12, we have shown the existence of a constrained-optimal policy for the discount or average criteria. The common feature in these two chapters is that there is a single constraint. In the case of one constraint, the Lagrange multiplier technique not only shows the existence of a constrained-optimal stationary policy; in fact, it establishes that such a constrained-optimal policy is a randomized stationary policy which differs in at most one state. However, such an approach may not apply directly to the case of multiple constraints, which has been dealt with by introducing occupation measures in Guo (2007b) [51].

On the other hand, the algorithm for computing constrained-optimal policies for the *finite* models of constrained continuous-time MDPs can be given by using the *uniformization technique* and the associated results in Puterman (1994) [129] and Sennott (1999) [141] for discrete-time MDPs. However, it is worthy and desirable to develop algorithms of computing constrained-optimal policies for the more general case of constrained continuous-time MDPs with unbounded transition or reward rates.

# Appendix A

Some results from real analysis and measure theory are repeatedly used throughout the book, and for ease of reference they are collected in this appendix. Since the proofs of these results are not required for an understanding of the text, they are omitted, but we give some related references.

## A.1 Limit Theorems

This section presents some useful theorems about limits.

**Proposition A.1** *Let $G$ be a finite (nonempty) set and $u(a,n)$ a real-valued function of $a \in G$ and $n \in \mathcal{N}_+ := \{0,1,2,\ldots\}$, the set of nonnegative integers.*

(a) $\liminf_{n\to\infty}[\min_{a\in G} u(a,n)] = \min_{a\in G}[\liminf_{n\to\infty} u(a,n)]$.
(b) *If* $\lim_{n\to\infty} u(a,n)$ *exists for all* $a \in G$, *then*

$$\lim_{n\to\infty}\Big[\min_{a\in G} u(a,n)\Big] = \min_{a\in G}\Big[\lim_{n\to\infty} u(a,n)\Big].$$

(c) $\liminf_{n\to\infty}\sum_{a\in G} u(a,n) \ge \sum_{a\in G}\liminf_{n\to\infty} u(a,n)$.
(d) *In particular (for $G = \{1,2\}$),*

$$\begin{aligned}\liminf_{n\to\infty} u(1,n) + \liminf_{n\to\infty} u(2,n) &\le \liminf_{n\to\infty}\big[u(1,n)+u(2,n)\big]\\ &\le \liminf_{n\to\infty} u(1,n) + \limsup_{n\to\infty} u(2,n)\\ &\le \limsup_{n\to\infty}\big[u(1,n)+u(2,n)\big]\\ &\le \limsup_{n\to\infty} u(1,n) + \limsup_{n\to\infty} u(2,n).\end{aligned}$$

*Proof* Parts (a)–(c) follow from Propositions A.1.3 and A.1.5 in Sennott (1999) [141], and (d) is from Klambauer (1973) [99, Sect. 4]. (It can be shown that Proposition A.1(a) does not hold when "min" is replaced by "max".) □

X. Guo, O. Hernández-Lerma, *Continuous-Time Markov Decision Processes*,
Stochastic Modelling and Applied Probability 62,
DOI 10.1007/978-3-642-02547-1, © Springer-Verlag Berlin Heidelberg 2009

The next result generalizes Proposition A.1 to the case of nonnegative functions and may be useful in the approximation of denumerable state models by finite state models.

**Proposition A.2** *Let $S$ be a countable set. Suppose that $\{E_n\}$ $(n \geq 0)$ is an increasing sequence of subsets of $S$ such that $\bigcup E_n = S$. Let $u(i,n)$ be a nonnegative function of $i \in E_n$ (or of $i \in S$) and $n \in \mathcal{N}_+$. Then,*

$$\liminf_{n\to\infty} \sum_{i\in E_n} u(i,n) \geq \sum_{i\in S} \liminf_{n\to\infty} u(i,n).$$

*Proof* See Proposition A.1.8 in Sennott (1999) [141]. □

We now present versions of Fatou's lemma and the dominated convergence theorem for the case in which the probability measure may also be a function of other parameters.

**Proposition A.3** (Generalized Fatou lemma) *Suppose that the following conditions hold*:

(i) *$S$ is a countable set with a probability measure $(p_i, i \in S)$ (i.e., $p_i \geq 0$ for all $i \in S$, and $\sum_{i\in S} p_i = 1$).*
(ii) *$\{E_n\}$ $(n \geq 0)$ is an increasing sequence of subsets of $S$ such that $\bigcup E_n = S$.*
(iii) *$(p_i(n), i \in E_n)$ is a probability measure on $S$ for each $n$, concentrated on $E_n$ (i.e., $\sum_{i\in E_n} p_i(n) = 1$) and such that $\lim_{n\to\infty} p_i(n) = p_i$ for all $i \in S$.*
(iv) *$u(i,n)$ is a nonnegative function of $i \in E_n$ and $n \in \mathcal{N}_+$.*

*Then* $\liminf_{n\to\infty} \sum_{i\in E_n} u(i,n) p_i(n) \geq \sum_{i\in S} [\liminf_{n\to\infty} u(i,n)] p_i$.

*Proof* See Proposition A.2.5 in Sennott (1999) [141]. □

As in the proof of the standard dominated convergence theorem, from Proposition A.3 above we obtain the following fact.

**Proposition A.4** (Generalized dominated convergence theorem) *Suppose that* (i)–(iii) *in Proposition* A.3 *hold, and in addition suppose that*:

(iv) *There exist functions $u$ and $w$ on $E_n \times \mathcal{N}_+$ and $S$, respectively, such that $|u(i,n)| \leq w(i)$ for all $i \in E_n$ and $n \in \mathcal{N}_+$, and $\sum_{i\in S} w(i) p_i < \infty$.*
(v) *The limits $\lim_{n\to\infty} u(i,n) =: \hat{u}(i)$ exist for all $i \in S$.*

*Then* $\lim_{n\to\infty} \sum_{i\in E_n} u(i,n) p_i(n) = \sum_{i\in S} \hat{u}(i) p_i$.

*Proof* The proposition follows from Theorem A.2.6 in Sennott (1999) [141] or Lemma 8.3.7 in Hernández-Lerma and Lasserre (1999) [74]. □

A crucial link between the infinite-horizon discounted-reward and the long-run average-reward criteria is provided by the following fact.

**Proposition A.5** (Tauberian theorem) *Suppose that $u(s)$ is a nonnegative real-valued measurable function of $s \geq 0$. Let $U(t) := \int_0^t u(s)\,ds$ for each $t \geq 0$, and $V(\alpha) := \int_0^\infty e^{-\alpha s} u(s)\,ds$ for each $\alpha > 0$. Then,*

$$\liminf_{t\to\infty} \frac{U(t)}{t} \leq \liminf_{\alpha\downarrow 0} \alpha V(\alpha) \leq \limsup_{\alpha\downarrow 0} \alpha V(\alpha) \leq \limsup_{t\to\infty} \frac{U(t)}{t}.$$

*Proof* See the Tauberian Theorem in Widder (1946) [156]. □

**Proposition A.6** *Suppose that $G$ is a finite or countable set and that $\{E_i, i \in G\}$ is a sequence of compact subsets of a metric space $(E, \rho)$ with metric $\rho$. Let $X := \prod_{i\in G} E_i$ be the product metric space of $\{E_i, i \in G\}$.*

*Then for any sequence $\{x_n, n = 1, 2, \ldots\}$ in $X$ (i.e., $x_n := (x_n(i), i \in G) \in X$ for all $n \geq 1$), there exist a subsequence $\{x_{n_k}\}$ of $\{x_n\}$ and a point $x \in X$ such that $\lim_{k\to\infty} x_{n_k}(i) = x(i)$ for all $i \in G$ (where the convergence is with respect to $\rho$).*

Proposition A.6 is a special case of Tikhonov's product theorem. As a particular case of Proposition A.6, we have the following fact.

**Proposition A.7** *Let $\{u_n\}$ $(n \geq 0)$ be a sequence of real-valued functions on a countable set $S$. Suppose that there exist two real-valued functions $v_k$ $(k = 1, 2)$ on $S$ such that $v_1(i) \leq u_n(i) \leq v_2(i)$ for all $i \in S$ and $n \geq 1$. Then there exist a subsequence $\{u_{n_k}\}$ and a function $u$ on $S$, with $v_1 \leq u \leq v_2$, such that $\lim_{k\to\infty} u_{n_k}(i) = u(i)$ for all $i \in S$.*

*Proof* This result follows from Proposition A.6 with $x_n(i) := u_n(i)$ and $E_i := [v_1(i), v_2(i)]$ for all $i \in G := S$. □

## A.2 Results from Measure Theory

In this section, we collect some important results from measure theory that are needed in some arguments in this book.

Let $\mathcal{F}$ be a collection of subsets of a nonempty set $X$. Then $\mathcal{F}$ is called an *algebra* (also known as a *field*) if and only if $X$ is in $\mathcal{F}$ and $\mathcal{F}$ is closed under complementation and finite unions, that is,

(a) $X \in \mathcal{F}$.
(b) If $A \in \mathcal{F}$, then $A^c \in \mathcal{F}$, where $A^c$ denotes the complement of $A$ (relative to $X$).
(c) If $A_1, A_2, \ldots, A_n$ $(n < \infty)$ are in $\mathcal{F}$, then $\bigcup_{k=1}^n A_k$ is in $\mathcal{F}$.

If in addition (c) is replaced by closure under *countable* unions, that is, (c′) If $A_1, A_2, \ldots$ are in $\mathcal{F}$, then $\bigcup_{k=1}^\infty A_k$ is in $\mathcal{F}$, then $\mathcal{F}$ is called a *$\sigma$-algebra* (also known as a *$\sigma$-field*).

A pair $(X, \mathcal{F})$ consisting of a (nonempty) set $X$ and a $\sigma$-algebra $\mathcal{F}$ of subsets of $X$ is called a *measurable space*.

If $\Gamma$ is a class of subsets of $X$, the smallest $\sigma$-algebra that contains $\Gamma$ will be denoted by $\sigma(\Gamma)$, and it is called the *minimal $\sigma$-algebra over $\Gamma$* (or the $\sigma$-algebra generated by $\Gamma$).

A function with values in the set of extended real numbers $\bar{\mathbb{R}} := \mathbb{R} \cup \{+\infty, -\infty\}$ is called an extended real-valued function. An extended real-valued function $\mu$ on an algebra $\mathcal{F}$ (not necessarily a $\sigma$-algebra) is called *countably additive* if and only if, for any finite or countable collection of *disjoint* sets $A_1, A_2, \dots$ in $\mathcal{F}$ whose union also belongs to $\mathcal{F}$, we have

$$\mu\left(\bigcup_n A_n\right) = \sum_n \mu(A_n).$$

If this requirement holds only for *finite* collections of disjoint sets in $\mathcal{F}$, then $\mu$ is said to be *finitely additive*. A nonnegative, finitely additive function $\mu$ on the algebra $\mathcal{F}$ is called *$\sigma$-finite* if and only if $X$ can be written as $\bigcup_{n=1}^{\infty} A_n$ with $A_n \in \mathcal{F}$ and $\mu(A_n) < \infty$ for all $n$.

**Definition A.8** Let $X$ be an arbitrary (nonempty) set, and $\mathcal{F}$ a family of subsets of $X$. If $\mathcal{F}$ is an algebra and $\mu$ is countably additive on $\mathcal{F}$, then $\mu$ is called a *signed measure*. If in addition $\mu(A) \geq 0$ for all $A \in \mathcal{F}$, then $\mu$ is called a *measure*, and it is said to be a probability measure if $\mu(X) = 1$. If $\mu$ is a measure on a $\sigma$-algebra $\mathcal{F}$, then the triplet $(X, \mathcal{F}, \mu)$ is called a *measure space*. If in addition $\mu(X) = 1$, then the measure space $(X, \mathcal{F}, \mu)$ is called a probability space.

In many cases, we can construct a measure on an algebra, which we then need to extend to a measure on a $\sigma$-algebra. To this end, we may use the following fact.

**Proposition A.9** (Carathéodory extension theorem) *Let $\mu$ be a $\sigma$-finite measure on an algebra $\mathcal{F}$. Then $\mu$ has a unique extension to a $\sigma$-finite measure on the minimal $\sigma$-algebra $\sigma(\mathcal{F})$. That is, there exists a unique $\sigma$-finite measure $\bar{\mu}$ on $\sigma(\mathcal{F})$ such that*

$$\bar{\mu}(A) = \mu(A) \quad \forall A \in \mathcal{F}.$$

*Proof* See Ash (2000) [6, p. 19], for instance. □

In the next two propositions, $(X, \mathcal{F})$ denotes a measurable space.

**Proposition A.10** (Jordan–Hahn decomposition theorem) *Let $\mu$ be a signed measure on $\mathcal{F}$. For every $A \in \mathcal{F}$, define*

$$\mu^+(A) := \sup\{\mu(B) : B \in \mathcal{F}, B \subset A\}, \tag{A.1}$$

$$\mu^-(A) := -\inf\{\mu(B) : B \in \mathcal{F}, B \subset A\}. \tag{A.2}$$

*Then $\mu^+$ and $\mu^-$ are measures on $\mathcal{F}$, and $\mu = \mu^+ - \mu^-$.*

*Proof* See Ash (2000) [6, p. 62], for instance. □

The measure $|\mu| := \mu^+ + \mu^-$ is called the *total variation* of $\mu$. A signed measure $\mu$ is said to be finite if $|\mu|(X) < \infty$.

A real-valued function $u$ on $X$ is called (Borel) *$\mathcal{F}$-measurable* if and only if the set $\{x : u(x) \leq t\}$ is in $\mathcal{F}$ for all $t \in (-\infty, \infty)$. For any $\mathcal{F}$-measurable function $u$ and a signed measure $\mu$ on $\mathcal{F}$, the integral of $u$ with respect to $\mu$ is denoted by $\int_X u\,d\mu := \int_X u\,d\mu^+ - \int_X u\,d\mu^-$, whenever one of the two integrals is finite.

Let $M(X)$ be the set of all finite signed measures on $\mathcal{F}$. Consider the *total variation norm* of $\mu \in M(X)$ given by

$$\|\mu\|_{TV} := |\mu|(X) = \sup_{|u|\leq 1} \left| \int_X u\,d\mu \right|.$$

By analogy, given a measurable function $w \geq 1$ on $X$, the *$w$-norm* of $\mu \in M(X)$ is defined as

$$\|\mu\|_w := \int_X w\,d|\mu| = \sup_{|u|\leq w} \left| \int_X u\,d\mu \right|.$$

Let $M_w(X) := \{\mu \in M(X) : \|\mu\|_w < \infty\}$. Since $w \geq 1$,

$$\|\mu\|_w \geq |\mu|(X) = \|\mu\|_{TV}.$$

This fact gives the second statement in the following proposition.

**Proposition A.11** *Fix a measurable function $w \geq 1$ on $X$. Then $M(X)$ and $M_w(X)$ are Banach spaces, and $M(X)$ contains $M_w(X)$.*

*Proof* See Proposition 7.2.1 in Hernández-Lerma and Lasserre (1999) [74]. □

The next result is on the classical Fubini's Theorem, which is used for the interchange of integrals.

**Proposition A.12** (Fubini's theorem) *Let $(X_1, \mathcal{F}_1, \mu_1)$ and $(X_2, \mathcal{F}_2, \mu_2)$ be measure spaces, and define the corresponding products*

$$X := X_1 \times X_2, \qquad \mathcal{F} := \sigma(\mathcal{F}_1 \times \mathcal{F}_2), \qquad \mu := \mu_1 \times \mu_2.$$

*Suppose that $\mu_1$ and $\mu_2$ are $\sigma$-finite measures. If $u$ is an $\mathcal{F}$-measurable function on $X$ such that $\int_X u\,d\mu$ exists, then*

$$\int_{X_1} \int_{X_2} u\,d\mu_2\,d\mu_1 = \int_{X_2} \int_{X_1} u\,d\mu_1\,d\mu_2.$$

*In particular, $\int_{X_1} \int_{X_2} u\,d\mu_2\,d\mu_1 = \int_{X_2} \int_{X_1} u\,d\mu_1\,d\mu_2$ for every nonnegative measurable $u$ on $X$.*

*Proof* See Ash (2000) [6, p. 108], for instance. □

To state the martingale stability theorem required in Chap. 8, we recall the concept of the conditional expectation of an extended random variable given a $\sigma$-algebra. (An *extended random variable* is a Borel-measurable function from $X$ to the set $\bar{\mathbb{R}}$ of extended real numbers.)

**Proposition A.13** (Existence of conditional expectations) *Let $Y$ be an extended random variable on a probability space $(X, \mathcal{F}, P)$, and let $\mathcal{F}_0$ be a sub-$\sigma$-algebra of $\mathcal{F}$. Suppose that the expectation of $Y$ exists. Then there is an $\mathcal{F}_0$-measurable function, denoted $E(Y|\mathcal{F}_0)$, such that*

$$\int_C Y\,dP = \int_C E(Y|\mathcal{F}_0)\,dP \quad \forall C \in \mathcal{F}_0,$$

*and the function $E(Y|\mathcal{F}_0)$ is unique $P$-a.s.*

*Proof* See Theorem 5.4.6 in Ash (2000) [6]. □

**Definition A.14** Let $Y$ be an extended random variable on $(X, \mathcal{F}, P)$ with finite expectation, and let $\mathcal{F}_0$ be a sub-$\sigma$-algebra of $\mathcal{F}$. Then the unique $\mathcal{F}_0$-measurable function $E(Y|\mathcal{F}_0)$ in Proposition A.13 is called the *conditional expectation* of $Y$ given $\mathcal{F}_0$.

In particular, when the sub-$\sigma$-algebra $\mathcal{F}_0$ is generated by a family of random variables, say $\{x(s) : s \in T\}$ with $T$ any subset of $[0, \infty)$, we write $E(Y|\mathcal{F}_0)$ simply as $E(Y|x(s), s \in T)$.

**Proposition A.15** (The Martingale stability theorem) *Let $\{Y_n, n = 1, 2, \ldots\}$ be a sequence of random variables on a probability space $(X, \mathcal{F}, P)$, and let $\{\mathcal{F}_n, n = 0, 1, \ldots\}$ a nondecreasing sequence of sub-$\sigma$-algebras of $\mathcal{F}$ such that $\{M_n, \mathcal{F}_n, n = 1, 2, \ldots\}$ with $M_n := \sum_{k=1}^n Y_k$ is a martingale (i.e., $E(M_{n+1}|\mathcal{F}_n) = M_n$ for all $n \geq 1$). If $1 < q \leq 2$ and*

$$\sum_{n=1}^{\infty} n^{-q} E\big(|Y_n|^q|\mathcal{F}_{n-1}\big) < \infty \quad P\text{-a.s.},$$

*then $\lim_{n\to\infty} \frac{1}{n} M_n = 0$ $P$-a.s.*

*Proof* See, for instance, Theorem 2.18 in Hall and Heyde (1980) [67]. □

Noting that $\sum_{n=1}^{\infty} n^{-q} E[|Y_n|^q] < \infty$ implies that $\sum_{n=1}^{\infty} n^{-q} E(|Y_n|^q|\mathcal{F}_{n-1}) < \infty$ $P$-a.s., as a consequence of Proposition A.15, we have the following fact:

**Proposition A.16** *Let* $\{Y_n, n=1,2,\ldots\}$, $\{\mathcal{F}_n, n=0,1,\ldots\}$, *and let* $M_n := \sum_{k=1}^{n} Y_k$ *be as in Proposition* A.15. *If* $1 < q \le 2$ *and*

$$\sum_{n=1}^{\infty} n^{-q} E\big[|Y_n|^q\big] < \infty,$$

*then* $\lim_{n\to\infty} \frac{1}{n} M_n = 0$ *P-a.s.*

We conclude this appendix with the following useful result.

**Proposition A.17** (Borel–Cantelli lemma) *Given a measure space* $(X, \mathcal{F}, \mu)$, *if* $A_n \in \mathcal{F}$ *for all* $n = 1, 2, \ldots$ *and* $\sum_{n=1}^{\infty} \mu(A_n) < \infty$, *then* $\mu(\limsup A_n) = 0$.

*Proof* See, for instance, Lemma 2.2.4 in Ash (2000) [6]. □

# Appendix B

In this appendix, we present some basic concepts and results about a continuous-time Markov chain with a *countable* state space $S$. Some of the results are attributed, some are proved, and some are stated without proof. It is not necessary to read these proofs to understand how the results in this appendix are applied in the text, but we mention some related references that may be consulted if necessary.

## B.1 Continuous-Time Markov Chains

Suppose that $x(t)$ (for each fixed $t \geq 0$) is a random variable on a probability space $(\Omega, \mathcal{F}, P)$ and takes values in the set $S$ (i.e., $x(t)$ is an $\mathcal{F}$-measurable function on $\Omega$ and takes values in $S$). Then we call the set $\{x(t), t \geq 0\}$ a *stochastic process* with the state space $S$.

**Definition B.1** A stochastic process $\{x(t), t \geq 0\}$ defined on a probability space $(\Omega, \mathcal{F}, P)$, with values in a countable set $S$ (to be called the *state space* of the process), is called a *continuous-time Markov chain* if, for any finite sequence of "times" $0 \leq t_1 < t_2 < \cdots < t_n < t_{n+1}$ and a corresponding set of states $i_1, i_2, \ldots, i_{n-1}, i, j$ in $S$, it holds that

$$\begin{aligned} &P\bigl(x(t_{n+1}) = j \,|\, x(t_1) = i_1, \ldots, x(t_{n-1}) = i_{n-1}, x(t_n) = i\bigr) \\ &\quad = P\bigl(x(t_{n+1}) = j \,|\, x(t_n) = i\bigr) \end{aligned} \tag{B.1}$$

whenever $P(x(t_1) = i_1, \ldots, x(t_{n-1}) = i_{n-1}, x(t_n) = i) > 0$.

Equation (B.1) is called the *Markov property*, and the probability

$$p(s, i, t, j) := P\bigl(x(t) = j \,|\, x(s) = i\bigr) \quad \text{for } 0 \leq s \leq t \text{ and } i, j \in S$$

is called the chain's *transition (probability) function*. Note that $p(s, i, t, j)$ denotes the transition probability of the process being in state $j$ at time $t$ starting from $i$ at time $s$.

X. Guo, O. Hernández-Lerma, *Continuous-Time Markov Decision Processes*, Stochastic Modelling and Applied Probability 62,
DOI 10.1007/978-3-642-02547-1, © Springer-Verlag Berlin Heidelberg 2009

It follows from the Markov property (B.1) that the transition function has the following properties.

**Proposition B.2** *Suppose that $p(s,i,t,j)$ is the transition function of a Markov chain. Then, for all $i, j \in S$ and $t \geq s \geq 0$:*

(a) $p(s,i,t,j) \geq 0$ *and* $\sum_{j\in S} p(s,i,t,j) \leq 1$.
(b) $p(s,i,s,j) = \delta_{ij}$ *(the Kronecker delta).*
(c) *(The Chapman–Kolmogorov equation)* $p(s,i,t,j) = \sum_{k\in S} p(s,i,v,k)p(v,k,t,j)$ *for all $i, j \in S$ and* $0 \leq s \leq v \leq t$.
(d) $p(s,i,t,j)$ *is continuous in $s \in [0,t]$ (right-continuous at 0, left-continuous at $t$).*
(e) $p(s,i,t,j)$ *is continuous in $t \in [s,+\infty)$ (right-continuous at $s$) and uniformly continuous in $j \in S$.*

*Proof* The proof of properties (a)–(c) follows from (B.1) directly. The proof of properties (d)–(e) can be found in Ye, Guo, and Hernández-Lerma (2008) [159], Theorem 1, for instance. □

**Definition B.3** A function $p(s,i,t,j)$ defined for $t \geq s \geq 0$ and $i, j \in S$ is called a *transition function* if it satisfies properties (a)–(c) in Proposition B.2 above. A transition function is called *standard* if it further satisfies that

$$\lim_{t\downarrow s} p(s,i,t,j) = \delta_{ij}$$

uniformly in $j \in S$ (for each fixed $i \in S$ and $s \geq 0$). A standard transition function $p(s,i,t,j)$ is called *regular* if $\sum_{j\in S} p(s,i,t,j) \equiv 1$, that is, $\sum_{j\in S} p(s,i,t,j) = 1$ for all $i \in S$ and $t \geq s \geq 0$.

**Proposition B.4** *Suppose that $p(s,i,t,j)$ is a standard transition function. Then, for any two states $i \neq j$ and $s \geq 0$:*

(a) *The limit* $q_{ii}(s) := \lim_{t\to s^+} \frac{p(s,i,t,i)-1}{t-s}$ *exists and is nonpositive (but it may be $-\infty$).*
(b) *The limit* $q_{ij}(s) := \lim_{t\to s^+} \frac{p(s,i,t,j)}{t-s}$ *exists, and it is nonnegative and finite.*
(c) $\sum_{j\neq i} q_{ij}(s) \leq q_i(s)$ *(hence $q_{ij}(s) \leq q_i(s)$), where* $q_i(s) := -q_{ii}(s) \geq 0$.

*Proof* See Theorem 1 in Ye, Guo, and Hernández-Lerma (2008) [159], for instance. □

**Definition B.5** A standard transition function $p(s,i,t,j)$ is called *stable* if $q_i(s)$ in Proposition B.4 is finite for every $i \in S$ and $s \geq 0$. Moreover, the $q_{ij}(s)$ are called the *transition rates* of $p(s,i,t,j)$ (or the transition rates of the associated Markov chain $\{x(t), t \geq 0\}$ with $p(s,i,t,j)$ as its transition function).

Noting that each variable $x(t)$ in a Markov process $\{x(t), t \geq 0\}$ is in fact a function of both $t \geq 0$ and $\omega \in \Omega$, we may write $x(t)$ as $x(t,\omega)$. In this case, the function

$x(t,\omega)$ of $t \geq 0$ (where $\omega$ is fixed) is called the *sample path* corresponding to $\omega$. A stochastic process $\{x(t), t \geq 0\}$ is called a *right process* if each sample path is right-continuous and has finite left-hand limits at every $t$.

It is well known (see, for instance, Anderson (1991) [4], Chung (1970, 1967) [29, 30], Gihman and Skorohod (1979) [47], or Williams (1979) [157]) that for a given stable standard transition function $p(s,i,t,j)$, there exists a right-process which has $p(s,i,t,j)$ as its transition function. Thus, without loss of generality, for a given stable and standard transition function, we may suppose that the corresponding Markov chain $\{x(t), t \geq 0\}$ is a right process, which in turn gives that $x(t,\omega)$ is measurable in $(t,\omega)$.

**Definition B.6** An extended random variable $\tau \geq 0$ on $(\Omega, \mathcal{F}, P)$ is called a *stopping time* (with respect to a given Markov chain $\{x(t), t \geq 0\}$ if $\{\tau \leq t\} \in \mathcal{F}_t$ for all $t \geq 0$, where $\mathcal{F}_t := \sigma(x(s) : 0 \leq s \leq t)$. A stopping time $\tau$ is said to be finite if $P(\tau < \infty) = 1$.

Obviously, a constant variable $\tau := t$ (for any fixed $t \geq 0$) is a stopping time, and if $\tau_1$ and $\tau_2$ are stopping times, then so are $\max\{\tau_1, \tau_2\}$, $\tau_1 + \tau_2$, and $\min\{\tau_1, \tau_2\}$.

For a given stopping time $\tau$, define the $\sigma$-algebra $\mathcal{F}_\tau := \{A \in \mathcal{F} : A \cap (\tau \leq t) \in \mathcal{F}_t \ \forall t \geq 0\}$.

Generalizing the Markov property to any stopping time $\tau$ yields the *strong Markov property*, that is, for all $t \geq 0$ and $j \in S$,

$$P\big(x(\tau+t) = j|\mathcal{F}_\tau\big) = P\big(x(\tau+t) = j|x(\tau)\big) \quad P\text{-a.s.} \tag{B.2}$$

We now wish to present an important interpretation of the transition rates $q_{ij}(s)$ in Proposition B.4. To do so, we first introduce some notation.

**Definition B.7** Suppose that $x(s) = i$ for some $s \geq 0$ and $i \in S$. Then the extended random variable

$$\tau_i(s) := \inf\big\{t > s|x(t) \neq i\big\} \tag{B.3}$$

is called the *holding* (or *sojourn*) *time* in state $i$ at time $s \geq 0$. (Note that $\tau_i(s) := +\infty$ if the set $\{t > s|x(t) \neq i\}$ is empty, which is due to the definition $\inf \emptyset := +\infty$.)

**Proposition B.8** *Consider a Markov chain* $\{x(t)\}$ $(t \geq 0)$ *with finite transition rates* $q_{ij}(t)$. *Then for all* $i, j \in S$,

(a) $P(\tau_i(s) > t|x(s) = i) = e^{-\int_s^t q_i(v)\,dv}$, *where* $q_i(v) := -q_{ii}(v)$.
(b) $P(x(\tau_i(s)) = j|x(s) = i) = \int_s^\infty e^{-\int_s^t q_i(v)\,dv} q_{ij}(t)\,dt$ *if* $j \neq i$.

*Proof* See Theorems 1.3 and 1.6 in Lothar (2003) [112] or Kitaev and Rykov (1995) [98, p. 150]. □

## B.2 Stationary Distributions and Ergodicity

If a continuous-time Markov chain $\{x(t), t \geq 0\}$ is such that, for $t \geq s \geq 0$, the transition probability $P(x(t) = j|x(s) = i)$ depends only on the difference $t - s$, but not on $s$ and $t$ individually, then we say that the Markov chain is time-homogeneous or simply *homogeneous*. In this case, $p(s, i, t, j) = P(x(t - s) = j|x(0) = i) = p(0, i, t - s, j)$, and we define $p_{ij}(t) := p(0, i, t, j)$ for all $i, j \in S$ and $t \geq 0$.

Thus, for any stopping time $\tau$, by the strong Markov property (B.2) we have

$$P\big[x(\tau + t) = j|A \cap \big(x(\tau) = i\big)\big] = p_{ij}(t)$$

for all $i, j \in S$, $t \geq 0$, and $A \in \mathcal{F}_\tau$ with $P(A \cap (x(\tau) = i)) > 0$.

**Definition B.9** The *transition function* of a homogeneous Markov chain is called a *stationary transition function* and is denoted by $p_{ij}(t)$.

In this subsection, we will limit ourselves to the case of a homogeneous Markov chain. Unless explicitly stated otherwise, $p_{ij}(t)$ is further assumed to be *standard* and *regular*, i.e.,

$$\lim_{t \downarrow 0} p_{ij}(t) = \delta_{ij} \quad \text{and} \quad \sum_{j \in S} p_{ij}(t) = 1 \quad \text{for all } i \in S \text{ and } t \geq 0,$$

respectively. Moreover, we see that the associated transition rates $q_{ij}(s)$ are independent of $s$, and they are denoted as $q_{ij}$ for simplicity. In particular, for a *stable* stationary transition function (i.e., $q_i := -q_{ii}$ is finite for every $i \in S$), we have the following basic properties.

**Proposition B.10** *Suppose that a given stationary transition function $p_{ij}(t)$ is stable. Then for each $i, j \in S$, the following hold:*

(a) *The derivative $p'_{ij}(t)$ exists and is finite and continuous on $[0, \infty)$, and for each $t \geq 0$, $\sum_{j \in S} |p'_{ij}(t)| \leq 2q_i$, where $q_i := -q_{ii}$.*
(b) *$p'_{ij}(s + t) = \sum_{k \in S} p'_{ik}(s) p_{kj}(t)$ for all $s > 0$ and $t \geq 0$, and $p'_{ij}(s + t) = \sum_{k \in S} p_{ik}(s) p'_{kj}(t)$ for all $t > 0$ and $s \geq 0$.*
(c) *$p'_{ij}(t) \geq \sum_{k \in S} q_{ik} p_{kj}(t)$ for all $t > 0$ (this is called the backward inequality), and $p'_{ij}(t) \geq \sum_{k \in S} p_{ik}(t) q_{kj}$ for all $t > 0$ (this is called the forward inequality).*
(d) *$p'_{ij}(t) = \sum_{k \in S} q_{ik} p_{kj}(t)$ if and only if $\sum_{j \in S} q_{ij} = 0$.*
(e) *The limits $p^*_{ij} := \lim_{t \to \infty} p_{ij}(t)$ exist, and $p^*_{ii} = (q_i m_{ii})^{-1}$, where $m_{ii} := E(\tau_i^+|x(0) = i)$ with $\tau_i^+ := \inf\{t > 0 : x(t) = i\}$. (We have $p^*_{ii} = 0$ when $q_i m_{ii} = \infty$.)*
(f) *$\lim_{t \to \infty} p'_{ij}(t) = 0$.*

*Proof* Parts (a) and (b) follow from Propositions 1.2.4 and 1.2.6, together with Corollary 1.2.5 in Anderson (1991) [4]. Parts (c) and (d) are from Proposition 1.2.7

in Anderson (1991) [4]. Part (e) comes from Theorem 5.1.3 and Proposition 6.2.1 in Anderson (1991) [4]. Finally, part (f) follows from the corollary in Chung (1967) [29, p. 211]. □

**Definition B.11** Given a *stationary transition function* $p_{ij}(t)$, a finite measure $\{\mu_i\}$ on $S$ (that is, $\mu_i \geq 0$ and $0 < \sum_{i\in S}\mu_i < \infty$) is called an *invariant measure* of $p_{ij}(t)$ if

$$\sum_{i\in S}\mu_i p_{ij}(t) = \mu_j \quad \text{for all } j \in S \text{ and } t \geq 0.$$

An invariant measure $\{\mu_i\}$ on $S$ is called an invariant probability measure (i.p.m.) if $\sum_{i\in S}\mu_i = 1$.

To give conditions for the existence of an i.p.m., we need to introduce some concepts.

**Definition B.12** Given two states $i, j \in S$, we say that $j$ can be reached from $i$, and write $i \hookrightarrow j$, if $p_{ij}(t) > 0$ for some $t > 0$. We say that $i$ and $j$ communicate, and write $i \leftrightarrow j$, if $i$ and $j$ can be reached from each other. A state $i$ is called recurrent if $\int_0^\infty p_{ii}(t)\,dt = \infty$, and transient if $\int_0^\infty p_{ii}(t)\,dt < \infty$.

Using the inequality $p_{ij}(s+t) \geq p_{ik}(s)p_{kj}(t)$ from Proposition B.2(c), it is easy to show that "$\leftrightarrow$" is an equivalence relation, partitioning $S$ into disjoint equivalence classes called *communicating classes*. We call $p_{ij}(t)$ (or the corresponding homogeneous Markov chain) *irreducible* if all the states in $S$ form the only one communicating class. In addition, a communicating class $C$ is called *closed* if $p_{ij}(t) = 0$ for all $t \geq 0$ whenever $i \in C$ and $j \notin C$.

**Proposition B.13** *Suppose that a given regular homogeneous transition function $p_{ij}(t)$ is irreducible. Then the following statements hold*:

(a) *The limits $\mu_j := \lim_{t\to\infty} p_{ij}(t)$ exist and are independent of $i$ (for all fixed $j \in S$).*
(b) *The set $\{\mu_j, j \in S\}$ in (a) is an invariant measure, and either*
   ($b_1$) *$\mu_j = 0$ for all $j \in S$; or*
   ($b_2$) *$\mu_j > 0$ for all $j \in S$, and $\sum_{j\in S}\mu_j = 1$.*

*Proof* See Theorem 5.1.6 in Anderson (1991) [4]. □

**Definition B.14** A recurrent state $i$ is called *positive* if $\lim_{t\to\infty} p_{ii}(t) > 0$, and *null* if $\lim_{t\to\infty} p_{ii}(t) = 0$.

From Definitions B.12 and B.14 it is clear that "positive" implies "recurrent."

**Proposition B.15** *Suppose that $i \leftrightarrow j$. Then,*

(a) *$i$ is transient (recurrent) if and only if $j$ is transient (recurrent).*
(b) *$i$ is recurrent positive (null) if and only if $j$ is recurrent positive (null).*

*Proof* See Propositions 5.1.1 and 5.1.4 in Anderson (1991) [4]. □

By Definition B.14 and Proposition B.13, it can be seen that being recurrent positive is equivalent to the following notion of ergodicity. An irreducible and regular transition function $p_{ij}(t)$ (or the associated Markov chain) is called *ergodic* if there is a probability measure $\{\mu_j, j \in S\}$ such that

$$\lim_{t\to\infty} p_{ij}(t) = \mu_j \quad \forall i, j \in S.$$

In the text, however, we will also use the following stronger notion of ergodicity.

*In the remainder of the appendix, we assume that $p_{ij}(t)$ is an irreducible regular transition function.*

**Definition B.16** An ergodic transition function $p_{ij}(t)$ with an i.p.m. $\{\mu_j, j \in S\}$ is said to be exponentially ergodic if there are constants $0 < \rho < 1$ and $C_{ij}$ such that

$$\left|p_{ij}(t) - \mu_j\right| \le C_{ij} e^{-\rho t} \quad \text{for all } i, j \in S \text{ and } t \ge 0, \tag{B.4}$$

in which case $\rho$ is called the convergence rate of the Markov chain.

**Proposition B.17** *The following statements are equivalent.*

(i) *$p_{ij}(t)$ is exponentially ergodic.*
(ii) *For some (and then for all) $i \in S$, there exists $\lambda > 0$ such that $E(e^{\lambda \tau_i^+} \mid x(0) = i) < \infty$, with $\tau_i^+$ as in Proposition* B.10(e).

*Proof* See Lemma 6.6.3 in Anderson (1991) [4]. □

The calculation of an i.p.m. and the verification of exponential ergodicity of a Markov chain are important problems in Markov chain theory. In the following appendix, we give conditions on the transition rates $q_{ij}$ under which such problems can be studied.

# Appendix C

For a given standard transition function $p(s,i,t,j)$, Proposition B.4 gives that the limits

$$q_{ij}(t) := \lim_{t' \to t^+} \frac{p(t,i,t',j) - \delta_{ij}}{t' - t}$$

exist. In this appendix, we consider a converse problem. Namely, given the transition rates $q_{ij}(t)$, can we construct a transition function $p(s,i,t,j)$ such that the limits

$$\lim_{t' \to t^+} \frac{p(t,i,t',j) - \delta_{ij}}{t' - t}$$

coincide with the given $q_{ij}(t)$?

*Remark* In the text, sometimes we write $q_{ij}(t)$ as $q(j|i,\pi_t)$ depending on a given policy $(\pi_t)$.

## C.1 The Construction of Transition Functions

We first introduce the definitions of a nonhomogeneous $Q(t)$-*matrix* and a nonhomogeneous $Q(t)$-*function*. Then we state some important properties of them.

**Definition C.1** For all $i, j \in S$, let $q_{ij}(t)$ be a real-valued function defined on $[0, +\infty)$. The matrix $Q(t) := [q_{ij}(t)]$ with $(i,j)$th element $q_{ij}(t)$ is said to be a *nonhomogeneous* $Q(t)$-*matrix* if for every $i, j \in S$ and $t \geq 0$:

(i) $q_{ij}(t) \geq 0$ if $i \neq j$, and $0 \leq q_i(t) := -q_{ii}(t) < \infty$.
(ii) $\sum_{j \in S} q_{ij}(t) \leq 0$.

If in addition $\sum_{j \in S} q_{ij}(t) = 0$ for all $i \in S$ and $t \geq 0$, then we say that $Q(t)$ is *conservative*.

If each component $q_{ij}(t)$ of $Q(t)$ is independent of $t$ (and let $q_{ij}(t) =: q_{ij}$), then such a $Q$-matrix is called *homogeneous*, where $Q := [q_{ij}]$.

X. Guo, O. Hernández-Lerma, *Continuous-Time Markov Decision Processes*, Stochastic Modelling and Applied Probability 62,
DOI 10.1007/978-3-642-02547-1, 

Obviously, by Proposition B.4, if a transition function $p(s,i,t,j)$ is stable, then its transition rates $q_{ij}(t)$ form a nonhomogeneous $Q(t)$-matrix. Conversely, for a given nonhomogeneous $Q(t)$-matrix $[q_{ij}(t)]$, can we construct a transition function $p(s,i,t,j)$ such that the limits $\lim_{t\to s^+}\frac{p(s,i,t,j)-\delta_{ij}}{t-s}$ exist and equal the given $q_{ij}(t)$? To answer this question, we now introduce a measurability–integrability assumption with respect to the Lebesgue measure on $[0,+\infty)$.

**Assumption C.2** Let $[q_{ij}(t)]$ be a nonhomogeneous $Q(t)$-matrix. Suppose that $q_{ij}(t)$ is Borel-measurable in $t\geq 0$ (for any fixed $i,j\in S$) and, moreover, $q_{ii}(t)$ is integrable on any *finite interval* in $[0,\infty)$.

Clearly, a homogeneous $Q$-matrix satisfies Assumption C.2.

The following definition relates a nonhomogeneous $Q(t)$-matrix and a transition function. The abbreviation *a.e.* (*almost every* or *almost everywhere*) refers to the Lebesgue measure on $[0,+\infty)$.

**Definition C.3** Let $[q_{ij}(t)]$ be a nonhomogeneous $Q(t)$-matrix satisfying Assumption C.2. If a transition function $p(s,i,t,j)$ satisfies that, for all $i,j\in S$ and a.e. $s\geq 0$, the partial derivatives $\frac{\partial p(s,i,t,j)}{\partial s}$ exist, and the limits

$$\lim_{t\to s^+}\frac{p(s,i,t,j)-\delta_{ij}}{t-s}$$

exist and equal $q_{ij}(s)$, then $p(s,i,t,j)$ is called a *$Q(t)$-function.*

Hence, the question above can be expressed as follows: Given a $Q(t)$-matrix, can we ensure the existence of a $Q(t)$-function? The following fact shows that this can be done if the $Q(t)$-matrix satisfies Assumption C.2.

**Proposition C.4** *Let $[q_{ij}(t)]$ be a nonhomogeneous $Q(t)$-matrix satisfying Assumption C.2. For all $i,j\in S$ and $0\leq s\leq t$, define recursively*

$$\begin{aligned} p_{ij}^{(0)}(s,t) &:= \delta_{ij}e^{-\int_s^t q_i(v)\,dv} \quad \textit{with } q_i(v):=-q_{ii}(v), \\ p_{ij}^{(n+1)}(s,t) &:= \int_s^t \sum_{k\neq i} e^{-\int_s^u q_i(v)\,dv} q_{ik}(u)p_{kj}^{(n)}(u,t)\,du \quad \forall n\geq 0. \end{aligned} \tag{C.1}$$

*Let $\hat{p}(s,i,t,j):=\sum_{n=0}^{\infty}p_{ij}^{(n)}(s,t)$. Then:*

(a) *$\hat{p}(s,i,t,j)$ is a $Q(t)$-function, and, moreover, it is standard and stable.*
(b) *$\hat{p}(s,i,t,j)$ is regular if and only if for all $i\in S$ and $t\geq s\geq 0$:*

$$\sum_{j\in S}\left[\int_s^t \sum_{k\in S}\hat{p}(s,i,v,k)q_{kj}(v)\,dv\right]\equiv 0.$$

(c) *The following Kolmogorov forward and backward equations hold*:

$$\frac{\partial \hat{p}(s,i,t,j)}{\partial t} = \sum_{k\in S} \hat{p}(s,i,t,k)q_{kj}(t),$$

$$\frac{\partial \hat{p}(s,i,t,j)}{\partial s} = -\sum_{k\in S} q_{ik}(s)\hat{p}(s,k,t,j)$$

*for all* $i, j \in S$ *and a.e.* $t \geq s \geq 0$.

(d) $\hat{p}(s,i,t,j) \leq p(s,i,t,j)$ *for any* $Q(t)$*-function* $p(s,i,t,j)$.

*Proof* See Theorems 2 and 3 in Ye, Guo, and Hernández-Lerma (2008) [159]. □

For a homogeneous $Q$-matrix, we can obtain a bit more results.

**Proposition C.5** *Let* $[q_{ij}]$ *be a homogeneous* $Q$*-matrix. For all* $i, j \in S$ *and* $t \geq 0$, *define recursively*

$$p_{ij}^{(0)}(t) = q_{ij}^{(0)}(t) := \delta_{ij} e^{-q_i t} \quad \textit{with } q_i := -q_{ii},$$

$$p_{ij}^{(n+1)}(t) := \int_0^t \sum_{k\neq i} e^{-q_i u} q_{ik} p_{kj}^{(n)}(t-u)\,du \quad \forall n \geq 0,$$

$$q_{ij}^{(n+1)}(t) := \delta_{ij} e^{-q_i t} + \int_0^t \sum_{k\neq i} e^{-q_i u} q_{ik} q_{kj}^{(n)}(t-u)\,du \quad \forall n \geq 0.$$

*Let* $\hat{p}_{ij}(t) := \sum_{n=0}^{\infty} p_{ij}^{(n)}(t)$ *and* $\hat{q}_{ij}(t) := \lim_{n\to\infty} q_{kj}^{(n)}(t)$. *Then*:

(a) $\hat{p}_{ij}(t) = \hat{q}_{ij}(t)$, *and* $\hat{p}_{ij}(t)$ *is a stationary* $Q$*-function, and, moreover, it is standard and stable.*

(b) $\hat{p}_{ij}(t)$ *is regular if and only if*

$$\sum_{j\in S}\left[\int_0^t \sum_{k\in S} \hat{p}_{ik}(v) q_{kj}\,dv\right] \equiv 0.$$

(c) *The following Kolmogorov forward and backward equations hold*:

$$\begin{aligned}\frac{d\hat{p}_{ij}(t)}{dt} &= \sum_{k\in S} \hat{p}_{ik}(t) q_{kj} \\ &= \sum_{k\in S} q_{ik} \hat{p}_{kj}(t) \quad \forall i, j \in S \textit{ and } t \geq 0.\end{aligned}$$

(d) $\hat{p}_{ij}(t) \leq p_{ij}(t)$ *for any* $Q$*-function* $p_{ij}(t)$.

*Proof* See Theorem 2.2.2 and the arguments in p. 71 by Anderson (1991) [4], for instance. □

Proposition C.5 gives a constructive method to obtain a homogeneous $Q$-function from a given $Q$-matrix. We next introduce another approach. (Recall from Proposition C.5 that $q_i := -q_{ii}$.)

**Proposition C.6** *For a given homogeneous $Q$-matrix $[q_{ij}]$ and any fixed $\alpha > 0$, define, for all $i, j \in S$ and $n \geq 1$,*

$$\varphi_{ij}^{(n)} := \begin{cases} \frac{\delta_{ij}}{\alpha+q_i} & \text{for } n = 1, \\ \frac{1}{\alpha+q_i}[\delta_{ij} + \sum_{k\neq i} q_{ik}\varphi_{kj}^{(n-1)}] & \text{for } n \geq 2. \end{cases}$$

*Then $\varphi_{ij}^{(n)}$ is nondecreasing in $n$, and*

$$\lim_{n\to\infty} \varphi_{ij}^{(n)} = \int_0^\infty e^{-\alpha t}\hat{p}_{ij}(t)\,dt,$$

*with $\hat{p}_{ij}(t)$ as in Proposition* C.5.

*Proof* See Anderson (1991) [4, pp. 121–122], for instance. □

For a nonhomogeneous $Q(t)$-matrix, Proposition C.4 shows the existence of a $Q(t)$-function $\hat{p}(s, i, t, j)$ and also that this $Q(t)$-function is the *minimum* of all the $Q(t)$-functions. But of course, in general, a $Q(t)$-function may *not* be unique, that is, there may be many different $Q(t)$-functions for the same given $Q(t)$-matrix; see Theorem 4.2.6 by Anderson (1991) [4] for details. To guarantee the uniqueness of a $Q(t)$-function for a given $Q(t)$-matrix, we have the following fact.

**Proposition C.7** *For a nonhomogeneous conservative $Q(t)$-matrix $[q_{ij}(t)]$ satisfying Assumption* C.2, *if*

$$\int_s^t \sum_{j\in S} \hat{p}(s, i, v, j)q_j(v)\,dv < \infty \quad \text{for all } t \geq s \geq 0 \text{ and } i \in S,$$

*then the $Q(t)$-function $\hat{p}(s, i, t, j)$ is unique and regular.*

*Proof* Since $q_{ij}(t) \geq 0$ for all $i \neq j$ and $t \geq 0$, and $\sum_{j\in S} q_{ij}(t) = 0$ for all $i \in S$ and $t \geq 0$, we have $\sum_{j\in S} |q_{ij}(t)| = 2q_i(t)$. Thus, by the present hypotheses, we have

$$\begin{aligned} \sum_{j\in S} \int_s^t \sum_{k\in S} \hat{p}(s, i, v, k)\big|q_{kj}(v)\big|\,dv &= \int_s^t \sum_{k\in S} \hat{p}(s, i, v, k)\Big[\sum_{j\in S} |q_{kj}|\Big]\,dv \\ &= 2\int_s^t \sum_{k\in S} \hat{p}(s, i, v, k)q_k(v)\,dv < \infty. \end{aligned}$$

Therefore, by Fubini's theorem (see, for instance, Proposition A.12) we have

$$\sum_{j\in S}\left[\int_s^t \sum_{k\in S}\hat{p}(s,i,v,k)q_{kj}(v)\,dv\right] = \int_s^t \sum_{k\in S}\hat{p}(s,i,v,k)\left[\sum_{j\in S}q_{kj}(v)\right]dv = 0.$$

Hence, by Proposition C.4(b), we obtain $\sum_{j\in S}\hat{p}(s,i,t,j) = 1$, and so the desired uniqueness result follows from Proposition C.4(d). □

A different approach to verify the regularity condition of the $Q(t)$-function $\hat{p}$ in Proposition C.7 is to use the following fact.

**Proposition C.8** *Let $W'$ be a nonnegative function on $S$, and let $c'$ be an arbitrary constant. Then the following statements are equivalent:*

(i) $\sum_{j\in S} W'(j)\hat{p}(s,i,t,j) \le e^{c'(t-s)}W'(i)$ *for all $i \in S$ and a.e. $s \in [0,t]$ and each $t \ge 0$ (with $\hat{p}(s,i,t,j)$ as in Proposition* C.4).

(ii) $\sum_{j\in S} W'(j)q_{ij}(t) \le c'W'(i)$ *for all $i \in S$ and a.e. $t \ge 0$.*

*Proof* (i) ⇒ (ii) For every $t \ge 0$, a.e. $s \in [0,t)$, and $i \in S$, we can rewrite the condition in (i) as

$$\sum_{j\ne i} W'(j)\frac{\hat{p}(s,i,t,j)}{(t-s)} \le \left[\frac{1-\hat{p}(s,i,t,i)}{t-s} - \frac{1-e^{c'(t-s)}}{t-s}\right]W'(i).$$

Then letting $t \downarrow s$, by the Fatou–Lebesgue lemma and Proposition C.4(a) we see that

$$\sum_{j\ne i} W'(j)q_{ij}(s) \le \left[-q_{ii}(s)+c'\right]W'(i) \quad \forall i \in S,$$

which gives (ii).

(ii) ⇒ (i) For each $t \ge 0$, a.e. $s \in [0,t]$, and $i \in S$, by Proposition C.4(a), it suffices to show that

$$\sum_{j\in S} W'(j)\left[\sum_{n=0}^{N} p_{ij}^{(n)}(s,t)\right] \le e^{c'(t-s)}W'(i) \quad \forall N \ge 0. \tag{C.2}$$

We are going to prove (C.2) by induction. In fact, the condition in (ii) gives $q_{ii}(t)W'(i) \le c'W'(i)$ for all $t \ge 0$, from which we obtain

$$e^{\int_s^t q_{ii}(u)\,du}W'(i) \le e^{c'(t-s)}W'(i) \quad \text{for all } t \ge s.$$

Hence, (C.2) holds for $N = 0$. Now suppose that (C.2) holds for some $N \ge 0$. Then, from (C.1) and the Fubini theorem, under the induction hypothesis and condition (ii), we obtain

$$\sum_{j\in S} W'(j)\left[\sum_{n=1}^{N+1} p_{ij}^{(n)}(s,t)\right]$$

$$= \int_s^t \sum_{k\in S} e^{-\int_s^u q_i(v)\,dv}\big(q_{ik}(u)+\delta_{ik}q_i(u)\big)\left[\sum_{j\in S} W'(j)\left(\sum_{n=0}^{N} p_{kj}^{(n)}(u,t)\right)\right] du$$

$$\le \int_s^t \left[\sum_{k\in S} e^{-\int_s^u q_i(v)\,dv}\big(q_{ik}(u)+\delta_{ik}q_i(u)\big)\right] e^{c'(t-u)} W'(k)\,du$$

$$\le c'W'(i)\int_s^t e^{-\int_s^u q_i(v)\,dv} e^{c'(t-u)}\,du + W'(i)\int_s^t e^{-\int_s^u q_i(v)\,dv} q_i(u) e^{c'(t-u)}\,du$$

$$= e^{c'(t-s)}W'(i) - e^{-\int_s^t q_i(v)\,dv} W'(i).$$

This gives that (C.2) holds for $N+1$, and so it does for all $N \ge 0$. □

From Propositions C.7 and C.8 we obtain the following result.

**Proposition C.9** *Let $Q(t) = [q_{ij}(t)]$ be as in Proposition C.7. If there exist constants $L_k > 0$ $(k = 1, 2)$ and a nonnegative real-valued function $w$ on $S$ such that*

$$\sum_{j\in S} w(j)q_{ij}(t) \le L_1 w(i) \quad \textit{and} \quad q_i(t) \le L_2 w(i) \quad \textit{for all } i \in S \textit{ and } t \ge 0,$$

*then the $Q(t)$-function $\hat{p}(s,i,t,j)$ is unique and regular.*

*Proof* By Proposition C.8, we have

$$\sum_{j\in S}\left[\int_s^t \hat{p}(s,i,v,j)q_j(v)\right] dv \le L_2 \sum_{j\in S}\int_s^t \hat{p}(s,i,v,j)w(j)\,dv$$

$$\le L_2 w(i)(t-s)e^{L_1(t-s)} < \infty.$$

Therefore, Proposition C.7 gives the desired result. □

## C.2 Ergodicity Based on the $Q$-Matrix

In Sect. B.2, we presented some results on the ergodicity of a stationary transition function $p_{ij}(t)$. These results are based on $p_{ij}(t)$ itself. In contrast, we now present related facts but based on the corresponding $Q$-matrix. Hence, throughout this subsection, we consider a fixed homogeneous $Q$-matrix $[q_{ij}]$, and $\hat{p}_{ij}(t)$ denotes the associated minimum $Q$-function obtained in Proposition C.5.

The next result shows that the irreducibility of $\hat{p}_{ij}(t)$ can be verified from the given $[q_{ij}]$.

**Proposition C.10** *The following statements are equivalent for $i, j \in S$ with $i \neq j$:*

(i) *$j$ can be reached from $i$ (i.e., $i \hookrightarrow j$).*
(ii) *Either $q_{ij} > 0$ or $q_{ii_1} q_{i_1 i_2} \cdots q_{i_n j} > 0$ for some distinct states $i_1, i_2, \ldots, i_n \in S$.*

*Proof* See Proposition 5.3.1 by Anderson (1991) [4], for instance. □

To verify the ergodicity of $\hat{p}_{ij}(t)$, we have the following fact.

**Proposition C.11** *For a given $Q$-matrix, suppose that the corresponding minimum $Q$-function $\hat{p}_{ij}(t)$ is irreducible. Then the following statements are equivalent:*

(i) *$\hat{p}_{ij}(t)$ is ergodic.*
(ii) *For some finite nonempty set $B \subset S$, the system*

$$\sum_{j \notin B, j \neq i} q_{ij} x_j \leq q_i x_i - 1 \quad \text{for } i \notin B, \qquad x_i = 0 \quad \text{for } i \in B,$$

*has a finite nonnegative solution $(x_i, i \in S)$ such that $\sum_{j \in S} q_{ij} x_j < \infty$ for $i \in B$.*

*Proof* See Theorem 6.2.3 in Anderson (1991) [4]. □

The following proposition states that the i.p.m. of $\hat{p}_{ij}(t)$ can also be determined by the given transition rates $q_{ij}$.

**Proposition C.12** *Fix any given $Q$-matrix $[q_{ij}]$.*

(a) *If the corresponding $\hat{p}_{ij}(t)$ is ergodic, then the i.p.m. $\{\mu_i, i \in S\}$ of $\hat{p}_{ij}(t)$ is the unique positive solution (that is, $\mu_i > 0$ for all $i \in S$) of the equations*

$$\sum_{i \in S} x_i = 1, \qquad \sum_{i \in S} x_i q_{ij} = 0 \quad \text{for all } j \in S.$$

(b) *If $[q_{ij}]$ is bounded in $i, j \in S$, then*

$$\lim_{t \to \infty} \hat{p}_{ij}(t) = \lim_{n \to \infty} \frac{1}{n+1} \sum_{m=0}^{n} \bar{p}_{ij}^{(m)},$$

*with $\hat{p}_{ij}(t)$ as in Proposition C.5, $\bar{p}_{ij} := \delta_{ij} + \frac{q_{ij}}{L}$, and $L$ such that $L \geq \sup_{i \in S} |q_{ii}|$.*

*Proof* (a) Follows from Theorem 5.4.5 and Proposition 5.4.6 in Anderson (1991) [4]. Part (b) follows from Proposition 5.4.11 in Anderson (1991) [4]. □

To verify the exponential ergodicity of $\hat{p}_{ij}(t)$, we may use the following fact.

**Proposition C.13** *For a given Q-matrix, suppose that the corresponding minimum Q-function $\hat{p}_{ij}(t)$ is irreducible. Then the following statements are equivalent:*

(i) *$\hat{p}_{ij}(t)$ is exponentially ergodic.*
(ii) *For some finite nonempty set $B \subset S$, there is $\delta$ with $0 < \delta < \inf_{i\in S} q_i$ for which the system*

$$\sum_{j\notin B, j\neq i} q_{ij}x_j \le (q_i - \delta)x_i - 1 \quad \text{for } i \notin B, \qquad x_i = 0 \quad \text{for } i \in B,$$

*has a finite nonnegative solution $(x_i, i \in S)$ such that $\sum_{j\in S} q_{ij}x_j < \infty$ for $i \in B$.*

*Proof* See Theorem 6.6.5 in Anderson (1991) [4]. □

We now apply Proposition C.13 to the case of birth-and-death processes.

**Proposition C.14** *Suppose that S is the set of nonnegative integers, and let $[q_{ij}]$ be a conservative birth-and-death process on S, that is, $q_{ii+1} =: \lambda_i > 0$ for all $i \ge 0$, $q_{ii-1} =: \mu_i > 0$ for all $i \ge 1$, $q_{ii} = -(\lambda_i + \mu_i)$ for $i \ge 0$, $\mu_0 = 0$, and $q_{ij} = 0$ otherwise. Let*

$$\sum_{i=1}^{\infty}\left[\frac{1}{\lambda_i} + \frac{\mu_i}{\lambda_i\lambda_{i-1}} + \frac{\mu_i\mu_{i-1}}{\lambda_i\lambda_{i-1}\lambda_{i-2}} + \cdots + \frac{\mu_i\cdots\mu_2}{\lambda_i\cdots\lambda_2\lambda_1}\right] = \infty,$$

$$\sum_{i=1}^{\infty}\frac{1}{\mu_{i+1}}\left[1 + \frac{\lambda_i}{\mu_i} + \frac{\lambda_i\lambda_{i-1}}{\mu_i\mu_{i-1}} + \cdots + \frac{\lambda_i\lambda_{i-1}\cdots\lambda_1}{\mu_i\mu_{i-1}\cdots\mu_1}\right] =: \bar{L} < \infty.$$

(a) *Take $\pi_0 := (1 + \sum_{i=1}^{\infty} \frac{\lambda_{i-1}\cdots\lambda_0}{\mu_i\cdots\mu_1})^{-1}$ and $\pi_i := \pi_0 \frac{\lambda_{i-1}\cdots\lambda_0}{\mu_i\cdots\mu_1}$ for $i \ge 1$. Then $\{\pi_i, i \in S\}$ is the unique i.p.m. of $\hat{p}_{ij}(t)$.*
(b) *Choose $\bar{n}$ (since $\bar{L} < \infty$) such that $\beta := \frac{1}{\mu_{\bar{n}}} + \sum_{i=\bar{n}}^{\infty} \frac{\lambda_i\lambda_{i-1}\cdots\lambda_{\bar{n}}}{\mu_{i+1}\mu_i\cdots\mu_{\bar{n}}} < 1$, and let $\delta$ be such that $0 < \delta < \min\{\inf_{i\in S} q_i, \frac{1-\beta}{\bar{L}+\mu_{\bar{n}}^{-1}}\}$. Take $B = \{0, 1, \ldots, \bar{n}-1\}$. Then there exists a sequence $(x_i, i \in S)$ such that*

$$\lambda_i x_{i+1} + \mu_i x_{i-1} = (\lambda_i + \mu_i - \delta)x_i - 1 \quad \text{for } i \ge \bar{n}, \qquad x_i = 0 \quad \text{for } i \in B,$$

*and $0 \le x_{i+1} - x_i \le \frac{\mu_i\cdots\mu_1}{\lambda_i\cdots\lambda_1}$ for all $i \ge \bar{n}$. Hence (by Proposition C.13), the corresponding $\hat{p}_{ij}(t)$ is exponentially ergodic.*

*Proof* Parts (a) and (b) follow from Anderson (1991) [4, pp. 197–198] and the proof of Proposition 6.6.6 in Anderson (1991) [4], respectively. □

Proposition C.13 gives conditions for the exponential ergodicity of $\hat{p}_{ij}(t)$. But then it is, of course, desirable to obtain an estimate of the convergence rate $\rho$ in (B.4) based on conditions on the given transition rates $q_{ij}$. To this end, we introduce the following concept.

*In the remainder of this subsection, we suppose, without loss of generality, that the state space $S$ is the set of nonnegative integers, i.e., $S = \{0, 1, 2, \ldots\}$.*

**Definition C.15** A stationary transition function $p_{ij}(t)$ is called *stochastically monotone* if, for any fixed $k \in S$ and $t \geq 0$, the sum $\sum_{j \geq k} p_{ij}(t)$ is a nondecreasing function of $i \in S$.

The next result shows that the monotonicity of $\hat{p}_{ij}(t)$ can be verified by using the transition rates $q_{ij}$.

**Proposition C.16** *Suppose that the homogeneous $Q$-matrix $[q_{ij}]$ is conservative, except possibly for the initial row (i.e., $\sum_{j \in S} q_{ij} = 0$ for all $i \geq 1$). Then the following statements are equivalent:*

(i) $\hat{p}_{ij}(t)$ *is stochastically monotone.*
(ii) $\sum_{j \geq k} q_{ij} \leq \sum_{j \geq k} q_{i+1,j}$ *for all $i, k \in S$ such that $k \neq i + 1$.*

*Proof* See Theorem 7.3.4 in Anderson (1991) [4]. □

We complete this subsection by using Propositions C.13 and C.16 to obtain a result on an estimate of the convergence rate $\rho$ in (B.4).

**Proposition C.17** *Suppose that $\hat{p}_{ij}(t)$ is ergodic (with i.p.m. $\mu$) and has conservative transition rates $q_{ij}$. In addition, suppose that:*

(i) *There exist a nondecreasing function $w \geq 1$ on $S$ and constants $\delta > 0$ and $b \geq 0$ such that $\sum_{j \in S} w(j) q_{ij} \leq -\delta w(i) + b\delta_{i0}$ $\forall i \in S$.*
(ii) $\sum_{j \geq k} q_{ij} \leq \sum_{j \geq k} q_{i+1,j}$ *for all $i, k \in S$ such that $k \neq i + 1$.*
(iii) *For all $j > i > 0$, there exist nonzero distinct states $i_1, i_2, \ldots, i_m \geq j$ such that $q_{ii_1} q_{i_1 i_2} \cdots q_{i_{m-1} i_m} > 0$.*

*Then,*

(a) $\hat{p}_{ij}(t)$ *is ergodic and has a unique i.p.m. $\mu$.*
(b) $|\sum_{j \in S} \hat{p}_{ij}(t) u(j) - \mu(u)| \leq 2\|u\|_w (1 + \frac{b}{\delta}) e^{-\delta t} w(i)$ *for all $i \in S, t \geq 0$, and real-valued functions $u$ on $S$, with $\mu(u) := \sum_{j \in S} u(j)\mu(j)$ and $\|u\|_w := \sup_{i \in S} \frac{|u(i)|}{w(i)} < \infty$.*

*Proof* The condition (i) also implies that the "drift" condition (2.4) by Lund, Meyn, and Tweedie (1996) [113] is satisfied with the drift function $w$. On the other hand, by (ii) and Proposition C.16 we see that the Markov chain $\{x(t)\}$ with transition function $\hat{p}_{ij}(t)$ is stochastically monotone. Furthermore, by condition (iii), this chain satisfies condition (2.1) in [113]. Thus, since $w \geq 1$, from Theorem 2.2 in [113] we obtain, for all $i \in S$ and $t \geq 0$,

$$\left|\sum_{j\in S} \hat{p}_{ij}(t)u(j) - \mu(u)\right| \leq 2\|u\|_w e^{-\delta t}\left[w(i) + \frac{b}{\delta}\right] \leq 2\|u\|_w \left(1 + \frac{b}{\delta}\right) e^{-\delta t} w(i).$$

This proves the proposition. □

## C.3 Dynkin's Formula

Let $\{x(t), t \geq 0\}$ be a continuous-time Markov chain with a denumerable state space $S$, transition matrix $P(t) = [p_{ij}(t)]$, and $Q$-matrix $Q = [q_{ij}]$. Given a real-valued function $u$ on $S$, we denote by $Qu$ the function on $S$ defined as

$$(Qu)(i) := \sum_{j\in S} u(j)q_{ij} \quad \text{for all } i \in S.$$

Furthermore, suppose that $u$ is such that, for all $i \in S$ and $t \geq 0$,

$$E_i\left|u\big(x(t)\big)\right| := \sum_{j\in S} |u(j)| p_{ij}(t) < \infty \tag{C.3}$$

and

$$E_i \int_0^t \left|Qu\big(x(s)\big)\right| ds := \sum_{j\in S} \int_0^t |Qu(j)| p_{ij}(s)\, ds < \infty. \tag{C.4}$$

Then, for all $i \in S$, $t \geq 0$, and $\lambda > 0$, we have *Dynkin's formula*

$$E_i e^{-\lambda t} u\big(x(t)\big) - u(i) = E_i \int_0^t e^{-\lambda s}\left[Qu\big(x(s)\big) - \lambda u\big(x(s)\big)\right] ds. \tag{C.5}$$

Letting $\lambda \downarrow 0$, we obtain

$$E_i u\big(x(t)\big) - u(i) = E_i \int_0^t Qu\big(x(s)\big)\, ds, \tag{C.6}$$

which is also known as *Dynkin's formula*.

A function $u : S \to \mathbb{R}$ that satisfies (C.3) and (C.4) is said to be in the domain of the *extended generator* of the Markov chain $\{x(t), t \geq 0\}$ with $p_{ij}(t)$ as its transition probability function.

Let $\tau$ be a stopping time with respect to the $\sigma$-algebras $\mathcal{F}_t$, $t \geq 0$. (See Definition B.6.) If $\tau$ is such that $E_i \tau < \infty$ for every $i \in S$, the (C.5) and (C.6) hold with $\tau$ in lieu of $t$.

There are several ways for obtaining Dynkin's formula. For instance, (C.6) follows from the fact that, for every $u$ that satisfies (C.3)–(C.4), the process

$$M_t(u) := u\big(x(t)\big) - u\big(x(0)\big) - \int_0^t (Qu)\big(x(s)\big)\,ds \quad \text{for } t \geq 0,$$

is a martingale with respect to the $\sigma$-algebras $\mathcal{F}_t$.

The results in this section are also valid for a nonhomogeneous Markov process $\{x(t), t \geq 0\}$, with the obvious notational changes.

# References


1. Albright, S. C., & Winston, W. (1979). A birth-death model of advertising and pricing. *Advances in Applied Probability*, *11*, 134–152.
2. Allen, L. J. S. (2003). *An introduction to stochastic processes with applications to biology*. Upper Saddle River: Pearson Education.
3. Altman, E. (1999). *Constrained Markov decision processes*. London/Boca Raton: Chapman & Hall/CRC Press.
4. Anderson, W. J. (1991). *Continuous-time Markov chains*. Berlin: Springer.
5. Arapostathis, A., Borkar, V. S., Fernandez-Gaucherand, E., Ghosh, M. K., & Marcus, S. I. (1993). Discrete-time controlled Markov processes with average cost criterion: a survey. *SIAM Journal on Control and Optimization*, *31*, 282–344.
6. Ash, R. B. (2000). *Probability and measure theory* (2nd ed.). San Diego: Harcourt/Academic Press.
7. Bailey, N. T. J. (1975). *The mathematical theory of infectious diseases and its applications*. Carimate: Griffin.
8. Bartholomew, D. J. (1973). *Stochastic models for social processes* (2nd ed.). New York: Wiley.
9. Bather, J. (1976). Optimal stationary policies for denumerable Markov chains in continuous time. *Advances in Applied Probability*, *8*, 144–158.
10. Bellman, R. (1957). *Dynamic programming*. Princeton: Princeton University Press.
11. Bertsekas, D. P. (1995). *Dynamic programming and optimal control, Vol. I, II*. Belmont: Athena Scientific.
12. Bertsekas, D. P. (2001). *Dynamic programming and optimal control, Vol. II* (2nd ed.). Belmont: Athena Scientific.
13. Beutler, F. J., & Ross, K. W. (1985). Optimal policies for controlled Markov chains with a constraint. *Journal of Mathematical Analysis and Applications*, *112*, 236–252.
14. Blackwell, D. (1962). Discrete dynamic programming. *Annals of Mathematical Statistics*, *33*, 719–726.
15. Blumenthal, R. M., & Getoor, R. K. (1968). *Markov processes and potential theory*. San Diego: Academic Press.
16. Borkar, V. S. (1989). *Pitman research notes in mathematics: Vol. 203. Optimal control of diffusion processes*. Harlow: Longman Scientific & Technical.
17. Borkar, V. S. (1991). *Pitman research notes in mathematics: Vol. 240. Topics in controlled Markov chains*. Harlow: Longman Scientific & Technical.
18. Cao, X. R. (1998). The relations among potentials, perturbation analysis, and Markov decision processes. *Discrete Event Dynamic Systems*, *8*, 71–87.
19. Cao, X. R. (2000). A unified approach to Markov decision problems and performance sensitivity analysis. *Automatica (Oxford)*, *36*, 771–774.



X. Guo, O. Hernández-Lerma, *Continuous-Time Markov Decision Processes*, Stochastic Modelling and Applied Probability 62, DOI 10.1007/978-3-642-02547-1, 

20. Cao, X. R. (2003). Semi-Markov decision problems and performance sensitivity analysis. *IEEE Transactions on Automatic Control, 48*, 758–769.
21. Cao, X. R. (2007). *Stochastic learning and optimization: a sensitivity-based approach*. Berlin: Springer.
22. Cao, X. R., & Chen, H. F. (1997). Potentials, perturbation realization and sensitivity analysis of Markov processes. *IEEE Transactions on Automatic Control, 42*, 1382–1397.
23. Cao, X. R., & Guo, X. P. (2004). A unified approach to Markov decision problems and performance sensitivity analysis with discounted and average criteria: multi-chain cases. *Automatica (Oxford), 40*, 1749–1759.
24. Cao, X. R., & Zhang, J. Y. (2008). The $n$th-order bias optimality for multi-chain Markov decision processes. *IEEE Transactions on Automatic Control, 53*, 496–508.
25. Cavazos-Cadena, R. (1991). A counterexample on the optimality equation in Markov decision chains with average cost criterion. *Systems and Control Letters, 16*, 387–392.
26. Cavazos-Cadena, R., & Fernandez-Gaucherand, E. (1995). Denumerable controlled Markov chains with average reward criterion: sample path optimality. *Mathematical Methods of Operations Research, 41*, 89–108.
27. Chen, M. F. (2000). Equivalence of exponential ergodicity and $L^2$-exponential convergence for Markov chains. *Stochastic Processes and Their Applications, 87*, 281–279.
28. Chen, M. F. (2004). *From Markov chains to non-equilibrium particle systems* (2nd ed.). Singapore: World Scientific.
29. Chung, K. L. (1967). *Markov chains with stationary transition probabilities* (2nd ed.). Berlin: Springer.
30. Chung, K. L. (1970). *Lectures on boundary theory for Markov chains*. Princeton: Princeton University Press.
31. Çinlar, E. (1975). *Introduction to stochastic processes*. New York: Prentice Hall.
32. Davis, M. H. A. (1993). *Markov models and optimization*. London: Chapman and Hall.
33. Dekker, R., & Hordijk, A. (1988). Average, sensitive and Blackwell optimal policies in denumerable Markov decision chains with unbounded rewards. *Mathematics of Operations Research, 13*, 395–420.
34. Dekker, R., & Hordijk, A. (1992). Recurrence conditions for average and Blackwell optimality in denumerable state Markov decision chains. *Mathematics of Operations Research, 17*, 271–289.
35. Dong, Z. Q. (1979). Continuous time Markov decision programming with average reward criterion—countable state and action space. *Scientia Sinica, SP ISS*(II), 141–148.
36. Doshi, B. T. (1976). Continuous-time control of Markov processes on an arbitrary state space: discounted rewards. *Annals of Statistics, 4*, 1219–1235.
37. Down, D., Meyn, S. P., & Tweedie, R. L. (1995). Exponential and uniform ergodicity of Markov processes. *Annals of Probability, 23*, 1671–1691.
38. Dynkin, E. B., & Yushkevich, A. A. (1979). *Controlled Markov processes*. Berlin: Springer.
39. Feinberg, E. A. (2004). Continuous-time jump Markov decision processes: a discrete-event approach. *Mathematics of Operations Research, 29*, 492–524.
40. Feinberg, E. A., & Shwartz, A. (1996). Constrained discounted dynamic programming. *Mathematics of Operations Research, 21*, 922–945.
41. Feinberg, E. A., & Shwartz, A. (Eds.) (2002). *Handbook of Markov decision processes*. Dordrecht: Kluwer Academic.
42. Feller, W. (1940). On the integro-differential equations of purely discontinuous Markoff processes. *Transactions of the American Mathematical Society, 48*, 488–515.
43. Fisher, L. (1968). On the recurrent denumerable decision process. *Annals of Mathematical Statistics, 39*, 424–434.
44. Fleming, W. H., & Soner, H. M. (1993). *Controlled Markov processes and viscosity solutions*. Berlin: Springer.
45. Gale, D. (1967). On optimal development in a multi-sector economy. *Review of Economic Studies, 34*, 1–19.
46. Gihman, I. I., & Skorohod, A. V. (1975). *The theory of stochastic processes II*. Berlin: Springer.

47. Gihman, I. I., & Skorohod, A. V. (1979). *Controlled stochastic processes*. Berlin: Springer.
48. Guo, X. P. (1999). Nonstationary denumerable state Markov decision processes with an average variance criterion. *Mathematical Methods of Operations Research*, *49*, 87–96.
49. Guo, X. P. (2000). Constrained nonhomogeneous Markov decision processes with expected total reward criterion. *Acta Applicandae Mathematicae Sinica (English Series)*, *23*, 230–235.
50. Guo, X. P. (2007a). Continuous-time Markov decision processes with discounted rewards: the case of Polish spaces. *Mathematics of Operations Research*, *32*, 73–87.
51. Guo, X. P. (2007b). Constrained optimality for average cost continuous-time Markov decision processes. *IEEE Transactions on Automatic Control*, *52*, 1139–1143.
52. Guo, X. P., & Cao, X. R. (2005). Optimal control of ergodic continuous-time Markov chains with average sample-path rewards. *SIAM Journal on Control and Optimization*, *44*, 29–48.
53. Guo, X. P., & Hernández-Lerma, O. (2003a). Continuous-time controlled Markov chains. *Annals of Applied Probability*, *13*, 363–388.
54. Guo, X. P., & Hernández-Lerma, O. (2003b). Continuous-time controlled Markov chains with discounted rewards. *Acta Applicandae Mathematicae*, *79*, 195–216.
55. Guo, X. P., & Hernández-Lerma, O. (2003c). Drift and monotonicity conditions for continuous-time controlled Markov chains with an average criterion. *IEEE Transactions on Automatic Control*, *48*, 236–245.
56. Guo, X. P., & Hernández-Lerma, O. (2003d). Zero-sum games for continuous-time Markov chains with unbounded transition and average payoff rates. *Journal of Applied Probability*, *40*, 327–345.
57. Guo, X. P., & Hernández-Lerma, O. (2003e). Constrained continuous-time Markov controlled processes with discounted criteria. *Stochastic Analysis and Applications*, *21*(2), 379–399.
58. Guo, X. P., & Liu, K. (2001). A note on optimality conditions for continuous-time Markov decision processes with average cost criterion. *IEEE Transactions on Automatic Control*, *46*, 1984–1989.
59. Guo, X. P., & Rieder, U. (2006). Average optimality for continuous-time Markov decision processes in Polish spaces. *Annals of Applied Probability*, *16*, 730–756.
60. Guo, X. P., & Shi, P. (2001). Limiting average criteria for nonstationary Markov decision processes. *SIAM Journal on Optimization*, *11*, 1037–1053.
61. Guo, X. P., & Yin, G. G. (2007). *Nonnegative models for continuous-time Markov decision processes with discounted and average costs*. Preprint.
62. Guo, X. P., & Zhu, W. P. (2002a). Denumerable state continuous-time Markov decision processes with unbounded cost and transition rates under the discounted criterion. *Journal of Applied Probability*, *39*, 233–250.
63. Guo, X. P., & Zhu, W. P. (2002b). Denumerable state continuous-time Markov decision processes with unbounded cost and transition rates under an average criterion. *ANZIAM Journal*, *34*, 541–557.
64. Guo, X. P., & Zhu, W. P. (2002c). Optimality conditions for continuous-time Markov decision processes with an average cost criterion. In Z. T. Hou, J. A. Filar, & A. Y. Chen (Eds.), *Markov processes and controlled Markov chains* (pp. 156–187). Dordrecht: Kluwer Academic. Chap. 10.
65. Guo, X. P., Hernández-Lerma, O., & Prieto-Rumeau, T. (2006). A survey of recent results on continuous-time Markov decision processes. *Top*, *14*(2), 177–246.
66. Guo, X. P., Song, X. Y., & Zhang, J. Y. (2009). Bias optimality for multichain continuous-time Markov decision processes. *Operations Research Letters*, *37*.
67. Hall, P., & Heyde, C. C. (1980). *Martingale limit theory and its applications*. San Diego: Academic Press.
68. Haviv, M., & Puterman, M. L. (1998). Bias optimality in controlled queuing systems. *Journal of Applied Probability*, *35*, 136–150.
69. Hernández-Lerma, O. (1989). *Adaptive Markov control processes*. Berlin: Springer.
70. Hernández-Lerma, O. (1994). *Aportaciones matemáticas: Vol. 3. Lectures on continuous-time Markov Control processes*. México: Sociedad Matemática Mexicana.

71. Hernández-Lerma, O., & González-Hernández, J. (2000). Constrained Markov controlled processes in Borel spaces: the discounted case. *Mathematical Methods of Operations Research*, *52*, 271–285.
72. Hernández-Lerma, O., & Govindan, T. E. (2001). Nonstationary continuous-time Markov control processes with discounted costs on infinite horizon. *Acta Applicandae Mathematicae*, *67*, 277–293.
73. Hernández-Lerma, O., & Lasserre, J. B. (1996). *Discrete-time Markov control processes: basic optimality criteria*. Berlin: Springer.
74. Hernández-Lerma, O., & Lasserre, J. B. (1999). *Further topics on discrete-time Markov control processes*. Berlin: Springer.
75. Hernández-Lerma, O., Vega-Amaya, O., & Carrasco, G. (1999). Sample-path optimality and variance-minimization of average cost Markov control processes. *SIAM Journal on Control and Optimization*, *38*, 79–93.
76. Hilgert, N., & Hernández-Lerma, O. (2003). Bias optimality versus strong 0-discount optimality in Markov control processes with unbounded costs. *Acta Applicandae Mathematicae*, *77*, 215–235.
77. Hitchcock, S. E. (1986). Extinction probabilities in predator–prey models. *Journal of Applied Probability*, *23*, 1–13.
78. Hiriart-Urruty, J. B., & Lemaréchal, C. (2001). *Fundamentals of convex analysis*. Berlin: Springer.
79. Hordijk, A., & Kallenberg, L. C. M. (1984). Constrained undiscounted stochastic dynamic programming. *Mathematics of Operations Research*, *9*, 276–289.
80. Hordijk, A., & Yushkevich, A. A. (1999a). Blackwell optimality in the class of stationary policies in Markov decision chains with a Borel state and unbounded rewards. *Mathematical Methods of Operations Research*, *49*, 1–39.
81. Hordijk, A., & Yushkevich, A. A. (1999b). Blackwell optimality in the class of all policies in Markov decision chains with a Borel state and unbounded rewards. *Mathematical Methods of Operations Research*, *50*, 421–448.
82. Hordijk, A., & Yushkevich, A. A. (2002). Blackwell optimality. In E. A. Feinberg & A. Shwartz (Eds.), *Handbook of Markov decision processes* (pp. 231–267). Dordrecht: Kluwer Academic.
83. Hou, Z. T., & Guo, Q. F. (1988). *Homogeneous denumerable Markov processes*. Berlin: Science Press and Springer.
84. Hou, Z. T., & Guo, X. P. (1998). *Markov decision processes*. Changsha: Science and Technology Press of Hunan (in Chinese).
85. Hou, Z. T. et al. (1994). *The Q-matrix problems for Markov processes*. Changsha: Science and Technology Press of Hunan (in Chinese).
86. Howard, R. A. (1960). *Dynamic programming and Markov processes*. New York: Wiley.
87. Hu, Q. Y. (1996). Continuous-time Markov decision processes with moment criterion. *Journal of Mathematical Analysis and Applications*, *203*, 1–12.
88. Iosifescu, M., & Tautu, P. (1973). *Stochastic processes and applications in biology and medicine, Vol. II: Models*. Berlin: Springer.
89. Jasso-Fuentes, H., & Hernández-Lerma, O. (2008). Characterizations of overtaking optimality for controlled diffusion processes. *Applied Mathematics and Optimization*, *21*, 349–369.
90. Jasso-Fuentes, H., & Hernández-Lerma, O. (2009a). Ergodic control, bias and sensitive discount optimality for Markov diffusion processes. *Stochastic Analysis and Applications*, *27*, 363–385.
91. Jasso-Fuentes, H., & Hernández-Lerma, O. (2009b). Blackwell optimality for controlled diffusion processes. *Journal of Applied Probability*, *46*, 372–391.
92. Kakumanu, P. (1971). Continuously discounted Markov decision models with countable state and action spaces. *Annals of Mathematical Statistics*, *42*, 919–926.
93. Kakumanu, P. (1972). Nondiscounted continuous-time Markov decision processes with countable state and action spaces. *SIAM Journal on Control*, *10*, 210–220.
94. Kakumanu, P. (1975). Continuous-time Markov decision processes with average return criterion. *Journal of Mathematical Analysis and Applications*, *52*, 173–188.

95. Kakumanu, P. (1977). Relation between continuous and discrete Markovian decision problems. *Naval Research Logistics Quarterly*, *24*, 431–439.
96. Kallenberg, L. C. M. (1983). *Mathematical centre tracts: Vol. 148. Linear programming and finite Markovian control problems*. Amsterdam: Mathematisch Centrum.
97. Kato, T. (1966). *Perturbation theory for linear operators*. Berlin: Springer.
98. Kitaev, M. Y., & Rykov, V. V. (1995). *Controlled queueing systems*. Boca Raton: CRC Press.
99. Klambauer, G. (1973). *Real analysis*. New York: American Elsevier.
100. Kermack, W. O., & McKendrick, A. G. (1927). Contributions to the mathematical theory of epidemics. *Proceedings of the Royal Society, A*, *115*, 700–721.
101. Kurano, M., Nakagami, J. I., & Huang, Y. (2002). Constrained Markov decision processes with compact state and action spaces: the average case. *Optimization*, *48*, 255–269.
102. Lasserre, J. B. (1988). Conditions for the existence of average and Blackwell optimal stationary policies in denumerable Markov decision processes. *Journal of Mathematical Analysis and Applications*, *136*, 479–490.
103. Lefèvre, C. (1979). Optimal control of the simple stochastic epidemic with variable recovery rates. *Mathematical Biosciences*, *44*, 209–219.
104. Lefèvre, C. (1981). Optimal control of a birth and death epidemic process. *Operations Research*, *29*, 971–982.
105. Leizarowitz, A. (1996). Overtaking and almost-sure optimality for infinite horizon Markov decision processes. *Mathematics of Operations Research*, *21*, 158–181.
106. Lembersky, M. R. (1974). On maximal rewards and $\varepsilon$-optimal policies in continuous time Markov chains. *Annals of Statistics*, *2*, 159–169.
107. Lewis, M. E., & Puterman, M. L. (2001). A note on bias optimality in controlled queueing systems. *Journal of Applied Probability*, *37*, 300–305.
108. Lewis, M. E., & Puterman, M. L. (2002a). A probabilistic analysis of bias optimality in unichain Markov decision processes. *IEEE Transactions on Automatic Control*, *46*, 96–100.
109. Lewis, M. E., & Puterman, M. L. (2002b). Bias optimality. In E. A. Feinberg & A. Shwartz (Eds.), *Handbook of Markov decision processes* (pp. 89–111). Dordrecht: Kluwer Academic.
110. Lippman, S. A. (1974/75). On dynamic programming with unbounded rewards. *Management Science*, *21*, 1225–1233.
111. Lippman, S. A. (1975). Applying a new device in the optimization of exponential queueing systems. *Operations Research*, *23*, 667–710.
112. Lothar, B. (2003). *From Markov jump processes to spatial queues*. Dordrecht: Kluwer Academic.
113. Lund, R. B., Meyn, S. P., & Tweedie, R. L. (1996). Computable exponential convergence rates for stochastically ordered Markov processes. *Annals of Applied Probability*, *6*, 218–237.
114. Mangel, M. (1985). *Decision and control in uncertain resource systems*. San Diego: Academic Press.
115. Massy, W. F., Montgomery, D. B., & Morrison, D. G. (1970). *Stochastic models of buying behavior*. Cambridge: MIT Press.
116. Meyn, S. P., & Tweedie, R. L. (1993). Stability of Markovian processes III: Foster–Lyapunov criteria for continuous-time processes. *Advances in Applied Probability*, *25*, 518–548.
117. Miller, B. L. (1968). Finite state continuous time Markov decision processes with an infinite planning horizon. *Journal of Mathematical Analysis and Applications*, *22*, 552–569.
118. Miller, B. L., & Veinott, A. F. (1969). Discrete dynamic programming with a small interest rate. *Annals of Mathematical Statistics*, *40*, 366–370.
119. Piunovskiy, A. B. (1997). *Optimal control of random sequences in problems with constraints*. Dordrecht: Kluwer Academic.
120. Piunovskiy, A. B. (2004). Multicriteria impulsive control of jump Markov processes. *Mathematical Methods of Operations Research*, *60*, 125–144.
121. Piunovskiy, A. B., & Mao, X. R. (2000). Constrained Markov decision processes: the dynamic programming approach. *Operations Research Letters*, *27*, 119–126.
122. Prieto-Rumeau, T. (2006). Blackwell optimality in the class of Markov policies for continuous-time controlled Markov chains. *Acta Applicandae Mathematicae*, *92*, 77–96.

123. Prieto-Rumeau, T., & Hernández-Lerma, O. (2005a). The Laurent series, sensitive discount and Blackwell optimality for continuous-time controlled Markov chains. *Mathematical Methods of Operations Research*, *61*, 123–145.
124. Prieto-Rumeau, T., & Hernández-Lerma, O. (2005b). Bias and overtaking equilibria for zero-sum continuous-time Markov games. *Mathematical Methods of Operations Research*, *61*, 437–454.
125. Prieto-Rumeau, T., & Hernández-Lerma, O. (2006a). Bias optimality for continuous-time controlled Markov chains. *SIAM Journal on Control and Optimization*, *45*, 51–73.
126. Prieto-Rumeau, T., & Hernández-Lerma, O. (2006b). A unified approach to continuous-time discounted Markov controlled processes. *Morfismos*, *10*, 1–40.
127. Prieto-Rumeau, T., & Hernández-Lerma, O. (2009, in press). Variance minimization and the overtaking optimality approach to continuous-time controlled Markov chains. *Mathematical Methods of Operations Research*.
128. Puterman, M. L. (1974). Sensitive discount optimality in controlled one-dimensional diffusions. *Annals of Probability*, *2*, 408–419.
129. Puterman, M. L. (1994). *Markov decision processes: discrete stochastic dynamic programming*. New York: Wiley.
130. Qiu, Q., Wu, Q., & Pedram, M. (2001). Stochastic modeling of a power-managed system: construction and optimization. *IEEE Transactions Computer Aided Design*, *20*, 1200–1217.
131. Rajarshi, M. B. (1981). Simple proofs of two threshold theorems for a general stochastic epidemic. *Journal of Applied Probability*, *18*, 721–724.
132. Ramsey, F. P. (1928). A mathematical theory of savings. *Econometrics Journal*, *38*, 543–559.
133. Rao, M. M. (1995). *Stochastic processes: general theory*. Dordrecht: Kluwer Academic.
134. Reuter, G. E. H. (1961). Competition processes. *Proceedings of the Fourth Berkeley Symposium on Mathematical Statistics and Probability*, *2*, 421–430.
135. Revuz, D., & Yor, M. (1999). *Continuous martingales and Brownian motion*. Berlin: Springer.
136. Ridler-Rowe, C. J. (1988). Extinction times for certain predator-prey models. *Journal of Applied Probability*, *25*, 612–616.
137. Ross, S. M. (1968). Non-discounted denumerable Markovian decision models. *Annals of Mathematical Statistics*, *39*, 412–423.
138. Ross, S. M. (1983). *Introduction to stochastic dynamic programming*. San Diego: Academic Press.
139. Roykov, V. V. (1966). Markov sequential decision processes with finite state and decision space. *Theory of Probability and Its Applications*, *11*, 302–311.
140. Sennott, L. I. (1991). Constrained discounted Markov chains. *Probability in the Engineering and Informational Sciences*, *5*, 463–476.
141. Sennott, L. I. (1999). *Stochastic dynamic programming and the control of queueing systems*. New York: Wiley.
142. Serfozo, R. F. (1979). An equivalence between continuous and discrete time Markov decision processes. *Operations Research*, *27*, 616–620.
143. Serfozo, R. F. (1981). Optimal control of random walks, birth and death processes, and queues. *Advances in Applied Probability*, *13*, 61–83.
144. Sladký, K. (1978). Sensitive optimality criteria for continuous time Markov processes. In *Transactions of the eighth Prague conference on information theory, statistical decision functions and random processes, Prague, 1978* (Vol. B, pp. 211–225).
145. Song, J. S. (1987). Continuous-time Markov decision programming with nonuniformly bounded transition rates. *Scientia Sinica*, *12*, 1258–1267 (in Chinese).
146. Tadj, L., & Choudhury, G. (2005). Optimal design and control of queues. *Top*, *13*, 359–412.
147. Tanaka, K. (1991). On discounted dynamic programming with constraints. *Journal of Mathematical Analysis and Applications*, *155*, 264–277.
148. Taylor, H. M. (1976). A Laurent series for the resolvent of a strongly continuous stochastic semi-group. *Mathematical Programming Studies*, *6*, 258–263.

149. Tweedie, R. L. (1981). Criteria for ergodicity, exponential ergodicity and strong ergodicity of Markov processes. *Journal of Applied Probability*, *18*, 122–130.
150. Veinott, A. F. (1966). On finding optimal policies in discrete dynamic programming with no discounting. *Annals of Mathematical Statistics*, *37*, 1284–1294.
151. Veinott, A. F. (1969). Discrete dynamic programming with sensitive discount optimality criteria. *Annals of Mathematical Statistics*, *40*, 1635–1660.
152. Vidale, M. L., & Wolfe, H. B. (1957). An operations research study of sales response to advertising. *Operations Research*, *5*, 370–381.
153. von Weizsäcker, C. C. (1965). Existence of optimal programs of accumulation for an infinite horizon. *Review of Economic Studies*, *32*, 85–104.
154. Wang, Z. K., & Yang, X. Q. (1992). *Birth and death processes and Markov chains*. Berlin: Springer, Science Press.
155. Wickwire, K. (1977). Mathematical models for the control of pests and infectious diseases: a survey. *Theoretical Population Biology*, *11*, 182–238.
156. Widder, D. V. (1946). *The Laplace transform*. Princeton: Princeton University Press.
157. Williams, D. (1979). *Diffusions, Markov processes, and martingales*. New York: Wiley.
158. Wu, C. B. (1997). Continuous-time Markov decision processes with unbounded reward and nonuniformly bounded transition rate under discounted criterion. *Acta Applicandae Mathematicae Sinica*, *20*, 196–208 (in Chinese).
159. Ye, L. E., Guo, X. P., & Hernández-Lerma, O. (2008). Existence and regularity of a nonhomogeneous transition matrix under measurability conditions. *Journal of Theoretical Probability*, *21*(3), 604–627.
160. Yin, G. G., & Zhang, Q. (1998). *Continuous-time Markov chains and applications*. Berlin: Springer.
161. Yushkevich, A. A. (1973). On a class of strategies in general Markov decision models. *Theory of Probability and Its Applications*, *18*, 777–779.
162. Yushkevich, A. A. (1977). Controlled Markov models with countable state and continuous time. *Theory of Probability and Its Applications*, *22*, 215–235.
163. Yushkevich, A. A. (1994). Blackwell optimal policies in a Markov decision process with a Borel state space. *Mathematical Methods of Operations Research*, *40*, 253–288.
164. Yushkevich, A. A. (1997). Blackwell optimality in continuous-in-action Markov decision processes. *SIAM Journal on Control and Optimization*, *35*, 2157–2182.
165. Yushkevich, A. A., & Feinberg, E. A. (1979). On homogeneous controlled Markov models with continuous time and finite or countable state space. *Theory of Probability and Its Applications*, *24*, 156–161.
166. Zhang, J. Y., & Cao, X.-R. (2009). Continuous-time Markov decision processes with $n$-bias optimality criteria. *Automatica*, *35*, 1628–1638.
167. Zhang, L. L., & Guo, X. P. (2008). Constrained continuous-time Markov decision processes with average criteria. *Mathematical Methods of Operations Research*, *67*, 323–340.
168. Zheng, S. H. (1991). Continuous-time Markov decision programming with average reward criterion and unbounded reward rate. *Acta Applicandae Mathematicae Sinica*, *7*, 6–16.

# Index



X. Guo, O. Hernández-Lerma, *Continuous-Time Markov Decision Processes*,
Stochastic Modelling and Applied Probability 62,
DOI 10.1007/978-3-642-02547-1, 

# Stochastic Modelling and Applied Probability
## formerly: Applications of Mathematics

1 Fleming/Rishel, **Deterministic and Stochastic Optimal Control** (1975)
2 Marchuk, **Methods of Numerical Mathematics** (1975, 2nd ed. 1982)
3 Balakrishnan, **Applied Functional Analysis** (1976, 2nd ed. 1981)
4 Borovkov, **Stochastic Processes in Queueing Theory** (1976)
5 Liptser/Shiryaev, **Statistics of Random Processes I: General Theory** (1977, 2nd ed. 2001)
6 Liptser/Shiryaev, **Statistics of Random Processes II: Applications** (1978, 2nd ed. 2001)
7 Vorob'ev, **Game Theory: Lectures for Economists and Systems Scientists** (1977)
8 Shiryaev, **Optimal Stopping Rules** (1978, 2008)
9 Ibragimov/Rozanov, **Gaussian Random Processes** (1978)
10 Wonham, **Linear Multivariable Control: A Geometric Approach** (1979, 2nd ed. 1985)
11 Hida, **Brownian Motion** (1980)
12 Hestenes, **Conjugate Direction Methods in Optimization** (1980)
13 Kallianpur, **Stochastic Filtering Theory** (1980)
14 Krylov, **Controlled Diffusion Processes** (1980, 2009)
15 Prabhu, **Stochastic Storage Processes: Queues, Insurance Risk, and Dams** (1980)
16 Ibragimov/Has'minskii, **Statistical Estimation: Asymptotic Theory** (1981)
17 Cesari, **Optimization: Theory and Applications** (1982)
18 Elliott, **Stochastic Calculus and Applications** (1982)
19 Marchuk/Shaidourov, **Difference Methods and Their Extrapolations** (1983)
20 Hijab, **Stabilization of Control Systems** (1986)
21 Protter, **Stochastic Integration and Differential Equations** (1990)
22 Benveniste/Métivier/Priouret, **Adaptive Algorithms and Stochastic Approximations** (1990)
23 Kloeden/Platen, **Numerical Solution of Stochastic Differential Equations** (1992, corr. 2nd printing 1999)
24 Kushner/Dupuis, **Numerical Methods for Stochastic Control Problems in Continuous Time** (1992)
25 Fleming/Soner, **Controlled Markov Processes and Viscosity Solutions** (1993)
26 Baccelli/Brémaud, **Elements of Queueing Theory** (1994, 2nd ed. 2003)
27 Winkler, **Image Analysis, Random Fields and Dynamic Monte Carlo Methods** (1995, 2nd ed. 2003)
28 Kalpazidou, **Cycle Representations of Markov Processes** (1995)
29 Elliott/Aggoun/Moore, **Hidden Markov Models: Estimation and Control** (1995, corr. 3rd printing 2008)
30 Hernández-Lerma/Lasserre, **Discrete-Time Markov Control Processes** (1995)
31 Devroye/Györfi/Lugosi, **A Probabilistic Theory of Pattern Recognition** (1996)
32 Maitra/Sudderth, **Discrete Gambling and Stochastic Games** (1996)
33 Embrechts/Klüppelberg/Mikosch, **Modelling Extremal Events for Insurance and Finance** (1997, corr. 4th printing 2008)
34 Duflo, **Random Iterative Models** (1997)
35 Kushner/Yin, **Stochastic Approximation Algorithms and Applications** (1997)
36 Musiela/Rutkowski, **Martingale Methods in Financial Modelling** (1997, 2nd ed. 2005, corr. 3rd printing 2009)
37 Yin, **Continuous-Time Markov Chains and Applications** (1998)
38 Dembo/Zeitouni, **Large Deviations Techniques and Applications** (1998, corr. 2nd printing 2009)
39 Karatzas, **Methods of Mathematical Finance** (1998)
40 Fayolle/Iasnogorodski/Malyshev, **Random Walks in the Quarter-Plane** (1999)
41 Aven/Jensen, **Stochastic Models in Reliability** (1999)
42 Hernandez-Lerma/Lasserre, **Further Topics on Discrete-Time Markov Control Processes** (1999)
43 Yong/Zhou, **Stochastic Controls. Hamiltonian Systems and HJB Equations** (1999)
44 Serfozo, **Introduction to Stochastic Networks** (1999)
45 Steele, **Stochastic Calculus and Financial Applications** (2001)
46 Chen/Yao, **Fundamentals of Queuing Networks: Performance, Asymptotics, and Optimization** (2001)

# Stochastic Modelling and Applied Probability
formerly: Applications of Mathematics

47 Kushner, **Heavy Traffic Analysis of Controlled Queueing and Communications Networks** (2001)
48 Fernholz, **Stochastic Portfolio Theory** (2002)
49 Kabanov/Pergamenshchikov, **Two-Scale Stochastic Systems** (2003)
50 Han, **Information-Spectrum Methods in Information Theory** (2003)
51 Asmussen, **Applied Probability and Queues** (2nd ed. 2003)
52 Robert, **Stochastic Networks and Queues** (2003)
53 Glasserman, **Monte Carlo Methods in Financial Engineering** (2004)
54 Sethi/Zhang/Zhang, **Average-Cost Control of Stochastic Manufacturing Systems** (2005)
55 Yin/Zhang, **Discrete-Time Markov Chains** (2005)
56 Fouque/Garnier/Papanicolaou/Sølna, **Wave Propagation and Time Reversal in Randomly Layered Media** (2007)
57 Asmussen/Glynn, **Stochastic Simulation: Algorithms and Analysis** (2007)
58 Kotelenez, **Stochastic Ordinary and Stochastic Partial Differential Equations: Transition from Microscopic to Macroscopic Equations** (2008)
59 Chang, **Stochastic Control of Hereditary Systems and Applications** (2008)
60 Bain/Crisan, **Fundamentals of Stochastic Filtering** (2009)
61 Pham, **Continuous-time Stochastic Control and Optimization with Financial Applications** (2009)
62 Guo/Hernández-Lerma, **Continuous-Time Markov Decision Processes** (2009)